Cognitive Technologies

Editor-in-Chief

Daniel Sonntag, German Research Center for AI, DFKI, Saarbrücken, Saarland, Germany

Titles in this series now included in the Thomson Reuters Book Citation Index and Scopus!

The Cognitive Technologies (CT) series is committed to the timely publishing of high-quality manuscripts that promote the development of cognitive technologies and systems on the basis of artificial intelligence, image processing and understanding, natural language processing, machine learning and human-computer interaction.

It brings together the latest developments in all areas of this multidisciplinary topic, ranging from theories and algorithms to various important applications. The intended readership includes research students and researchers in computer science, computer engineering, cognitive science, electrical engineering, data science and related fields seeking a convenient way to track the latest findings on the foundations, methodologies and key applications of cognitive technologies.

The series provides a publishing and communication platform for all cognitive technologies topics, including but not limited to these most recent examples:

- Interactive machine learning, interactive deep learning, machine teaching
- Explainability (XAI), transparency, robustness of AI and trustworthy AI
- Knowledge representation, automated reasoning, multiagent systems
- Common sense modelling, context-based interpretation, hybrid cognitive technologies
- Human-centered design, socio-technical systems, human-robot interaction, cognitive robotics
- Learning with small datasets, never-ending learning, metacognition and introspection
- Intelligent decision support systems, prediction systems and warning systems
- Special transfer topics such as CT for computational sustainability, CT in business applications and CT in mobile robotic systems

The series includes monographs, introductory and advanced textbooks, state-of-the-art collections, and handbooks. In addition, it supports publishing in Open Access mode.

András Kornai

Vector Semantics

 Springer

András Kornai
HLT
SZTAKI Computer Science Research Institute
Budapest, Hungary

ISSN 1611-2482   ISSN 2197-6635   (electronic)
Cognitive Technologies
ISBN 978-981-19-5609-6   ISBN 978-981-19-5607-2   (eBook)
https://doi.org/10.1007/978-981-19-5607-2

This Springer imprint is published by the registered company Springer Nature Singapore Pte Ltd.
The registered company address is: 152 Beach Road, #21-01/04 Gateway East, Singapore 189721, Singapore

To Ágnes

There is nothing as practical as a good theory (Lewin, 1943)

Mathematics is the art of reducing any problem to linear algebra (William Stein, quoted in Kapitula (2015))

Algebra is the offer made by the devil to the mathematician. The devil says: I will give you this powerful machine, it will answer any question you like. All you need to do is give me your soul: give up geometry and you will have this marvelous machine (Atiyah, 2001)

Preface

This book is a direct continuation of (Kornai, 2019), but unlike its predecessor, it is no longer a textbook. The earlier volume, henceforth abbreviated S19, mostly covered material that is well known in the field, whereas the current volume is a research monograph, dominated by the author's own research centering on the 4lang system.

S19 attempted to cater to students of four disciplines, linguistics; computer science; cognitive science; and philosophy. As Hinrich Schütze wrote at the time: "This textbook distinguishes itself from other books on semantics by its interdisciplinarity: it presents the perspectives of linguistics, computer science, philosophy and cognitive science. I expect big changes in the field in coming years, so that a broad coverage of foundations is the right approach to equipping students with the knowledge they need to tackle semantics now and in the future."

The big changes were actually already under way, in no small part due to Schütze, 1993, who took the fundamental step in modeling word meaning by vectors in ordinary Euclidean space. S19:2.7 discusses some of the mathematical underpinnings. This material is now standard, so much so that the main natural language processing (NLP) textbook, Jurafsky and Martin (2022) is already incorporating it in its new edition (our references will be to this new version). But for now, vectorial semantics has relatively few contact points with mainstream linguistic semantics, so little that the most comprehensive (five volumes) contemporary summary, Gutzmann et al. (2021), has not devoted a single chapter to the subject. Sixty years ago, McCarthy (1963) urged:

> Mathematical linguists are making a serious mistake in their concentration on syntax and, even more specially, on the grammar of natural languages. It is even more important to develop a mathematical understanding and a formalization of the kinds of information conveyed in natural language

and here we continue with the original plan by trying to use not just word vectors, both static and contextual, but the broader machinery of linear and multilinear algebra to describe meaning representations that make sense both to the linguist and to the computer scientist. In this process, we will reassess the word vectors themselves, arguing that in most cases words correspond not to vectors, but to polytopes in n-space, and we will offer novel models for many traditional concerns of linguistic semantics from presuppositions to indexicals, from rigid designators to variable binding. In Kornai, 2007 we wrote:

> Perhaps the most captivating aspect of mathematical linguistics is not just the existence of discrete mesoscopic structures but the fact that these come embedded, in ways we do not fully understand, in continuous signals

and vector semantics makes a virtue of necessity: whether we fully understand it or not, by embedding words, obviously discrete, in continuous Euclidean space, we are accounting for an essential feature of their internal organization. In the meantime, similar changes are taking place in speech recognition, see e.g. Bohnstingl et al., 2021. Obviously, we cannot discuss speech in any detail here, but it seems clear that the early goals of neural modeling, greatly frustrated at the time by insufficient computing power, are finally coming in view. The recent move to *dynamic* or *contextual* embeddings, by now an entrenched standard in computational linguistics (CL) and NLP, has left a key question unanswered, that of compositionality (S19:1.1): how we represent the meaning of larger expressions. The importance of the issue has been realized early on (Allauzen et al., 2013), but so far no proposed solution such as Purver et al., 2021 has gained wider acceptance. In fact, within the CL/NLP community the issue has largely receded from view, owing to the influence of what Noah Smith called "converts to e2e religion and the cult of differentiability" (see LeCun, Bengio, and Hinton, 2015 and Goldberg, 2017 for a clear summary of the end-to-end differentiable paradigm).

For now, the schism between linguists, who set store by intermediary structures built from units of analysis ranging from the morpheme to the paragraph and beyond, and the computational linguists, who are increasingly in favor of end-to-end (e2e) systems that emphatically do not rely on intermediary units or structures, not even the basic similarity structure of the lexicon that was brought to light by static word vectors, seems unresolvable. Yet it seems clear that both parties want the same thing, *learnable* models of linguistic behavior, and the difference is a matter of strategy: linguists are looking for explainable, modular systems whose learnability can be studied as we go along, whereas computational linguists insist on models that are learnable right now, often at the expense of issues like one-shot and zero-shot learning which occupy a more central place in theoretical linguistics, where the phenomenon is known as productivity. Also, CL/NLP is perfectly happy with using multi-gigaword training sets, while linguists want an algorithm that is responsive to the primary linguistic data, unlikely to exceed a few million words total.

In this book, we try to make both sides happy, by (i) using intermediate representations, and (ii) providing learning algorithms for these. In Chapter 1 we begin by defining the formal system we will use to assign meanings to words by means of symbolic techniques. As Gérard Huet observed at the time, in S19:4,5.8 we used the "elegant formalism of Eilenberg machines" instead of "kludgy imperative devices with tapes and reading heads". But this really just kicked the learnability can down the road, especially

as it is well known (Angluin, 1981; Angluin, 1987) that finite state (FS) devices are not at all trivial to learn. The frontier of FS learnability work is now in phonology (Rogers et al., 2013; Yli-Jyrä, 2015; Chandlee and Jardine, 2019; Rawski and Dolatian, 2020), where the data has significant temporal structure. It remains to be seen how much of this can be transferred to semantics, where memory is typically random access (see 7.4) and temporal structure, the succession of words, can be largely irrelevant (in free word order languages). Therefore, the main thrust of the current volume is to link the linguistic theory of semantics to continuous vector spaces, rather than Eilenberg machines, while trying to preserve as much of the elegance of relational thinking as possible.

Our approach is formal, and has the express goal of making the formalism useful for computational linguists. Yet it owes a great deal to a decidedly informal theory, cognitive linguistics. In fact, the volume could be called *Formal Lexical Semantics*, were it not for the now entrenched terminology that presents *formal* and *lexical* as direct opposites. The influence of 'cognitive' work (Jackendoff, 1972; Jackendoff, 1983; Jackendoff, 1990; Lakoff, 1987; Langacker, 1987; Talmy, 2000) will be visible throughout. Many of these cognitive theories are presented informally (indeed, most exponents of cognitive grammar, with the notable exception of Jackendoff, are positively anti-formal); and others, both in AI and in cognitive science proper, remain silent on word meaning, with Fodor (1998) being quite explicit that words are atomic. In Chapter 2 we present a formal theory of non-compositional semantics that is suitable for morphology, i.e. for describing the semantics of clitics and bound affixes as well, and extends smoothly to the compositional domain. That something of the sort is really required is evident from cross-linguistic considerations, since the same meaning that is expressed by morphology in one language will often be expressed by syntactic means in another.

As Kurt Lewin famously said, "there is nothing as practical as a good theory". We will illustrate this thesis by presenting a highly formal reconstruction of much of cognitive grammar, albeit one cast in algebraic terms rather than the generative machinery preferred by Jackendoff. We take on board several thorny issues such as temporal and spatial semantics in Chapter 3; negation in Chapter 4; probabilistic reasoning in Chapter 5; modals and counterfactuality in Chapter 6; implicature and gradient adjectives in Chapter 7; proper names and the integration of real-world knowledge in Chapter 8; and some applications in Chapter 9.

Perhaps more significant, we take on board the *entire* lexicon, both in terms of breadth and depth. The 4lang computational project aims at reducing the entire vocabulary to a core defining set. For breadth, we will discuss some representatives of all the standard (Buck, 1949) semantic fields from "Physical World" to "Religion and Beliefs" (see S19:6.4 and 5.3). We aim at exhaustivity at the class level, but not at the individual level: for example we do not undertake to systematically catalogue all 50+ "Body Parts and Functions" considered by Buck. 4lang uses only a couple dozen of these, and we see no need to go beyond representative examples: once the reader sees how these are treated, the general idea will be clear. As for the rest (e.g. *navel* is outside 4lang) we rely on general purpose dictionaries, LDOCE (Procter, 1978) in particular, and consider 'the small hollow or raised place in the middle of your stomach' satisfactory as long as the words appearing in the definition, and their manner of combination, are defined. The reader is helped by the Appendix starting on p. 253, where each of the 4lang defining words are listed with a pointer to the main body of the text where the entry is discussed, and by many cross-references for those who prefer to drill down rather than follow along the in-breadth order of exposition necessitated by the subject matter.

In terms of depth, we often go below the word level, considering bound morphemes, both roots and suffixes, as lexical entries (see 2.2). Unlike many of its predecessors, 4lang doesn't stop at a set of primitives, but defines these as well, in terms of the other primitives, wherever possible. There remain a handful of truly irreducible elements, such as the question morpheme *wh*, but more interesting are the 99% of cases like *judge* defined as `human, part_of court/3124, decide, make official(opinion)` (see 1.3 for the syntax of the formal language used in definitions) where we can trace the constituent parts to any depth.

The key observation here is that true undefinability is more an anomaly than the norm. We simply cannot hang the rest of the vocabulary on the few undefinable elements, because we encounter irreducible circularity in the definitions long before we could reduce everything else to these. Through this book, we embrace this circularity and raise it to the level of a heuristic method: once sufficient machinery is in place (especially in Chapter 6 and beyond), we will spend considerable time on chasing various chains of definitions by means of repeat substitutions.

Consider the days of the week. The Longman Defining Vocabulary bites the bullet, and lists all of Sunday, Monday, ..., Saturday as primitives. But clearly, as soon as one of these is primitive, the others are definable. Rather than arbitrarily designating one of them as the basic one, 4lang treats each definition as an equation, and the entire lexicon as a set of equations mutually constraining all meanings. How this is done is the subject of the book. The impatient reader may jump ahead to 9.5 where the algorithm, built bottom-up throughout the volume, is summarized in a top-down fashion.

Who should read this book

Semantics studies how meaning is conveyed from one person to another. This is a big question, and there are several academic disciplines that want a piece of the action. The list includes linguistics; logic; computer science; artificial intelligence; philosophy; psychology; cognitive science; and semiotics. Many practitioners in these disciplines would tell the student that semantics *only* makes sense if studied from the viewpoint of their discipline. Here we take a syncretic view and welcome any development that seems to make a contribution to the big question.

As with S19, the ideal reader is a hacker, 'a person who delights in having an intimate understanding of the internal workings of a system'. But this time we aim at the graduate student, and assume not just S19 as a prerequisite, but also a willingness to read research papers. A central element of the Zeitgeist is to bring Artificial General Intelligence (AGI) to this world. This is tricky, in particular in terms of making sure that AGIs are not endangering humanity (see Kornai, 2014, now superseded by Fuenmayor and Benzmüller, 2019, and S19:9 for the author's take on this). Clearly, a key aspect of AGI is the ability to communicate with humans, and the book is designed to help create a way for doing so (as opposed to helping with the sensory system, the motor capabilities, etc). This is an undertaking involving a large number of people most of whom operate not just without central direction, but often without knowledge of each other. Even though only 9.4 address the issue directly, the book is recommended to all people interested in the linguistic aspects of AGI.

At the same time, it is our express goal to get linguists and cognitive scientists, who may or may not be skeptical about the AGI goal, back in the game. The enormous predictive success of deep models, transformers in particular, in producing fluent text of impeccable grammaticality makes clear that syntax is, in the learning sense, easier than semantics. The current frontier of this work is AlphaCode (Li et al., 2022), which generates software of remarkable semantic understanding from programming problems stated in English, much as earlier generations of computational models produced systems of equations from MCAS-level word problems (Kushman et al., 2014). In our view, such systems bypass the 'fast thinking' cognitive competence that characterizes human language understanding, and model 'slow thinking', the Type 2 processes of Kahneman, 2011. Our interest here is with the former, in particular with the *naive* world-view that predates our contemporary scientific world-view both ontogenically and phylogenically.

How to read it

Again, the book is primarily designed to be read on a computer. We make heavy use of inline references, typeset in blue, particularly to Wikipedia (WP) and the Stanford Encyclopedia of Philosophy (SEP), especially for concepts and ideas that we feel the reader will already know but may want to refresh. Because following these links greatly improves the reading experience, readers of the paper version are advised to have a cellphone on hand so that they can scan the hyperlinks which are also rendered as QR codes on the margin.

The current volume also comes with an external index starting at page 249 and also accessible at http://hlt.bme.hu/semantics/external2 that collects a frozen copy of the external references to protect the reader against dead links. A traditional index, with several hundred index terms, is also provided, but the reader is encouraged to search both indexes and, as a last resort, the file itself, if a term is missing from these. Within the Appendix (Chapter 9.5), those definitions that are explained in the text are also indexed. These words are highlighted on the margin where the definition is found.

Linguistic examples are normally given in *italics*, and if a meaning (paraphrase) is provided, this appears in single quotes. Italics are also used for technical terms appearing the first time and for emphasis. The 4lang computational system contains a concept dictionary, which initially had bindings in four languages, representative samples of the major language families spoken in Europe, Germanic (English), Slavic (Polish),

Romance (Latin), and Finno-Ugric (Hungarian). Today, bindings exist in over 40 languages (Ács, Pajkossy, and Kornai, 2013; Hamerlik, 2022), but the printed version staring on 253 is restricted to English. In the text, dictionary entries, definitions, and other computationally pertinent material, will be given in `typewriter font`.

As is common with large-scale research programs with many ingredients, many of the specifics of `4lang` have changed since the initial papers were published. This issue was largely covered up in S19 by the conscious effort to put the work of others front and center, as befits a textbook, and minimize direct discussion of `4lang`. The problem of slow drift (there were no major conceptual upheavals, even the shift from Eilenberg machines to vector semantics could be accomplished by deprecating one branch and adding another) is now addressed by versioning: S19 corresponds to Release V1 of `4lang`, and the current volume corresponds to V2. Unless specifically stated otherwise, all definitions, formulas, and statistics discussed here are from V2, see 9.5 for release notes. A great deal of work remains for further releases. This is noted occasionally in the text, always as an invitation to join the great free software hive mind, and discussed more systematically in Chapter 9.

Acknowledgments

Some of the material presented here appeared first in papers, some of which are joint work with others whose contributions are highly significant: here we single out Marcus Kracht Bielefeld, (Kornai and Kracht, 2015; Borbély et al., 2016), and Zalán Gyenis Jagellonian University, Krakow (Gyenis and Kornai, 2019) for their generosity of letting me reuse some of this work.

Much of the heavy lifting, especially on the computational side, but also in terms of conceptual clarifications and new ideas, was done by current and former HLT students, including Judit Ács SZTAKI, Gábor Borbély BME, Márton Makrai Institute of Cognitive Neuroscience and Psychology, Ádám Kovács TU Wien, Dániel Lévai Upright Oy, Dávid Márk Nemeskey Digital Heritage Lab, Gábor Recski TU Wien, and Attila Zséder Lensa.

Some of the material was taught in the fall semester of 2020/21 at BME and at ESSLLI 2021. I owe special thanks to the students at these courses and to all readers of the early versions of this book, who caught many typos and stylistic infelicities, suggested excellent references, and helped with the flow of the presentation: Judit Ács BME AUT, Miklós Eper (BME TTK), Kinga Gémes (BME AUT), Tamás Havas (BME TTK), Máté L. Juhász (ELTE), Máté Koncz (BME TTK), Ádám Kovács (BME AUT), and Boglárka Tauber (BME TTK).

The help of those who commented on some parts of the manuscript, offering penetrating advice on many points evident only to someone with their expertise, in particular Avery Andrews (Australian National University), Cleo Condoravdi (Stanford), András Cser PPKE, Hans-Martin Gärtner Research Institute for Linguistics, András Máté ELTE Logic, Richard Rhodes (Berkeley), András Simonyi PPKE, Anna Szabolcsi (NYU), and Madeleine Thompson (OpenAI) is gratefully acknowledged. The newly (V2) created

Japanese and Chinese bindings reflect the expertise and generosity of László Cseres-nyési Shikoku Gakuin University and Huba Bartos Research Institute for Linguistics, and the updated Polish bindings, originally due to Anna Cieślik Cambridge, benefited from the help of Małgorzata Suszczynska University of Szeged. I am particularly grateful to my colleague Ferenc Wettl BME, and my Doktorvater, Paul Kiparsky (Stanford), who both provided detailed comments. Needless to say, they do not agree with everything in the book, the views expressed here are not those of the funding agencies, and all errors and omissions remain my own.

The work was partially supported by 2018-1.2.1-NKP-00008: Exploring the Mathematical Foundations of Artificial Intelligence; by the Hungarian Scientific Research Found (OTKA), contract number 120145; the Ministry of Innovation and Technology NRDI Office within the framework of the Artificial Intelligence National Laboratory Program, and MILAB, the Hungarian artificial intelligence national laboratory. The writing was done at the Algebra department of the Budapest University of Technology and Economics (BME) and at the Computer Science Institute (SZTAKI).

I am grateful for the continuing professionalism and painstaking support of my editors at Springer, Celine Chang and Alexandru Ciolan. Open Access was made possible by grant MEC_K 141539 from NKFIH, the Hungarian National Research, Development and Innovation Office.

Contents

1

Foundations of non-compositionality

Contents

For the past half century, linguistic semantics was dominated by issues of compositionality to such an extent that the meaning of the atomic units (which were generally assumed to be words or their stems) received scant attention. Here we will put word meaning front and center, and base the entire plan of the book on beginning with the lowest meaningful units, morphemes, and building upward. In 1.1 we set the stage by considering the three major approaches to semantics that can be distinguished by their formal apparatus: formulaic, geometric, and algebraic. In 1.2 we summarize some of the lexicographic principles that we will apply throughout: universality, reductivity, and keeping the lexicon free of encyclopedic knowledge. In 1.3 we describe the formulaic theory of lexical meaning. This is linked to the geometric theory in 1.4, and to the algebraic theory in 1.5. The links between the algebraic and the geometric theory are discussed in 1.6, where we investigate the possibility of a meta-formalism that could link all three approaches together.

1.1 Background

The **formulaic** (logic-based) theory of semantics (S19:3.7), Montague Grammar (MG) and its lineal descendants such as Discourse Representation Theory and Dynamic Semantics reigned supreme in linguistic semantics until the 21st century in spite of its well known failings because it was, and in some respects still is, the only game in town: the alternative 'cognitive' theory went largely unformalized, and was deemed 'markerese' (Lewis, 1970) by the logic-based school. Here we will attempt to formalize many, though

A. Kornai, *Vector Semantics*, Cognitive Technologies, https://doi.org/10.1007/978-981-19-5607-2_1

by no means all, insights of the cognitive theory, an undertaking made all the more necessary by the fact that MG has little to offer on the nature of atomic units (Zimmermann, 1999).

Starting perhaps with (Schütze, 1993; Schütze, 1998) and propelled to universal success by (Collobert and Weston, 2008; Collobert et al., 2011) an entirely new, **geometric** theory, mapping meanings to vectors in low-dimensional Euclidean space, became standard in computational linguistics (S19:2.7 Example 2.3 et seqq). Subjects central to semantics such as compositionality, or the relation of syntactic to semantic representations, hitherto discussed entirely in a logic-based framework, became the focus of attention (Allauzen et al., 2013) for the geometric theory, but there is still no widely accepted solution to these problems. One unforeseen development of the geometric theory was that morphology, syntax, and semantics are to some extent located in different layers of the multilayer models that take word vectors as input (Belinkov et al., 2017b; Belinkov et al., 2017a) but 'probing' the models is still an art, see (Karpathy, Johnson, and Fei-Fei, 2015; Greff et al., 2015) for some of the early work in this direction, and (Clark et al., 2019; Hewitt and Manning, 2019) for more recent work on contextual embeddings.

At the same time, the **algebraic** theory of semantics (S19:Def 4.5 et seqq) explored in Artificial Intelligence since the 1960s (Quillian, 1969; Minsky, 1975; Sondheimer, Weischedel, and Bobrow, 1984), which used (hyper)graphs for representing the meaning of sentences and larger units, was given new impetus by Google's efforts to build a large repository of real-world knowledge by finding named entities in text and anchoring these to a large external knowledge base, the KnowledgeGraph, which currently has over 500m entities linked by 170b relations or 'facts' (Pereira, 2012). More linguistically motivated algebraic theories (Kornai, 2010a; Abend and Rappoport, 2013; Banarescu et al., 2013), coupled with a renewed interest in dependency parsing (Nivre et al., 2016), are contributing to a larger reappraisal of the role of background knowledge and the use of hypergraphs in semantics (Koller and Kuhlmann, 2011).

Through this book, we will try to link these three approaches, giving mathematical form to the belief that they are just the trunk, leg, and tail of the same elephant. This is not to say that these are 'notational variants' (Johnson, 2015), to the contrary, each of them make predictions that the others lack. A better analogy would be the algebraic (matrix) and the geometrical (transformation) view of linear algebra: both are equally valid, but they are not equally useful in every situation.

One word of caution is in order: the formulas we will study in 1.3 are not the formulas of higher order intensional logic familiar to students of MG, but rather the basic building blocks of a much simpler proto-logic, well below first order language in complexity. The graphs that we will start studying in 1.5 are hypergraphs, very similar to the notational devices of cognitive linguistics, DG, LFG, HPSG and those of AI, but not letter-identical to any of the broad variety of earlier proposals. Only the geometry is the same n-dimensional Euclidean geometry that everyone else is using, but even here there will be some twists, see 1.4.

1.2 Lexicographic principles

Universality `4lang` is a concept dictionary, intended to be universal in a sense made more precise below. To take the first tentative steps towards language-independence, the system was set up with bindings in four languages, representative samples of the major language families spoken in Europe: Germanic (English), Slavic (Polish), Romance (Latin), and Finno-Ugric (Hungarian). In Version 1, automatically created bindings exist in over 40 languages (Ács, Pajkossy, and Kornai, 2013), but the user should keep in mind that these bindings provide only rough semantic correspondence to the intended concept. In the current Version 2 (see 9.5) two Oriental languages, Japanese and Chinese, were added manually by László Cseresnyési and Huba Bartos respectively, and further automatic binding were created (Hamerlik, 2022).

The experience of parallel development of `4lang` in four languages reinforces a simple point that lexicographers have always considered self-evident: words or word senses don't match up across languages, not even in the case of these four languages that share a common European cultural/civilizational background. It's not just that some concepts are simply missing in some languages (a frequent cause of borrowing), but the whole conceptual space (see 1.4) can be partitioned differently.

For example, English tends to make no distinction between verbs that describe actions that affect their subjects and their objects the same way: compare *John turns, John bends* to *John turns the lever, John bends the pipe*. In Polish, we need a reflexive object pronoun *się* 'self' to express the fact that it is John who is turning/bending in the first case. The semantics is identical, yet in English *??John turns/bends himself* would sound strange. In Hungarian, we must use different verbs derived from the same root: 'turn self' is *ford-ul* whereas 'turn something' is *ford-ít*, and similarly for *haj-ol* 'bend self' and *hajl-ít* 'bend something', akin to Latin *versor/verso, flector/flecto*, but Latin also offers the option of using a pronoun *me flecto/verso*.

Where does this leave us in regards to the lofty goal of universality? At one extreme, we find the strong Sapir-Whorf hypothesis that language determines thought. This would mean that a speaker of English cannot share the concept of bending with a speaker of Hungarian, being restricted to one word for two different kinds of situations that Hungarian has two different words for. At the other extreme, we find the methodology followed here: we resort to highly abstract units (core lexemes) which we assume to be shared across languages, but permit larger units to be built from these in ways that differ from language to language. Here the key notions we must countenance include *self*, `self` which is defined as `=pat[=agt]`, `=agt[=pat]` (see also 3.3), and *bend*, which we take to be basic in the intransitive form, see 2.4. We turn to the issue of how in general transitives can be defined by their objectless counterparts in 3.1.

How formulas such as these are to be created, manipulated, and understood will be discussed in 1.3, here we begin with high-level formatting. The main `4lang` file is divided into 11 tab-separated fields, of which the last is reserved for comments (these begin with a percent sign). A typical entry, written as one line in the file but here in the text generally broken up in two for legibility, would be

```
water víz aqua woda mizu 水 shui3 水 2622 u N
        liquid, lack colour, lack taste, lack smell, life need
```

As can be seen, the first four columns are the 4 original language bindings given in EN HU LA PL order. In Version 1, all extended Latin characters were replaced by their base plus a number, e.g. o3 for ő, o2 for ö, and o1 for ó. This was to keep the behavior of standard unix utilities like `grep` constant across platforms (scripts for conversion to/from utf8 were available). In Version 2, two new columns are added after the fourth for JA ZH (see 9.5), and utf8-encoded accented characters are used throughout. The seventh column (in V1, the fifth) is a unique number per concept, most important when the English bindings coincide:

```
cook főz     coquo gotować 825 V
     =agt make <food>, ins_ heat
cook szakács coquus kucharz 2152 N
     person, <profession>, make food
```

The eighth (in V1, sixth) column is an estimate of reducibility status and can take only four values: p means primitive, an entry that seems impossible to reduce to other entries.

wh   An example would be the question morpheme *wh,* here given as `wh ki/mi/hogy quo kto/co/jak 3636 p G wh`. Note that the definiendum (column 1) appears in the definiens (column 10), making the irreducibility of this entry evident. At the other end we find entries marked by e, which means eliminable. An example would be

three   *three* `three három tres trzy 2970 e A number, follow two`. In between we find entries marked by c, which are candidates for core vocabulary: and exam-

see   ple would be *see* `see lát video widzieć 1476 c V perceive, ins_ eye`; and u, unknown reducibility status.

The ninth (in V1, seventh) column is a rough lexical category symbol, see 2.1 for further discussion. Our main subject here is the 10th (in V1, eighth) column, which gives the `4lang` definition. We defer the formal syntax of definitions to 1.3, after we discussed some further lexicographic principles, and use the opportunity to introduce some of the notation informally first. Many technical devices such as `=agt`, `=pat`, `wh`, `gen`, … make their first appearance here, but will be fully explained only in subsequent chapters. Very often, we will have reason to present lexical entries in an abbreviated form, showing only the headword and the definition (with the index, reducibility, and lexical category shown or suppressed as needed):

```
bend 975 e V has form[change], after(lack straight/563)
```

drunk   Where such abbreviated entries appear in running text, as *drunk* here, `drunk ittas potus pijany 1165 c A quality, person has quality, alcohol cause_, lack control` the headword is highlighted on the margin. For human readability, the concept number is omitted whenever the English binding is unique, so we have `person` in the above definition rather than `person/2185`, but we would spell out `man/659` 'homo' to disambiguate from `man férfi vir mężczyzna 744 man e N person, male`. In running text we generally omit the Japanese and Chinese equivalents for ease of typesetting.

Generally, we take examples from V2/700.tsv, but on occasion we find it necessary to go outside the `700.tsv` set to illustrate a point, and (very rarely) even outside the V1 file.

Reductivity In many ways, `4lang` is a logical outgrowth of modern, computationally oriented lexicographic work beginning with Collins-COBUILD (Sinclair, 1987), the Longman Dictionary of Contemporary English (LDOCE) (Boguraev and Briscoe, 1989), WordNet (Miller, 1995), FrameNet (Fillmore and Atkins, 1998), and VerbNet (Kipper, Dang, and Palmer, 2000). The main motivation for systematic reductivity was spelled out in (Kornai, 2010a) as follows:

"In creating a formal model of the lexicon the key difficulty is the circularity of traditional dictionary definitions – the first English dictionary, Cawdrey, 1604 already defines *heathen* as `gentile` and *gentile* as `heathen`. The problem has already been noted by Leibniz (quoted in Wierzbicka, 1985):

> Suppose I make you a gift of a large sum of money saying you can collect it from Titius; Titius sends you to Caius; and Caius, to Maevius; if you continue to be sent like this from one person to another you will never receive anything.

One way out of this problem is to come up with a small list of primitives, and define everything else in terms of these."

The key step in minimizing circularity was taken in LDOCE, where a small (about 2,200 words) defining vocabulary called LDV, Longman Defining Vocabulary was created, and strictly adhered to in the definitions with one trivial exception: words that often appear in definitions (e.g. the word *planet* is common to the definition of Mercury, Mars, Venus, ...) can be used as long as their definition is strictly in terms of the LDV. Since *planet* is defined 'a large body in space that moves around a star' and *Jupiter* is defined as 'the largest planet of the Sun' it is easy to substitute one definition in the other to obtain for Jupiter the definition 'the largest body in space that moves around the Sun'.

`4lang` generalizes this process, starting with a core list of defining elements, defining a larger set in terms of these, a yet larger set in terms of these, and so on until the entire vocabulary is in scope. As a practical matter we started from the opposite direction, with a seed list of approximately 3,500 entries composed of the LDV (2,200 entries), the most frequent 2,000 words according to the Google unigram count (Brants and Franz, 2006) and the BNC (Burnard and Aston, 1998), as well as the most frequent 2,000 words from Polish (Halácsy et al., 2008) and Hungarian (Kornai et al., 2006). Since Latin is one of the four languages supported by `4lang`, we added the classic Diederich, 1939 list and Whitney, 1885.

Based on these 3,500 words, we reduced the defining vocabulary by means of a heuristic graph search algorithm (Ács, Pajkossy, and Kornai, 2013) that eliminated all words that were definable in terms of the remaining ones. The end-stage is a vocabulary with the *uroboros property*, i.e. one that is minimal wrt this elimination process. This list (1,200 words, not counting different senses with multiplicity) was published as Appendix 4.8 of S19 and was used in several subsequent studies including (Nemeskey and Kornai, 2018). (The last remnant of the fact that we started with over 3k words is that numbers in the 5th column are still in the 1-3,999 range, as we decided against renumbering the set.) This '1200' list is part of Release V1 of 4lang on github, and has bindings to Release 2.5 of Concepticon (List, Cysouw, and Forkel, 2016).

By now (Release V2), this list has shrunk considerably, because improvements in the heuristic search algorithm (see Ács, Nemeskey, and Recski (2019) and uroboros.py) and a systematic tightening of 4lang definitions by means of def_ply_parser.py made further reductions possible. The name of the '700' list is somewhat aspirational (the Version 2 file has 739 words in 776 senses) but we believe the majority of the 359 senses marked e are indeed eliminable, and the eventual uroboros core (p and c entries) will be below 200 senses. With every substitution, we decrease the sparseness of the system. In the limiting case, with a truly uroboros set of maybe 120 elements, we expect the definitions to become much longer and more convoluted. This phenomenon is very observable in the Natural Semantic Metalanguage (NSM) of (Wierzbicka, 1992; Wierzbicka, 1996; Goddard, 2002), which in many ways served as an inspiration for 4lang.

The two theories, while clearly motivated by the same goal of searching for a common universal semantic core, differ in two main respects. First, by using English definitions rather than a formal language, NSM brings many subtle syntactic problems in tow (see Kornai (2021) for a discussion of some of these). Second, NSM is missing the reduction algorithm that 4lang provides. In brief, for any sense of any word we can look up the definition in a dictionary, convert this definition to a 4lang graph that contains only words from the LDV, and for any LDV word we can follow its reduction to V1, and further, to V2 terms. Preliminary work on V3 suggests that it will still have about twice as many primitives than the 63 primes currently used in NSM.

Indeed, just by looking at an ordinary English word such as *random* (see S19:Ex.° 4.21) we are at a complete loss how to define it in terms of the NSM system beyond the vague sense that the prime MAYBE may be involved. With 4lang , we start with 'aimlessly, without any plan' (LDOCE). We know (see 6.4) that *-ly* is semantically empty, and that *-less* is to be translated as lack stem_. Further, from 4.5 we know that *any* is defined as <one>, =agt is_a, so that *any plan* is defined as <one> plan. Since here neither the presence of *one* not its absence (see Rule 6 of 1.6 that the ⟨⟩ signify optionality) adds information, we have lack aim, lack plan. At this point, all defining terms are there in the (V2) core vocabulary, we are done.

Perhaps someone with deeper familiarity with NSM could concoct a definition using only the primes, though it appears that none of the 63 primes except WANT seem related to aims, goals, plans, or any notion of purposive action. To the extent that Gewirth, 1978 includes 'capability for voluntary purposive action' as part of the definition of what defines a human as a 'prospective purposive agent', this lack of defining NSM terms is highly problematic, placing the people whose language is describable in purely NSM terms on the level of infants with clear wants but no agency to plan. But our issue is a more general one: it is not this particular example that throws down the gauntlet, it is the lack of a general reduction algorithm.

In contrast, since at any stage the uroboros vocabulary is obtained by systematic reduction of a superset of the LDV, it is still guaranteed that every sense of every word listed in LDOCE (over 82k entries) are definable in terms of these. Since the defining vocabularies of even larger dictionaries such as Webster's 3rd (Gove, 1961) are generally included in LDOCE, we have every reason to believe that the entire vocabulary of English, indeed the entire vocabulary of any language, is still definable in terms of the uroboros concepts.

Redefinition generally requires more than string substitution. Take again PLANET, a word LDOCE uses in the same manner as NSM uses semantic molecules, and defines as 'a large body in space that moves around a star'. If we mechaically substitute this in the definition of *Jupiter*, 'the largest __ of the Sun' we obtain 'the largest a large body in space that moves around a star of the Sun'. It takes a great deal of sophistication for the substitution algorithm to realize that *a large* is subsumed by *the largest* or that *a star* is instantiated by *the Sun*. People perform these operations with ease, without conscious effort, but for now we lack parsers of the requisite syntactic and semantic sophistication to do this automatically. Part of our goal with the strict definition syntax that replaces English syntax on the right-hand side (rhs) of definitions is to study the mechanisms required by an automated parser for doing this, see Chapter 2.

Encyclopedic knowledge In light of the foregoing, the overall principle of keeping linguistic (lexicographic) knowledge separate from real-world (encyclopedic) knowledge is already well motivated. First, universality demands a common lexical base, whereas it is evident that real-world knowledge differs from culture to culture, and thus from language to language – in the limiting case, it differs within the same culture and the same language from period to period. Since the completion of the Human Genome Project in 2003, our knowledge of genes and genomes have exploded: at the time of this writing the Cancer Genome Atlas holds over 2.5 petabytes of data, yet the English language is pretty much the same as it was 20 years ago. The need to keep two so differently growing sources of knowledge separate is obvious.

Second, reductivity demands that knowledge be expressed in words. This may have made sense for biology two hundred years ago (indeed, biological taxa are traditionally defined by means of the same Aristotelian technology of *genus* and *differentia specifica* (S19:2.7) that we rely on), but clearly makes vanishingly little sense in chemistry, physics, and elsewhere in the sciences where knowledge is often expressed by a completely different language, that of mathematics. As we shall see in Chapter 8, trivia like *Who won the World Series in 1967?* are within scope for the 4lang Knowledge Representation (KR) system. But core scientific statements, from the Peano Axioms (see 3.4) to Gauss' Law of Magnetism, $\nabla \cdot \mathbf{B} = 0$, are out of scope.

How are the lines to be drawn between lexical and encyclopedic, verbally express-ible and mathematics-intense knowledge? This is a much debated isse (see Peeters, 2000 for a broad range of views) and `4lang` clearly falls at the Aristotelian end of the dual-ist/monist spectrum introduced in Cabrera, 2001. We begin our discussion with a simple item. The first edition of LDOCE (Procter, 1978) defines *caramel* as 'burnt sugar used for giving food a special taste and colour'. In `4lang` this could be recast as

```
caramel sugar[burnt], cause_ {food has {taste[special],
colour[specal], <taste[sweet]>, <colour[brown]>}}
```

where quite a bit of the syntax is implicit, such as the fact that `caramel` is the subject of `cause_`, see Section 1.3, and we sneaked in some real world knowledge that the special taste is (in the default case) sweet, and the special color is brown.

`special`   As the preceding make clear, we could track further *special* (defined in `4lang` as `lack common`), or *food*, or *burnt*, or any term, but here we will concentrate on *sugar* 'a sweet white or brown substance that is obtained from plants and used to make food and drinks sweet'. Remarkably, this definition would also cover xylitol ($CH_2OH(CHOH)_3CH_2OH$) or stevia ($C_{20}H_{30}O_3$) which are used increasingly as replacements for common household sugar ($C_6H_{12}O_6$).

This is not to say that the editors should have been aware in 1978 that a few decades later their definition will no longer be specific enough to distinguish sugar from other sweeteners. Yet the clause 'obtained from plants' is indicative of awareness about sac-charine ($C_7H_5NO_3S$) which is also sweet, but is not obtained from plants.

`4lang` takes the line that encyclopedic knowledge has no place in the lexicon. In-stead of worrying about how to write clever definitions that will distinguish sugar not just from saccharine but also from xylitol, stevia, and whatever new sweeteners the fu-ture may bring, it embraces simplicity and provides definitions like the following:

```
rottweiler dog
greyhound dog
```

This means that we fail to fully characterize the competent adult speaker's ability to use the word *rottweiler* or *greyhound*, but this does not seem to be a critical point of language use, especially as many adult speakers seem to get along just fine without a detailed knowledge of dog breeds. To quote Kornai, 2010a:

> So far we discussed the *lexicon*, the repository of linguistic knowledge about words. Here we must say a few words about the *encyclopedia*, the repository of world knowledge. While our goal is to create a formal theory of lexical defini-tions, it must be acknowledged that such definitions can often elude the grasp of the linguist and slide into a description of world knowledge of various sorts. Lexicographic practice acknowledges this fact by providing, somewhat begrudg-ingly, little pictures of flora, fauna, or plumbers' tools. A well-known method of avoiding the shame of publishing a picture of the yak is to make reference to

`Bos grunniens` and thereby point the dictionary user explicitly to some encyclopedia where better information can be found. We will collect such pointers in a set **E**

Today, we use Wikipedia for our encyclopedia, and denote pointers to it by a prefixed @ sign, see Section 1.3. Our definitions are

```
sugar cukor saccharum cukier  440 N
     material, sweet, <white>, in food, in drink
sweet eldes dulcis    sllodki 495 A
     taste, good, pleasant, sugar has taste, honey has taste
```

Instead of sophisticated scientific taxonomies, `4lang` supports a naive world-view (Hayes, 1979; Dahlgren, 1988; Gordon and Hobbs, 2017). We learn that *sugar* is sweet, and *sweet* is_a taste – the system actually makes no distinction between predicative (is) and attributive (is_a) usage. We learn that sugar is to be found in food and drink, but not where exactly. In general, the lexicon is restricted to the core premises of the naive theory. When in doubt about a particular piece of knowledge, the overriding principle is not whether it is true. In fact the lexicon preserves many factually untrue propositions, see e.g. the discussion in 3.1 of how the heart is the seat of love. The key issue is whether a meaning component is learnable by the methods we suggest in 5.3 and, since these methods rely on embodiment, a good methodological guideline is 'when in doubt, assign it to the encyclopedia'.

One place where the naive view is very evident is the treatment of high-level abstractions. For example, the definition of *color* has nothing to do with photons, frequency ranges in the electromagnetic spectrum, or anything of the sort – what we have instead is `sensation, light/739, red is_a, green is_a, blue is_a` and `colour` when we turn to e.g. *red* we find `colour, warm, fire has colour, blood` `red` `has colour`. Another field where we support only a naive theory is grammar, see 2.5.

As with *sugar* and *sweet*, we posit something approaching a mutual defining relation between *red* and *blood*, but this is not entirely like Titius and Caius sending you further on: actually *blood* gets eliminated early in the uroboros search as we iteratively narrow the defining set, while *red* stays on. Eventually, we have to have some primitives, and we consider *red*, a Stage II color in the (Berlin and Kay, 1969) hierarchy, a very reasonable candidate for a cross-linguistic primitive. In fact, `uroboros.py` is of the same opinion (in no run does *red* get eliminated, hence the marking c (core) in column 7).

So far, we have discussed the fact that separating the encyclopedia from the lexicon leaves us with a clear class of lexical entries, exemplified so far by colors and flavors, where the commonly understood meaning is anchored entirely outside the lexicon. There are also cases where this anchoring is partial, such as the suffix *-shaped*. The meaning of *guitar-shaped, C-shaped, U-shaped,* ... is clearly compositional, and relies on cultural primitives such as *guitar, C, U,* ... that will remain at least partially outside the lexicon. According to Rosch (1975), lexical entries may contain pointers to non-verbal material, not just primary perceptions like color or taste, but also prototypical images. We can say that *guitar* is a stringed musical instrument, or that *C* and *U* are letters of the alphabet, and this is certainly part of the meaning of these words, but it is precisely for the image aspect highlighted by *-shaped* that words fail us. Again anticipating notation that we will fully define only in 2.2, we can define *guitar-shaped* as `has shape, guitar has shape` and in general

-shaped `has shape, stem_ has shape, "_-shaped" mark_ stem_`

and leave it to the general unification mechanism we will discuss in 1.5 and 8.3 to guarantee that it is the same shape that the stem and the denotation of the compound adjective will share.

1.3 The syntax of definitions

Here we discuss, somewhat informally, the major steps in the formal analysis of `4lang` definitions. A standard lex-yacc parser, `def_ply_parser.py` is available on github. The syntax is geared towards *human* readability, so that plaintext lexical entries where the definiens (usually a complex formula) is given after the definiendum (usually an atomic formula) are reasonably understandable to those working with `4lang`. In 1.5 we will discuss in more detail the omission of overt subjects and objects, an *anuvṛtti*-like device, that greatly enhances readability. Here we present a simple example:

```
April    month, follow march/1563, may/1560 follow
bank     institution, money in
```

The intended graph for April will have a 0 link from the definiendum to month, a 1 link to march/1563 and a 2 link to may/1560. Strictly speaking, anuvṛtti removes redundancies across stanzas (sūtras) whereas our method operates within the same stanza across the left- and right-hand sides, but the functional goal of compression is the same.

Often, what is at the other side of the binary is unspecified, in which case we use the gen symbol "plugged up". Examples:

```
vegetable  plant, gen eat
sign gen perceive, information, show, has meaning
```

Thus, *vegetable* is a plant that someone (not specified who) can eat (it is the object of eating, subject unspecified), and *sign* is_a information, is the object of perception, is_a show (nominal, something that is or can be shown) and has meaning.

Starting with 'disambiguated language' (S19:3.7), semanticists generally give themselves the freedom to depart from many syntactic details of natural language. For example Cresswell, 1976 uses

λ-deep structures that look as though they could become English sentences with a bit of tinkering. In this particular work I am concerned more with the underlying semantic structure than with the tinkering.

By aiming at a universal semantic representation we are practically forced to follow the same method, since the details of the 'tinkering' change from language to language, but we try to be very explicit about this, using the `mark_` primitive that connects words to their meanings (see 2.5). One particular piece of tinkering both Cresswell and I are guilty of is permitting semantics to cross-cut syntax and morphology, such as by reliance on a comparative morpheme `er_` (called **er than** in Cresswell, 1976) but really, what can we do? The comparative *-er* is a morpheme used in about 5% of the definitions, and there is no reason to assume it means different things following different adjectival stems.

`mark_`

`er_`

Coordination A `4lang` definition always contains one or more clauses (hypergraph nodes, see 1.5) in a comma-separated list. The first of these is distinguished as the *head* (related to, but not exactly the same as the *root* in dependency graphs). In 1.5 the top-level nodes will be interpreted so as to include graph edges with label 0 running from the definiendum to the definiens. The simplest definitions are therefore of the form x, where x is a single atomic clause. Example

```
aim cell finis cel 363 N
     purpose
```

that is, the word *aim* is defined as *purpose*. Somewhat more complex definitions are given by a comma-separated list:

```
board lap    tabula tablica 456 N
        artefact, long, flat
boat  hajo1 navis  11o1dz1 976 N
        ship, small, open/1814
```

(The number following the '/', if present, serves to disambiguate among various definitions, in this case adjectival *open* 'apertus' from verbal *open* 'aperio'. These numbers are in column 7 of the `4lang` file.) In 1.4 we will discuss the appropriate vector space semantics for coordination of defining properties in more detail, but as a first approximation it is best to think of these as strictly intersective.

Subordination Deefinitions can have dependent clauses e.g. *protect* `=agt cause_ {=pat[safe]}` 'what *X protects Y* means is that *X* causes *Y* to be safe'. Of particular interes are relative clauses, which are handled by unification, without an overt *that* morpheme, e.g. 'red is the color that blood has' is expressed by a conjunction `red is_a color, blood has color` where the two tokens of `color` are automatically unified, see 8.3.

`protect`

External pointers Sometimes (42 cases in the 1,200 concepts published in S19:4.8) a concept doesn't fully belong in the lexicon, but rather in the encyclopedia. In the formal language defined here, such *external pointers* are marked by a prefixed @. Examples:

```
Africa land, @Africa
London city, @London
Muhammad  man/744, @Muhammad
U letter/278, @U
```

These examples, typically less than 5% of any dictionary, are but a tiny sample from millions of person names, geographic locations, and various other proper names. We will discuss such 'named entities' in greater detail in Chapter 8.

Subjects and objects In earlier work, staring with Kornai, 2010a, we linked `4lang` to the kind of graphical knowledge representation schemas commonly used in AI. Such (hyper)graphs have (hyper)edges roughly corresponding to concepts, and *links* connecting the concepts. `4lang` has only three kinds of links marked 0,1, and 2.

0 links cover both predicative *is*, cf. the definition of *sugar* as `sweet, in food, in drink` above, and subsumptive `is_a` which obtains both between hyponyms and hypernyms and between instances and classes. 1 links cover subjects, and 2 links cover objects. We will discuss hypergraphs further in 1.5 and the link inventory in 2.3.

In addition to 0 links, definitions often explain the definiendum in terms of it being the subject or object of some binary relation. In some cases, these relations are highly grammatical, as `for_`, known as "the dative of purpose":

```
handle 834 u N part_of object, for_ hold(object in hand)
```

while in other cases the relation has a meaning that is sufficiently close to the ordinary English meaning that we make no distinction. An example of the latter would be `for` used to mark the price in an exchange as in *He sold the book for $10*, or `has` used to mark possession as in *John has a new dog*. When we use a word in the sense of grammar, we mark this with an underscore, as in `for 2824` versus `for_ 2782`. We defer discussing the distinction between "ordinary" and "grammatical" terms to 2.5, but note here that the English syntax of such terms can be very different from their `4lang` syntax. Compare `-er 14` which is a suffix attaching to a single argument, the stem (which makes it a unary relation), to `er_ 3272` which has two obligatory arguments (making it a binary relation).

Direct predication In a formula `A[B]` means that there is a 0-link from A to B. This is used only to make the notation more compact. The notation B(A) means the same thing, it is also just syntactic sugar. Both brackets and parens can contain full subgraphs.

```
tree plant, has material[wood], has trunk/2759, has crown
```

That trees also have roots is not part of the definition, not because it is inessential, but because trees are defined as plants, and plants all have roots, so the property of having roots will be inherited.

Defaults In principle, all definitional elements are strict (can be defeased only under exceptional circumstances) but time and again we find it expedient to collapse strongly related entries by means of defaults that appear in angled brackets.

```
ride travel, =agt on <horse>, ins_ <horse>
```

These days, a more generalized *ride* is common (*riding the bus, catching a ride, …* so the definition `travel` should be sufficient as is. The historically prevalent mode of traveling, on horseback, is kept as a default. Note that these two entries often get translated by different words: for example Hungarian distinguishes *utazik* 'travel' and *lovagol* 'rides a horse', a verb that cannot appear with an object or instrument the same way as English *ride a bike* can. Defaults are further discussed in 6.4.

Agents, patients The relationship between horseback riding (which is, as exemplified above, just a form of traveling) and its defining element, the horse, is indirect. The horse is neither the subject, not the object of travel. Rather, it is the rider who is the subject of the definiendum and the definiens alike, corresponding to a graph node that has a 1 arrow leading to it from both. This node is labeled by `=agt`, so when we wish to express the semantic fact that Hungarian *lovagol* means 'travel on a horse' we write

```
lovagol travel, =agt on horse
```

Note that the horse is not optional for this verb in Hungarian: it is syntactically forbidden (*lovagol* is intransitive) and semantically obligatory. (Morphologically it is already expressed, as the verb is derived from the stem *ló* 'horse' though this derivation is not by productive suffixation.) Remarkably, when the object is_a horse (e.g. a colt is a young horse, or a specific horse like Kincsem) we can still use *lovagol* as in *János a csikót lovagolta meg* or *Elijah Madden Kincsemet lovagolta*.

For the patient role, consider the word *know*, defined as 'has information about'. For this to work, the expression `x know y` has to be equivalent to `x has information about y` i.e. we need to express the fact that the subject of has is the same as the subject of *know* (this is done by the `=agt` placeholder) and that the object of `about` is the same as the object of knowing – this will be done by the `=pat` placeholder.

As discussed in Kornai, 2012 in greater detail, these two placeholders (or *thematic roles*, as they are often called) will be sufficient, but given the extraordinary importance of these notions in grammatical theory, we will discuss the strongly related notions of thematic relations, deep cases, and *kārakas* in 2.4 further.

More complex notation When using [] or (), both can contain not just single nodes but entire subgraphs. For subgraphs we also use { }, see 1.6.

```
stock re1szve1ny syngrapha papier_wartos1ciowy 3626 N
    document, company has, {person has stock} prove
    {person has part_of company}
```

'stocks are documents that companies have, if a person has stock it proves that a person owns a part of the company'.

1.4 The geometry of definitions

Computational linguistics increasingly relies on *word embeddings* which assign to each word in the lexicon a vector in n-dimensional Euclidean space $\mathbb{R}^n$, generally with $150 \leqslant$

$n \leqslant 800$ (typically, 300). These embeddings come in two main varieties: *static*, where the same vector $\mathbf{v}(w)$ is used for each occurrence of a string w, and *dynamic* (also called *context-sensitive*) where the output depends on the context x_y in which w appears in text. On the whole, dynamic embeddings such as BERT (Devlin et al., 2019) work much better, but here we will concentrate on the static case, with an important caveat: we permit *multi-sense* embeddings where a single string such as *free* may correspond to multiple vectors such as for 'gratis' and 'liber'. Our working hypothesis is that dynamic embeddings just select the appropriate sense based on the context.

Embeddings, both static and dynamic, are typically obtained from large text corpora (billions of words) by various training methods we shall return to in Chapter 8, though other sources (such as dictionaries or paraphrase databases) have also been used (Wieting et al., 2015; Ács, Nemeskey, and Recski, 2019). Most of the action in a word embedding takes place on the unit sphere: the length of the vector roughly corresponds to the log frequency of the word in the data (Arora et al., 2015), and similarity between two word vectors is measured by cosine distance. Words of a similar nature, e.g. first names *John, Peter,...* tend to be close to one another. Remarkably, analogies tend to translate to simple vector addition: $\mathbf{v}(\text{king}) - \mathbf{v}(\text{man}) + \mathbf{v}(\text{woman}) \approx \mathbf{v}(\text{queen})$ (Mikolov, Yih, and Zweig, 2013), a matter we shall return to in 2.3.

For cleaner notation, we reverse the multi-sense embeddings and speak of vectors (in the unit ball) of $\mathbb{R}^n$ that can carry *labels* from a finitely generated set D^* and consider the one-to-many mapping $l : \mathbb{R}^n \to D^*$. We note that the degree of non-uniqueness (e.g. a vector getting labeled both *faucet* and *tap*) is much lower on the average than in the other direction, and we feel comfortable treating l, at least as a first approximation, as a function.

Definition 1. *A voronoid $V = \langle \mathcal{P}, P \rangle$ is a pairwise disjoint set of polytopes $\mathcal{P} = \{Y_i\}$ in $\mathbb{R}^n$ together with exactly one point p_i in the inside of each Y_i.*

In contrast to standard Voronoi diagrams, which are already in use psychological classification (see in particular Gärdenfors, 2000 3.9), here there is no requirement for the p_i to be at the center of the Y_i, and we don't require facets of the polytopes to lie equidistant from to labeled points. Further, there is no requirement for the union of the Y_i to cover the space almost everywhere, there can be entire regions missing (not containing a distinguished point as required by the definition). Given a label function l, if $p_i \in Y_i$ carries the label $w_i \in D^*$ we can say that the entire Y_i is labeled by w_i, written $l(Y_i) = w_i$.

Now we turn to learning. As in PAC learning (Valiant, 1984), we assume that each concept c corresponds to a probability distribution π_c over $\mathbb{R}^n$, and we assume that level sets for increasingly high probabilities bound the prototypical instance increasingly tightly, as happens with the Gaussians often used to model the π_c. An equally valid view is to consider the polytopes themselves as already defining a probability distribution, with sharp contours only if the softmax temperature is low.

It is often assumed in cognitive psychology that concepts such as *candle* are associated not just to other verbal descriptors (e.g. that it is roughly cylindrical, has a wick at the axis, is made of wax, is used on festive occasions, etc.) but also to nonverbal ones, such as a picture of 'the candle' or even the characteristic smell of burning candles. In fact, image labeling algorithms such as YOLO9000 (Redmon et al., 2016) have considerable success in finding things in pictures and naming them, but generating prototypical images remains a research goal even for human faces, where the state of the art is most developed.

Definition 2. *A* linear voronoid *is a voronoid defined by hyperplanes h_j such that every facet of every polytope lies in one of these.*

By adding a hyperplane for each facet of every polytope, every voronoid can be made into a linear one, but our interest is with the sparse case, when many facets, not just those for adjacent polytopes, are on the same hyperplane. Thus we have two objectives: first, to enclose the bulk of each concept set c in some Y_i so that $\pi_c(Y_i)$ is sufficiently close to 1, and second, to reduce the cardinality of the hyperplane set. Each half-space is defined by a normal vector $\mathbf{f}$ and an offset (called the *bias*), and we call these *features* (rather than half-spaces) in keeping with standard terminology in machine learning.

Definition 3. *A* vector $\mathbf{v}$ satisfies *a feature* $\mathbf{f}$ *iff* $\langle \mathbf{v}, \mathbf{f} \rangle > b$

Since our central interest is with just one half-space to the exclusion of the other (see Chapter 4), we orient the normal vector so that a feature takes positive value in this affine half-space. Note that a normed vector has $n-1$ free parameters and the bias adds the nth, so feature vectors are not qualitatively different from word vectors. So that we don't have to move to a dual space we will also call the positive half-spaces features, and denote them by F_j.

Now we can restate our sparsity goal as finding features $F_1, \ldots, F_k$ so that all polytopes can be defined by the intersection of a few of these. We leave open the possibility $k > n$, i.e. that the system of features is *overcomplete*. As a practical matter, models with $n = 300$ work reasonably well, while we expect k to be in the 500–1200 range. What we are looking for is a finite system $\mathcal{F} = \{F_1, \ldots, F_k\}$ such that each of the Y_i is expressible as a sparse vector with nonzero (positive) elements only on a few (in practice, less than 10) coordinates.

Remarkably, these simple (and in case of Def. 3, completely standard) definitions are already sufficient for a rudimentary theory of communication. Assume two parties, a speaker and a hearer. They both have *mental spaces*, a place where they store not just words and other linguistic expressions, but also concepts, sensory memories, things that philosophers of language would generally treat as sortally different. The term is chosen to express our indebtedness to (Fauconnier, 1985; Talmy, 2000) and the entire loosely connected school of Cognitive Linguistics, but we don't use 'mental space' in exactly the same way as Fauconnier, especially as we are modeling it by ordinary n-dimensional Euclidean space.

Ideally, the speaker and the hearer share the same voronoids, and simple ideas or sensations can simply be communicated by uttering the label of the polytope where it falls: I see a candle, and say *candle*. This is sufficient for the hearer to know which polytope was meant, and thereby gain some rough understanding of my mental activity. In reality, both speakers and hearers are aware that their mental spaces are not identical: my notion of a candle can differ from yours in ways that may be significant. But day-to-day communication is seldom hindered by this, by asking for a fork I'm unlikely to be handed a spoon. This is not because our $Y_{\texttt{fork}}$ polytopes have identical boundaries, but rather because the boundaries cover so much of the $\pi(\texttt{fork})$ probability mass that the symmetric difference between the polytopes of speaker and hearer is negligible.

The same logic extends to the vexing cases of hyperintensionals (Cresswell, 1975), phrases that describe contents that are not instantiated at all. I can speak of a pink elephant, and anybody who understands English understands what I mean with the same degree of (im)precision as they understand 'pink' and 'elephant'. Putting these two polytopes together just gives us their intersection, which works quite well even though in the real world this intersection happens to be empty. Note that the intersection can be empty even where there is no counterfactuality involved: a *former president* is by definition not a president, and at any rate it seems hard to maintain a subset of the space that contains former things. Since *former x* means 'was x, no longer x' i.e. a change of its *x*-ness, the point under discussion is one that has left Y_x.

In logical semantics it is a standard assumption that *extensions* of words, here modeled by polytope volumes, are changing with time. If I decide to paint a formerly black wall white, the meanings of *black* and *white* (standardly modeled by an indexed set of extensions e_λ, with the indexes running over the class of 'possible worlds' and called the *intension* of a word) remain constant, it is just their extensions that change with λ. We will assume a discrete time index t and require only three values 'before', 'now', and 'after'. We will discuss temporal semantics in greater detain in 3.2 – here we will simply assume three voronoids V_b, V_n, V_a and consider *former* an operator that effects a change from the identically labeled polytope, say Y for 'president' that somehow moves a point corresponding to the subject, say *Obama*, from the interior of Y in V_b to the exterior in V_n.

We have in both of these models a vector $\mathbf{p}$ corresponding to *president* and a vector $\mathbf{O}$ corresponding to *Obama*. The key insight is that not only do these vectors remain static, but the polytope Y that surrounds $\mathbf{p}$ also remains unchanged. What changes is the scalar product: in V_b we had $\langle \mathbf{O}, \mathbf{p} \rangle > b$ and in V_n we have $\langle \mathbf{O}, \mathbf{p} \rangle < b$. It is not that the threshold for presidency b has changed: what changed is the definition of the scalar product. We will assume the standard basis for V_n, but some B (before) basis in V_b, some A (after) basis in V_a, and use $\langle \mathbf{O}B, \mathbf{p}B \rangle > b$ conjoined with $\langle \mathbf{O}, \mathbf{p} \rangle < b$ to express the meaning of *former*. We return to scalar products in Chapter 2.3, but note in advance that we follow the literature in being a bit more loose in terminology than is common in mathematics: we will use *basis* also for generating systems that are not necessarily linearly independent, and *scalar product* also for bilinear forms that are not necessarily symmetrical.

The geometric model offers its own sortal types: vectors, half-spaces, polytopes, matrices, and so on. We will link these up to the lexical categories of 4lang in 2.1, but to build intuition we list some of the key correspondences here. Proper names are points, a matter we will discuss in greater detail in Chapter 8. This doesn't mean that all points p in concept space receive a label l that is a proper name, but by and large, all things can be named (have a proper name assigned to them), not just people, pets, or boats. Adjectives are typically half-spaces, with gradient effects modeled by the bias term, whereas common nouns are often polytopes (finite intersections of half-spaces) or projections thereof. Verbs, including the copula, carry time information, and their description often involves not just V_n, but also V_a and/or V_b as well.

Fig. 1.1: Dependence of voronoids on metric chosen

Note that any set of vectors defines its own voronoid, but the boundaries of the cells depend on the metric chosen. This is illustrated in Fig. 1.1, which was generated using http://yunzhishi.github.io/voronoi.html. Since the probability mass is near the center, exact placement of the boundaries is of little interest.

We will use voronoids to represent the nominal aspects of *conceptual schemas*, compact configurations of knowledge pertinent to some domain. With the addition of verbal information (in particular, timing, see 3.2) these schemas become a linear algebraic version of Schankian scripts. As an example, consider the exchange_ schema, roughly depicted in Fig. 1.2.

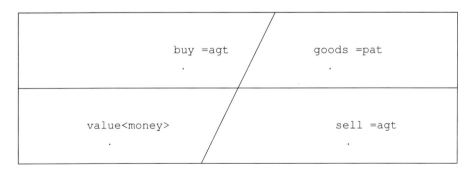

Fig. 1.2: exchange_

The words used are highly evocative: if we hear *sell* we automatically typecast the subject in the seller role, and the object in the 'goods' role. If we choose *buy*, the subject is bound to the buyer role, and again the object of buying is treated as the goods. Before the exchange, the seller has the goods, and the buyer the money, afterwards the buyer has the object and the seller the money. This analysis (similar to the one proposed in Hovav and Levin, 2008) is easily implemented as hypergraph unification (see 1.5), and also in the vector calculus we are using, but we defer the details to 3.2, where we discuss handling the temporal aspects `before` and `after`.

While unification proceeding from the keywords *buy* or *sell* proceeds naturally, the word *goods*, rarely used outside the context of shipping/insurance contracts, is quite a bit less evocative in English, and is really used just for want of a better term. The same can be said for the word *money*, even though the association is strong, buying is what money is for, and selling is what earns money. (Also, in the full lexical entries for *buy/sell*, `money` is merely a default: clearly goods can exchanged for services and other things of value.) Typically, we invoke the schema from the perspective of the controlling participants, the potential agents, though alternatives like promoting the money to the agent role, *In this village, ten thousand will buy you a beautiful house*, are often feasible.

In Fig. 1.3 we depict the two simplest schemas. The left panel shows a voronoid with a single region labeled, for want of a better name, `one`. Since this encompasses everything, we could have called it `all` or `whole` just as well. The ambiguity between *one* and *all*, reminiscent of the first basic principle of Plotinus "the One" (or "the Good"), will not play the same generative role here as with Plotinus, and we will also refrain from entertaining analogies between the right panel and Gnostic thought.

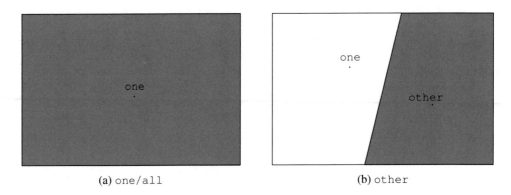

(a) `one`/`all`   (b) `other`

Fig. 1.3: `one` and `other`

The type difference between our first quantifier, `gen`, defined simply as a vector with the same value $1/d$ on each component, and `all`, is very clear. `gen` is simply a nominal, whereas `all` is a schema that requires implicit or explicit typecasting: as in *all books are for sale*, where *all* is already limited to *in this store* (Kornai, 2010b).

The same difficulty of naming certain regions of the voronoid, a problem we already encountered with `goods`, is manifest on the white side of the right panel of Fig. 1.3. The blue side directly defines `other`, but whatever is on the white side is typecast to one. (Numerical *one* is defined in opposition to *more*, and has little to do with the `one` in either part of the figure.) To keep such technical names distinct from the vocabulary, we will suffix them with _. The names of schemas (voronoids defined by unordered sets of vectors) will be enclosed in the same curly braces { } as more ad hoc statements like `{person has stock}` in 1.3 above. The graph-theoretic view, where schemas are simply hypernodes, does not do full justice to schemas as information objects – we will discuss this problem in 1.6.

To summarize the key geometric ideas, most words (proper names in particular, but also common nouns, adjectives, and verbs) correspond to vectors or polytopes with distinguished vectors. We can also compute vectors for adpositions and other function words like *be* that we will shortly turn to, but we actually consider these to be matrices (relations between vectors). Consider *bark*. For us, this is a pair of vectors $bark_1$ 'cortex' and $bark_2$ 'latrat(us)' indistinguishable without context, be it morphological (*barked, barking* can only refer to dog bark) or larger (*birch bark* can only refer to tree bark). When we assign just one vector to this, this is just the log frequency weighted sum of the two vectors corresponding to the two senses, sitting in polytopes Y_1 and Y_2 that are not even adjacent in concept space.

In terms of the distinguished points it is easy to tell them apart: Y_2 falls in the `sound` subset since $bark_2$ is defined as `sound[short, loud]`, `<dog>` make, whereas Y_1 is not a sound. The separating surface is not unique, $bark_1$ is some kind of covering that trees have, and as such, it is an *object* defined by the cluster of properties that physical objects have: `thing`, `<has colour>`, `has shape`, `has weight`, `<has surface>`, `has position`, `<lack life>`, and clearly dog barks have none of these, so any of these surfaces can be used to separate the two polytopes.

`bark/2517`

`object`

The one vector for *bark* that we obtain from running GloVe (Pennington, Socher, and Manning, 2014), word2vec (Mikolov et al., 2013), or any of the other algorithms must be related to the $bark_1$ and $bark_2$ vectors by addition, weighted by log frequency (Arora et al., 2015). How is this differentiated from cases like `boat` being defined as `ship, small, open`? In other words, how do we know that *bark* contains two vectors corresponding to two distinct senses, while *boat* contains only one, corresponding to a single unified sense? The answer is that this fact can't be read off of the vectors themselves, but can be read off the polytopes: in the *bark* case we have two, but in the *boat* case only one polytope. This is actually a key distinguisher between the more common variety of vector semantics that relies on word vectors directly and the variety that is presented here, since without polytopes the 'raw' vectors for homonymous and polysemous cases are indistinguishable.

A related question is how to distinguish the head from the subordinate elements in a definition: how would the definition `ship, small, open` differ from `open, small, ship`? Here we could rely on the fact that addition, after softmax, is not associative: $\sigma(\mathbf{a} + \sigma(\mathbf{b} + \mathbf{c})) \neq \sigma(\sigma(\mathbf{a} + \mathbf{b}) + \mathbf{c})$, in fact it is the term added last that would receive the greatest weight. More important is the observation that in this definition, `open` really means it lacks a deck (while an *open bottle* lacks a cork, and an *open letter* lacks the privacy protection offered by an envelope) so we have 'open in the way ships can be open' so a more pedantic definition would be 'ship, open as ship' and of course 'small as ship' for a boat is quite large on a default (human) scale. This gives `ship` a weight of 3, `open` and `small` a weight of 1 each. After softmax ($\beta = 1$) this becomes (0.787,0.106,0.106).

`London`   In general, we will assume that the head carries larger weight than the modifiers. This is especially clear in definitions such as *London* as `city,` `@London`. People, not just readers of the Wikipedia article that `@London` points to, but all competent speakers of English, have a wealth of information about London. Much of this information (e.g. images of Tower Bridge, Beefeaters, Parliament, ...) is non-linguistic (not pertinent to grammar), and the projection on the subspace L is dominated by one component (one-hot) on `city`.

The key link type in the algebraic (hypergraph) description we now turn to is the type 0 (is, is_a) link, which simply corresponds to set-theoretic containment: if A, as subset of $\mathbb{R}^n$ is contained in B, we say that

$$l(A) \text{ is\_a } l(B) \tag{1.1}$$

In the algebraic representation lexemes, and larger sentence representations, are hypergraphs, hypergraph unification is a well-defined symbol-manipulation operation, and such symbol manipulation can be performed by neural nets (Smolensky, 1990). In 2.3 we will present a more direct, geometric description in terms of a simple eigenspace model, keeping in effect only the linear and the quadratic terms from the full generality of the tensor model. This will answer a whole set of vexing problems, such as defining the meaning of *be*, where even the magnificent LDOCE resorts to circularity, offering the following senses:

1. used with a present participle to form the tenses of verbs
2. used with past participles to form the passive
3. used in sentences about an imagined situation
4. used in sentences to introduce an aim when you are saying what must be done in order to achieve it
5. used instead of 'have' to form the tense of some verbs
6. used to say that someone or something is the same as the subject of the sentence
7. used to say where something or someone is
8. used to say when something happens
9. used to describe someone or something, or say what group or type they belong to
10. to behave in a particular way
11. used to say how old someone is
12. used to say who something belongs to
13. used to talk about the price of something
14. to be equal to a particular number or amount
15. to exist

We emphasize that we are not singling out LDOCE for unfair treatment here. The online Cambridge Dictionary has a very similar assortment of 'used to' definitions:

1. used to say something about a person, thing, or state, to show a permanent or temporary quality, state, job, etc. *He is rich. It's cold today. I'm Andy. That's all for now. What do you want to be (= what job do you want to do) when you grow up? These books are (= cost) $3 each. Being afraid of the dark, she always slept with the light on. Never having been sick himself, he wasn't a sympathetic listener. Be quiet! The problem is deciding what to do. The hardest part will be to find a replacement. The general feeling is that she should be asked to leave. It's not that I don't like her - it's just that we rarely agree on anything!*

2. used to show the position of a person or thing in space or time *The food was already on the table. Is anyone there? The meeting is now (= will happen) next Tuesday. There's a hair in my soup.*

3. used to show what something is made of *Is this plate pure gold? Don't be so cheeky! Our lawyers have advised that the costs could be enormous. You have to go to college for a lot of years if you want to be a doctor. Come along - we don't want to be late! Oranges, lemons, limes and grapefruit are types of citrus fruit.*

4. used to say that someone should or must do something *You're to sit in the corner and keep quiet. Their mother said they were not to (= not allowed to) play near the river. There's no money left - what are we to do?*

5. used to show that something will happen in the future *We are to (= we are going to) visit Australia in the spring. She was never to see (= she never saw) her brother again.*

6. used in conditional sentences to say what might happen *If I were to refuse they'd be very annoyed. (formal) Were I to refuse they'd be very annoyed.*

7. used to say what can happen *The exhibition of modern prints is currently to be seen at the City Gallery.*

8. to exist or live *(formal) Such terrible suffering should never be. (old use or literary) By the time the letter reached them their sister had ceased to be (= had died).*

More traditional dictionaries, such as *Webster's New World* (Guralnik, 1958), use even more vague terms in the definition, such as 'used to express futurity, possibility, obligation, intention, etc'; *The Concise Oxford* (McIntosh, 1951) has, distributed among several senses, 'exist, occur, live, remain, continue, occupy such a position, experience such a condition, have gone to such a place, busy oneself so, hold such a view, be bound for such a place, belong under such a description, coincide in identity with, amount to, cost, signify'. A more unified treatment seems warranted, and will in fact be provided in 2.3.

1.5 The algebra of definitions

The method of capturing meaning by definitions is at the heart of our undertaking: each definition (line in the dictionary) corresponds to an equation or inequality in the overall system that determines the meaning of each part. Of the three methods discussed here, compositional semantics has long been dominated by the formulaic approach. This approach would have to be coupled to a theory of *grounding* and a theory of *meaning postulates* to fulfill its promise (see S19:3.7-8 for details) and we will not spend any time trying to turn our algebraic formulas into formulas of logic.

The use of (hyper)graphs is an algebraic method on its own, one that can be matched to the compositional manner in which we build the formulas by means of parallel synchronous rewriting. When it comes to detaching meaning representations from linear ordering, graphs are particularly useful, but to take full advantage of them we will need a workable definition of a 'well-formed hypergraph'. To this end, let us first recapitulate the syntax of the definitions we surveyed 1.3 in context-free rules.

1. Definition → Definiendum Definiens (% Comment)
2. Definiendum → Atom
3. Definiens → MarkedClause (',' MarkedClause)*
4. Comment → (ArbitraryString)
5. MarkedClause →DefaultClause|PositionClause|ComplexClause|Clause
6. DefaultClause →'⟨'Clause'⟩'|λ
7. PositionClause → PositionMarker mark_ UnaryAtom
8. ComplexClause → {Definiens}
9. PositionMarker → '"'SuffixMarker|PrefixMarker|InfixMarker'"'
10. Atom → PlainAtom|NumberedAtom|ExternalAtom|PositionMarker
11. NumberedAtom → PlainAtom'/'Number
12. ExternalAtom → '@'WikipediaPointer
13. PlainAtom → UnaryAtom|BinaryAtom
14. UnaryAtom → `Asia|acid|...|yellow|young|=agt|=pat`
15. BinaryAtom → `at|between|cause_|er_|follow|for_|from|has|in| ins_|is_a|lack|mark_|on|part_of|under`
16. Clause → 0Clause|1Clause|2Clause|FullClause
17. 0Clause → Atom'['Definiens']'|Atom'('Definiens')'|Atom
18. 1Clause → BinaryAtom Clause
19. 2Clause → Clause BinaryAtom
20. FullClause → ComplexClause BinaryAtom ComplexClause

As usual in syntax definitions, | in a rule indicates choice and () indicates optionality. (This is the metalanguage: in the language itself we use angled brackets to denote optional parts of definitions, see Rule 6.) This way, 1. abbreviates two rules, one containing no comment and the other containing a Comment after the % sign, which can be expanded to an arbitrary string by Rule 4. Needless to say, comments are irrelevant for the emerging representations, and in the system of parallel synchronized rewriting that we will turn to in 1.6 rules governing the comments will be discarded.

In regards to Rule 2, it should be noted that Atom is intended in the sense of 'dictionary entry' and may include expressions such as *I beg your pardon* which have a unitary meaning 'please repeat what you just said' quite disinct from their compositional sense. The intuition is the same as with lexemes (cf. S19:3.8,4.5) both in linguistics and lexicographic practice: different senses e.g. for *chrome*$_1$ 'hard and shiny metal' and *chrome*$_2$ 'eye-catching but ultimately useless ornamentation, especially for cars and software' often correspond to different words in another language (and when they systematically fail to, we begin to suspect that the purported senses are not distinct after all).

The right-hand side of a definition, the Definiens, is given as one or more marked clauses. The marking can be for defaults, marked by $\langle\rangle$, see Rule 6; for position (within word, or more rarely, among words), marked by doublequoted material, see Rule 7 and 2.2; for complexity, set-theoretical comprehension of several elements, marked by {}; or it may not be marked for any of these, yielding a clause. (As a practical matter, less than 20% of `4lang` defining clauses are marked.)

In a similar manner, we differentiate between ordinary (plain) atoms, and those that are numbered for disambiguation using Rule 11. NumberedAtoms are there simply to provide the same kind of sense disambiguation that lexicographers generally do by subscript numbering, except that we find it expedient to keep the index set 1–3,999 fixed, rather than restarting indexing for each (English) word. For example, we define *set/2746* somewhat similarly to mathematical sets as `group, has many(item),` `together, unit, item has common(characteristic)` but *set/2375* as `=agt cause_ {=pat at position[<stable>,<pro-per>]}`.   set/2746 set/2375

In a more hardcore system we could keep only the numbers: the words are there only to help with human readability. It is a historical accident that English uses the same syllable for both `2746` and `2375`, but from the Hungarian-Latin-Polish bindings it is evident that `kollekcio1 classis kolekcja` and `tesz pono kl1as1c1` are not the same thing. (In this particular case this would also follow from their lexical categories, see 2.1, but these are never used for disambiguation.) Generally we suppress the disambiguation indexes, but note that the ambiguity of English *set* cannot be expressed by making these optional: whenever there is more than one lexical entry with the same English printname, disambiguation numbers are obligatory (as the true heads of the NumberedAtom construction, they are the only obligatory part).

Another kind of specially marked atom is provided in Rule 12 by pointers to the encyclopedia. These are given in abbreviated style: for example the *Asia* in `@Asia` corresponds to `https://en.wikipedia.org/wiki/Asia`. Finally, the use of position markers, doublequoted strings with an explicit insertion locus marker __ that shows whether the definiendum is prefixed, suffixed, or infixed (Rule 9) is no more than a simple workaround to make sure semantics doesn't get entangled in all the technical issues of morphophonology (see 2.2 and 2.5 for further discussion).

Rule 3 is the one we started out with: a definition is the comma-separated conjunction of one or more (marked or unmarked) Clauses. Only a quarter of the definitions have four or more conjunct clauses, and over a quarter have only one, the average number of clauses is 2.68. To understand the internal structure of a clause, we need to look more closely at the alternatives in Rule 16. 0Clauses are elementary predicates, and ComplexClauses can be pretty much anything a definiens can be. The 1Clause and 2Clause constructs serve to help make the syntax human-readable, at least for those humans who are comfortable with SVO word order. Take something like

```
blood ve1r sanguis krew 2599 N
        liquid, in body, red
```

The first clause simply says that blood is a `liquid`, and the third one says it is (or is_a, `4lang` makes no distinction) `red`. In the middle we find a 1Clause (subject clause, Rule 18) that puts `blood` to the left, and `body` to the right of a relational predicate `in`, guaranteeing that `blood in body` is part of the definition of `blood`, without making it appear in the definiens. 2Clauses (object clauses, Rule 19) behave similarly:

```
mud sa1r lutum bl1oto 2056 N
    substance, wet, earth, soft, sticky, water in
```

abbreviating `water in mud` which makes clear that it is mud that contains water, not the other way around. Relational elements are discussed further in 2.3, but Rule 15 makes clear that they come from a small, closed list containing only 16 elements. Similarly, unary atoms come from the closed list given in the Appendix, which represents considerable reduction compared to the list of 1,200 elements in S19:4.8.

The method of having implicit elements in a rule harkens back to the Pāṇinian device of *anuvṛtti* (see Kornai, 2007 7.3.1 for a brief description, and Joshi and Bhate, 1984 for a full treatment). For Pāṇini the goal of anuvṛtti is to enhance brevity in order to lessen the effort to memorize (improve human recallability), while here the shortening of definitions enhances human readability. For simplicity, Release V2 of `4lang` provides both a more machine-readable expanded version, and a more human-readable compacted one, with software to create each from the other, see 9.5.

Definition 4 An (edge-labeled, finite) *hypergraph* with an alphabet (label set) Σ, a (finite) vertex set V, and (finite) hyperedge set E is defined by a mapping att:$E \rightarrow V^*$ that assigns a sequence of pairwise distinct attachment nodes att(e) to each $e \in E$ and a mapping lab:$E \rightarrow \Sigma$ that labels each hyperedge. The size of the sequence att(e) is called the *type* or *arity* of the label lab(e). As Eilenberg machines (S19:Def.4.4) come with input and output mappings, hypergraphs come with a sequence of pairwise distinct *external nodes* denoted 'ext'. This sequence may be empty, a choice that makes the more standard notion of hypergraphs a special case of our definition.

While the definition of hypergraphs stated above is reasonably standard, and it enables hooking up our machinery with that of s-graph grammars (Courcelle and Engelfriet, 2012; Koller, 2015) by means of synchronized string and hypergraph rewriting in 1.6, in `4lang` we concentrate on a simpler class of (hyper)graphs we will call *hypernode graphs* or *RDF graphs* or just `4lang` graphs.

Definition 5 A `4lang` graph or *hypernode graph* contains only ordinary directed edges (arrows) between a starting and an endpoint, these can be labeled 0,1, or 2, no other edge labels (colors) are countenanced. The hypernodes are ordered triples (`x,y,z`) where `x` or `z` may remain empty, As in the Resource Description Framework, members of the triple are called the 'subject', 'predicate', and 'object' of the triple. Subjects and objects (but not predicates) can themselves be `4lang` graphs.

This definition is again supported by a series of syntactic conventions to support human readability. Edge type 0 is used both for attribution *John is brave* and for IS_A indiscriminately. In larger graphs, we will write dashed arrows, $-\rightarrow$ instead of $\xrightarrow{0}$. Edge type 1 has the type number suppressed, we write $\rightarrow$ rather than $\xrightarrow{1}$. Finally edge type 2 will be depicted by a dotted arrow $\cdots\!\!>$ rather than $\xrightarrow{2}$.

In triple notation, $x \leftarrow y$ can be written as [`x,y,`], and $y \cdots\!\!> z$ can be written as [`,y,z`]. A full triple [`x,y,z`] could be depicted as $x \leftarrow y \cdots\!\!> z$. For ease of presentation, we introduce a special symbol @ (not to be confused with the external pointer delimiter of Rule 12 above) that will be placed in the middle of edges that should *in their entirety* be the terminal point of some other edge. Consider the sentence *video patrem venire* traditionally analyzed in Latin grammar with an infinitival object, meaning that the object of seeing is neither the father, nor his coming, but rather the entire 'coming of father'. An English translation could be *I see father's coming* or even *I see father coming*.

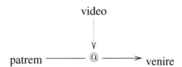

Fig. 1.4: *Video patrem venire*

In Chapter 5 this kind of graph structure will be further enriched by mappings from graph (hyper)nodes and (hyper)edges to small discrete partially or fully ordered sets.

Definition 6 A *valuation* is a partial mapping from some elements (both nodes and edges) of a hypergraph to a finite poset.

We will see in Chapter 2 in far greater detail how morphology and syntax are handled by the same mechanism, and here we omit the details of how syntax and morphological analysis of Latin sentences ordinarily proceeds hand in hand (S19:5.3).

Semantically, we have two units `father`, and `come`, the former being the subject of the latter. This is expressed in `4lang` syntax by `father[come]` or `come(father)`, keeping alive both function-argument alternatives explored in early Montague Grammar. Since this entire clause is the object of seeing, the whole sentence can be written as (1stsg) see father[come]. We parenthesized the 1st singular pronoun, not overt in the Latin original, but inferable for the conjugated form of the verb, in anticipation of a fuller discussion of pronouns in 3.3. In RDF-style triple notation we have [I,see,[father,come,]].

In terms of hypergraphs, we can consider father a single (atomic) vertex, but in light of the 4lang definition

```
father apa pater ojciec 173 N
        parent, male
```

we are equally free to consider it a small hyperedge containing two vertices parent and male. We do not fully explore the hypergraph connection here (see S19:4.1, Nemeskey et al., 2013, Ács and Recski, 2018, and 7.4 for further discussion) but we note that our concept of "doing grammar by spreading activation" is almost identical to that of Jackendoff and Audring, 2020 7.2.3. This is not at all surprising, as they both go back to the same ideas (Quillian, 1969; Collins and Loftus, 1975), but it is worth emphasizing that this view comes hand in hand with obliterating the usual distinction between rules and representations. In effect, all the work is done by the representations and there are only a few generic rules that apply to all representations, primitive and derived, intermediary or final, the same way. This uniformity, characteristic of early combinatorial system like the untyped lambda calculus (Church, 1936) and categorial grammar (Ajdukiewicz, 1935) is maintained in all implementations of 4lang, be they by Eilenberg machines (which directly formalize spreading activation), by (hyper)graph kernel methods (Ghosh et al., 2018), or by direct linear algebraic manipulation.

1.6 Parallel description

So far, we have three main approaches to endowing natural language expressions with semantics: the formulaic, the geometric, and the algebraic approaches discussed in 1.3, 1.4, and 1.5 respectively. All three have a long tradition going back to the 1960s, with many current variants. No doubt other approaches, such as the (now deprecated) automata-theoretic work, are feasible. The view we take here is that all these approaches are algebras of their own, and as such they can be connected by a parallel hyperedge rewriting system with as many branches as there are contenders for the notion 'semantic representation'. For example, the Abstract Meaning Representation (AMR) theory Banarescu et al. (2013) could be added as another branch, and for those content with the rough semantics encoded in explicit marking of head-dependent relations, Universal Dependencies could be added as yet another branch. In fact, some of the applied work discussed in 9.1 already transduces UD to 4lang.

The idea of *syntax-directed translation*, going back to Aho and Ullman, 1971, is standard both in compiler design and in semantics, where it is considered to implement the Fregean principle of compositionality (see S19:1.1) by two systems operating in parallel: a syntax that, proceeding from the bottom (leaf) nodes gradually collects these together, and a semantics that computes at each step a formula based on the formulas associated to the leaves and associates it to the parent node, using only *synthesized* attributes in the sense of Knuth, 1968. The basic idea has been fruitfully generalized for more powerful rewriting methods (Rambow and Satta, 1994; Shieber, 2004), and here we suggest, with implementation planned for Release V3, a hyperedge replacement framework (see Drewes, Kreowski, and Habel (1997) for a detailed overview) for two reasons: first, because it offers great clarity in regards to separating the metalanguage from the language, the tools from the objects themselves, and second, because it has an efficient implementation, the Algebraic Language Toolkit (Alto).

Alto (Gontrum et al., 2017) is an open-source parser which implements a variety of algebras for use with Interpreted Regular Tree Grammars (Koller and Kuhlmann, 2011; Koller, 2015) to simultaneously encode transformations between strings, trees, and `4lang` graphs. Alto has been used for semantic parsing both in Groschwitz, Koller, and Teichmann, 2015 and in the applied work we will discuss in Chapter 9, but a full Alto implementation of `4lang` is still in the planning stage. While this is hard to guarantee in advance, early experience suggests Alto will work well as the computational substratum, the kind of abstract machine the calculus is implemented on. S19:Def 5.8 used Eilenberg machines for implementing spreading activation, an approach we still consider viable for theoretical clarity, but one that has not gained traction beyond a small group of devotees. As Maler and Pnueli, 1994 already warned

> Another sociological problem associated with Eilenberg's construction is the elegant, concise, and motivationless algebraic style in which it is written, which makes it virtually inaccessible to many contemporary theoretical computer scientists.

This time we go with the flow, and take to heart William Stein's maxim: *Mathematics is the art of reducing any problem to linear algebra.* But much of the linear algebraic development has to wait until Chapter 6 and beyond, and in the meantime we assume a different, still algebraic but perhaps better motivated, system built on the hypernode graphs of Definition 5. To prevent any confusion, we emphasize that the machinery we propose, *hyperedge* replacement, uses a metalanguage that relies on a different notion of hypergraphs (Definition 4) than the object language. That the metalanguage is not the same as the object language should come as no surprise to students of logic or computer science: a well known example is regular expressions which describe finite state object languages but use a context-free metalanguage.

One particular semantic representation that we shall pay attention to is the *translational* approach whereby the semantics of one natural language is explicated in terms of another natural language. For this to work, we need to consider each natural language a kind of string algebra, operating on semantic atoms, morphemes. For the sake of simplicity, we will consider only one string operation, concatenation, even though more complex nonconcatenative operations are present in many languages. To the extent syntactic structure explicates semantic relations (e.g. the head-dependent relation that plays a central role in dependency grammar), we may even decorate the nodes with the appropriate graph structure links (see Chapter 9).

The atomic components of all algebras are the morphemes and words (including multi-word expressions that contain orthographic word boundaries (whitespaces). These are conceptualized as small, and individually rather limited nodes loosely connected by an is_a network. This network is a DAG but not necessarily a tree: undirected cycles are common, as in the classic Nixon diamond (Reiter and Criscuolo, 1983). Edges of this network are labeled 0. There are two other networks, with edges labeled 1 and 2. In these, no undirected or directed cycles have been found, but confluences (directed edges originating in different nodes but terminating in the same node) are not rare. Rough translational equivalents are provided across the 4 languages of 4lang and in principle pivot-based translation across these using the synchronous rewrite mechanism is possible.

This is not to say that the elementary components (nodes) are devoid of non-linguistic content: they may contain pointers pointing to all kinds of encyclopedic (verbal) knowledge as well as non-verbal memory: sounds, images, smell. Further, activation of such may bring activation of the nodes, so these pointers (associative links) are often bidirectional, or better yet, directionless. The entire set of nodes is viewed as adiabatically changing: new nodes are added as the individual, whose linguistic capabilities are being modeled, is acquiring new words/morphemes.

In addition to these static node-like structures, we permit the building of more dynamic structures, hypernodes, by a process of *grouping*. In the simplest case, this is just coordinating a few elementary nodes: instead of *Tom, Dick, and Harry* we can refer to the collective entity they form as *the boys*. Typically, hypernodes are nonce elements: *boys* may very well refer to other groups, say *Bill and Dave*, depending on context. Such temporary configurations, best thought of as the meanings of constituents, are denoted in the syntax by curly brackets. On rare but important occasions we will also encounter strongly lexicalized groupings we call *schemas*. For example, we will distinguish `place`, defined as `point, gen at` from {place}, a complex schema we will discuss in great detail in 3.1.

place

One conceptual difficulty we already touched upon in 1.5 is that nodes and hyperedges are not that different. In fact, when we define *fight* as `person want {harm at other(person)}, ins_ weapon` this means that we can at any time replace the node `fight` by the hypernode {`person want {harm at other (person), ins_ weapon}` *salva veritate*. This kind of substitution plays a major role in the low-level deduction process that takes place synchronous with text comprehension: when we hear *John fought the coyote with his bare hands* we automatically put *bare hands* in the ins_ slot and typecast it as a weapon.

fight

Complex deduction like this will have to be built from more elementary operations. The nodes (in what follows, we well refer to hypernodes also as nodes, unless there is a specific reason to distinguish the two) are capable of (i) activating themselves and adjacent edges to various degrees; (ii) copying themselves (triggered by the keyword `other`); (iii) unifying subnodes. This unification, which is automatic for nodes named identically (or for the element `gen`, which is capable of unification with anything), is not to be confused with coercion (see 3.3), though the effects are somewhat similar.

1. Definition → Definiendum Definiens (% Comment)

Unlike in generative theories of the lexicon (Pustejovsky, 1995), where the process of enumerating senses is assumed to start from some start symbol S, we see our system of definitions as a network (hypergraph). This is a large structure with tens, if not hundreds of thousands of hypernodes characterizing the lexical component of adult linguistic competence, and there is no starting point as such. Even developmentally, the first words learned will often correspond to rather complex sensory units (*mama* is a great deal more complex than *light*) as long as they are motivationally salient. As diary studies of early vocabularies clearly demonstrate, new words are often completely unattached to the existing inventory: before a child learns *peepee* or *doodoo* (apparently equally applicable for toilets, people on the toilet, or hearing the toilet flush) there is not one word related to excretion that could be used to describe the meaning (Rescorla, 1980).

For us, this rule plays a key role in *expansion*, the operation whereby we substitute the definiendum by the definiens. We emphasize that this is not a generative operation, but a deductive one that replaces one hypergraph, in which the definiendum appears as a node, by another one, where this node is replaced by the entire definiens, typically resulting in a more complex hypergraph. For example, in *John appears drunk* we may replace *appear* by its definition `gen think {=agt is_a =pat}` to obtain `gen think John is_a drunk`. As we shall see in Chapter 9, expansion, now implemented using the `GraphMatcher` class of the NetworkX library, plays a key role in analyzing lexical entailment (Kovács et al., 2022a). We return to this operation, our model of *spreading activation*, in 7.4.

appear

In terms of vector representations, substitution doesn't change the actual system of vector space objects described, but may bring to light a view of these objects from another basis. Consider for example *crime*, defined as `action, illegal` and trace *illegal* through the system by expanding it as `bad for_ law` to obtain `action, bad for_ law`. By tracing further *bad* as `cause_ hurt` we end up with an even more compact definition of crime: `action, hurt law` – this has the advantage that we don't have to get sidetracked with the issues of experiencer subjects (see 2.4) that the use of `for_` would bring in tow. At the same time, by highlighting the fact that crimes are actions, this definition makes evident that crime has a temporal dimension (and an agent, given that *action* is defined as `person do`). A noun like *tree* which is defined by `plant, has material[wood], has trunk/2759, has many(branch)` will have neither of these implications.

crime
illegal
bad

action
tree

2. Definiendum → Atom

Definienda are always numbered atoms. (The numbering is generally omitted for ease of presentation.) Semicompositional definienda, where a great deal (but not all) of the meaning can be inferred from the parts will have to be adjoined as atoms. We discuss the key technique, *subdirect decomposition*, in 2.2, but offer a simple, and from the lexicographic standpoint easy to defend, example here.

Consider *preferred stock* 'stock that entitles the holder to a fixed dividend, whose payment takes priority over that of ordinary share dividends' (Oxford) 'has a higher claim on assets and earnings than common stock has' (Investopedia). The defintion `stock, preferred` captures most of the meaning, both that preferred stock is a kind of stock, `preferred` and that it is in some sense *preferred*, a notion defined in `4lang` as `{gen like/3382 =pat}` `er_ {gen like/3382 other}, =agt choose =pat`. However, this does not say under what circumstances will this preference be manifest. Clearly it not the preference of the buyer that is relevant here, for if it were, nobody would ever buy common stock. The technical definition makes clear that it is for dividends, and in case of the division of assets, that preferred stock has an advantage, and this fact is external to (cannot be inferred from) the meaning of *prefer, preferred,* or *preference.*

Semicompositional expressions are spread over a continuum with fully compositional expressions at one end, and entirely non-compositional ones at the other. For a multi-word example consider *go Dutch* 'split the bill after a meal' and for a single word consider *went* which will mean, under any analysis, the past tense of *go, *go-ed.* If we assign meaning representation f to expression F and g to G, no case where the meaning of FG involves some extra element h beyond f and g can be considered fully compositional. A great deal depends on the lexicographic purpose: the same FG will be considered compositional if for some reason we consider the h element negligible, and non-compositional if we must make substantive use of it. For example the difference between *hold* and *give* is generally quite clear, yet in the expressions *hold/give a lecture* they are fully interchangeable, acting as *light verbs* (Jespersen, 1965) that contribute little beyond adding a verbal aspect to *lecture* which, in isolation, is ambiguous between noun and verb.

3. Definiens → MarkedClause (',' MarkedClause)*

In terms of graphs, each of the defining clauses are linked to the definiendum by type 0 links. In terms of the vectorial representation, the polytopes corresponding to the clauses are intersected. Noncompositionality arises precisely in those cases where the intersection of the clause polytopes is a superset of the definiendum polytope.

4. Comment → (ArbitraryString)

Comments are restricted to a separate column of the file. Since the comments themselves only benefit the human reader of the file, the rule is a no-op as far as its effect on meaning is concerned. Most of the comments list potentially interesting cross-linguistic tidbits, e.g. that the hand of an English person has four fingers and a thumb, while the hand of a Hungarian has five fingers, as the thumb is called *nagyujj* 'big finger'. Phenomena such as this are common (indeed, typical) and they served as motivating examples for taking the abstract, algebraic view.

5. MarkedClause →DefaultClause|PositionClause|ComplexClause|Clause

Unless overriden, default clauses are carried (credulous inference). This hides a great deal of complexity, both in terms of the deontical status of default existents (see 6.2) and the default logic overall (see 6.4). When the default fails, we use rewrite rule 6.

Position clauses, just as `mark_`, are language-specific. They are used in a rudimentary fashion throughout the book, mostly to indicate whether a form is free-standing or affixal, and if an affix, is it a prefix or a suffix, and sometimes to describe slightly more complex situations (infixes, circonfixes, tripartite constructions like `er_`). Of necessity, we abstract away from a great deal of micro-syntax, since most of the 'tinkering' is both highly syntactic and highly language-specific, while our focus is with the semantic and the universal.

Complex clauses are typically used in subordinate position. As an example, take *attract* `=agt cause_ {=pat want {=pat near =agt}}`. What is being caused   attract is itself a complex state of affairs, the patient wanting something, and that something again is a complex state, the patient being near the agent.

Rule 5 groups all these together with simple clauses, but this is only for the convenience of the formula parser. There are no deep similarities between default clauses and complex clauses, but one is surrounded by ⟨⟩ and the other by {} so the notation brings them close.

6. DefaultClause →'⟨'Clause'⟩'|λ

In expansion, the second alternative means we do override i.e. we omit the default for some reason. Consider *sugar* defined as `material, sweet, <white>, in`   sugar `food, in drink`. We still have to deal with *brown sugar* and not get entangled in some sophistry about how brown is really a kind of white, or how brown sugar is both brown *and* white, etc., see 6.4.

7. PositionClause → PositionMarker mark_ UnaryAtom

`mark_`, as opposed to the non-technical *mark* `sign, visible`, is a semi-technical   mark term, the closest we will get to the Sausserean sign: its agent is a sign, its patient is a meaning, and it itself means 'represent': *mark_* `=agt[sign], =pat[meaning],`   mark_ `represent`. A typical example would be in the last clause defining the English word *buy* we discussed in 1.4: `=agt receive =pat, =agt pay seller, "from`   buy `_" mark_ seller`. Whatever follows the string "from" is the seller in English – in Hungarian it would be whatever precedes the ablative case marker.

8. ComplexClause → {Definiens}

The key distinction between simplex and complex clauses is that the former appear in intersective situations, while the latter are unions, both in graphs and in vectors. Consider

defend =agt cause_ {=pat[safe]}. The agent doesn't cause the patient, or the defen
safety, what the agent causes is the safety of the patient, a complex situation with two
components. In our example of *attract* above, what the agent causes is also a complex
situation, one that has another complex situation as one of its components.

Here it is perhaps worth emphasizing that there cannot be two agents, or two patients,
or indeed, two of anything, unless this is signalled by the other keyword. Unification
is an automatic low-level process that we have not incorporated in these rewrite rules in
order to keep them simple, but are used in the IRTG/Alto system under development.

9. PositionMarker → ""'SuffixMarker|PrefixMarker|InfixMarker'""

Position clauses are language-dependent, and 4lang only gives them for English. They
are primarily used in morphology, where the underscore _ is written together with the
stem, and in the rare cases where English uses positional marking (e.g. subjects in prever-
bal, objects in postverbal position) they are separated by whitespace. The reader should
not take this simple notation as some profound statement about proto-syntax – position
markers appear in less than 5% of the dictionary, and English syntax offers many constructions that are inconvenient to describe by this mechanism (see 2.1 on the autonomy
of syntax).

The system gives a good indication of what is what, e.g. that in *buy* "from _"
mark_ seller, but without more developed morphophonological machinery this is
generally insufficient to drive a parser. This is because the quoted strings rarely stay
invariant: there can be all kinds of changes both to the stem and to the affix (e.g. in
the Hungarian ablative, *-tOl* the choice of realizing *O* as *ó* or *ő* depends on the vowel
harmonic properties of the stem), linking vowels or consonants may appear, material
may get truncated, there are suppletive forms, etc etc.

10. Atom → PlainAtom|NumberedAtom|ExternalAtom|PositionMarker

Atoms, just as clauses, are grouped here together for ease of parsing. Loosely speaking,
an Atom is a minimal entry in 4lang – a PlainAtom is just a word or morpheme,
signifying a unique concept. Non- and semi-compositional entries get their own atoms
(see discussion of Rule 2 above). We emphasize that the presence of compositionally
non-derivabable meaning is insufficient for us to declare the entry non-compositional,
for example, the Battle of Jena is just that, a battle that took place at Jena. We may
very well be aware that Clausewitz was captured by the French in this battle, but such
knowledge belongs in the encyclopedia, not the lexicon. Such knowledge is *inessential*
for understanding what this battle was, even a graduate student of history can get an A
on an exam or paper that doesn't mention this fact. This is in sharp contrast to the case
of preferred stock: not knowing how it is preferred amounts to not understanding the
MWE.

11. NumberedAtom → PlainAtom'/'Number

The numbering of the Atoms, effected by a slash followed by a serial number below 4,000, is just the standard disambiguation device to get around homonymy. A more human-friendly dictionary would use subscripts for different word senses. At the core level we are most interested in (Kornai, 2021) the numbering carries very little load: over 95% of the English headwords has only one sense in `4lang`. An interesting counterexample would be `place/1026` 'locus' versus `place/2326` 'spatium', see 3.1 for discussion.

12. ExternalAtom → '@'WikipediaPointer

ExternalAtoms are pointers to Wikipedia. They refer to concepts about which a great deal is known, such as the Battle of Jena, where this knowledge is properly considered a part of history, or Tulip, where the knowledge is really part of biology. As we discussed in 1.2, linguistic semantics is a weak theory that cannot serve as the foundation for all this kind of knowledge amassed by the sciences over the centuries.

13. PlainAtom → UnaryAtom|BinaryAtom

Almost all our atoms are unary. Binary atoms are a small, closed subset (see Rules 14-15), and we do not permit atoms of higher arity (Kornai, 2012).

14. UnaryAtom → `Asia|acid|...|yellow|young|=agt|=pat`

There can be millions of unary atoms such as pointers to the encyclopedia (see Chapter 8). `4lang` concentrates on the defining set, where we already know that less than a thousand items are sufficient. However, these are not defined uniquely. In linear algebraic terms, it is just the dimension of the basis that is given, the basis vectors can be chosen in many ways. A handful of elements like `=agt, =pat, wh, ...` are reasonable candidates from a universal standpoint, but many others, including natural kinds, are not. In (Kornai, 2010a) we wrote

> The biggest reason for the inclusion of natural kinds in the LDV is not conceptual structure but rather the eurocentric viewpoint of LDOCE: for the English speaker it is reasonable to define the yak as ox-like, but for a Tibetan defining the ox as yak-like would make more sense. There is nothing wrong with being eurocentric in a dictionary of an Indoeuropean language, but for our purposes neither of these terms can be truly treated as primitive.

More important than the actual selection of defining words is the method we employ in proving that the set so selected is actually capable of defining everything else. Once this is demonstrated, the issue of *which* elements are chosen is seen to be equivalent to deciding which equations to simplify by substituting the definiens for the definiendum.

How do we define words in general? Our method is akin to the use of multi-stage rockets in lifting a payload. In Stage 1, we simply look up the word in the dictionary, typically LDOCE. For example, at *intrude* we find 'interrupt someone or become involved in their private affairs in an annoying and unwanted way'. In Stage 2, those familiar with the system will translate this to =agt cause_[pause in =pat], after(=agt part_of =pat), =agt cause_ [=pat[angry]] manually. In the implementation we use the Stanza NLP package[1] to create a UD parse of the definition, and the dict_to_4lang system (Recski, 2018) to transform this to 4lang syntax.

One can be far more faithful to the original definition than we were here: clearly *annoying/unwanted* is not exactly the same as *make angry*. If this is significant for some purpose, we may trace the LDOCE definition of *annoy* 'make someone feel slightly angry and unhappy'; that of *slightly* to 'a little'; and adjust the last clause of the above definition to =agt cause_[=pat[angry[little]]]. The claim here is that there is no shade of meaning that is inexpressible by these methods, not that the automatic system can already create perfectly faithful definitions for each and every word in each and every context for each and every langauge. As is typical in NLP, the automated systems are somewhat inferior to the best human-achievable performance. We return to the matter of contextual disambiguation, whether to choose fall/2694 'cado' or fall/1883 'autumnus' in 6.4.

For other languages, we need to begin (Stage 0) with a bilingual dictionary translating the word into English, and proceed from there. Let us consider a word that is often claimed to have no English equaivalent, *schadenfreude* 'pleasure derived by someone from another person's misfortune' (Oxford). In Stage 1, we consult LDOCE to find that *pleasure* can be replaced by *joy*. This is not to say that these two words are perfect synonyms, but whatever shades of meaning distinguish the two appear irrelevant in the definition of schadenfreude. In Stage 2, we can go even further, and replace *joy* by

joy   its 4lang definition sensation, good to obtain 'good sensation caused by other person's harm' which becomes in the formal language of definitions sensation, good, {other(person) has harm} cause_. In this step we switched from *misfortune* to *harm* manually, because the former specifically implies bad luck (and thereby absolves the experiencer of responsibility) while the latter stands neutral on whether the person is the cause of their own bad situation or not. Since schadenfreude is appropriate for both cases, we need to revise the Oxford definition a bit.

This last step of emending a definition may look at first blush as something beyond the powers of any automated dictionary builder algorithm. But keep in mind that we already have several systems that assign vectors to words purely on the basis of corpora, and we may resort to these in refining any definition. Even more important, the addition of a new definition will bring in one more unknown, the definiendum, and one more equation, the definition itself. Therefore, if the original system was solvable, the new one will also be solvable.

---

[1] https://stanfordnlp.github.io/stanza

15. BinaryAtom → `at|between|cause_|er_|follow|for_|from|has|in|`

`ins_|is_a|lack|mark_|on|part_of|under`

Unlike unaries, which come from a large open list, binaries are restricted to a small closed set. We represent unaries by vectors, as standard, or by polytopes surrounding these, a slight extension of the standard. For binaries we use matrices, which are much more expensive, n^2 parameters for n-dimensional vectors. By far the largest group are spatial (or, in the sense of Anderson (2006), 'local') cases and adpositions which we will discuss in 3.1. These are the prototypical ones, and we will see how temporal, and even more abstract cases such as the instrumental, can be brought under the same formal umbrella (see 6.2 for instruments, and 2.4 for causation).

16. Clause → 0Clause|1Clause|2Clause|FullClause

For ease of parsing we group together a variety of Clauses subject to different expansion in an anuvṛtti-like process, as explained below.

17. 0Clause → Atom'['Definiens']'|Atom'('Definiens')'|Atom

0Clauses are defining clauses linked to the definiendum by a 0 link. A typical example would be *below* defined as `under`, or *fast* defined as `quick` – these are to be understood as 'below is a (kind of) under' or 'fast is a (kind of) quick'. When there are several defining 0Clauses, as is typical, the definiens is in 0 'is/is_a' relation to each of them: *dot* `mark`, `small`, `round` means 'a dot is a mark, a dot is small, a dot is round'. The square brackets are also abbreviating is/is_a in `A[B]` constructions, as in *energy* `work[physical]` which means 'energy is work (that) is physical' or, for even better conformity with English syntax, 'energy is physical work'. (We exhort the reader not to get bogged down in high school physics where energy is *capacity* for work. Our definitions, intended to capture a naive world-view, will rarely stand up to scrutiny from the contemporary scientific standpoint.)

below
fast

dot
energy

Constructions involving parentheses, `B(A)` are strictly equivalent to `A[B]` and are used only when this order sounds more natural. Example: *powder* `substance`, `more(particle)`. There is nothing in the system of definitions that strictly requires this: we are catering to English syntax where adjectives are preceeding the noun but can be reversed as in *blue box, the box is blue* but numerals and similar quantifiers don't really tolerate the same reversal *four legs, ??the legs are four.*

powder

18. 1Clause → BinaryAtom Clause

1Clauses are used whenever the definiendum should occupy the subject (1) slot in the definiens. Example: *bee* `insect`, `has wing`, `sting`, `make honey`. The implicit 0Clause links are bee is_a insect, bee is_a sting (yes, and dog is_a bark, a design

bee

decision that makes a great deal of sense within the larger system of unaries and binaries we will discuss in 2.1) but we will not say that a bee is_a 'has wing' or a 'make honey'. Rather, `has` is a BinaryAtom, and `make` is a non-atomic binary (an obligatory transitive as its definition contains an `=agt` and a `=pat`). When a clause begins with a binary, we automatically put the definiendum in its subject slot 'bee has wing', 'bee make honey'.

19. 2Clause → Clause BinaryAtom

`food`   2Clauses are similar to 1Clauses, except the definiendum fills the object slot of the defining clause. Example: *food* `substance, gen eat` 'food is what people eat' (see 4.5 for the treatment of the generic quantifier `gen`). In parsing, each clause needs to be inspected whether it has a binary, and if so, whether the binary has both valences filled

`make`   in, as in *make* `=agt cause_ {=pat[exist]}`. If the pre-binary position is empty, we are dealing with a 1Clause, if the post-binary position is empty, we are dealing with a 2Clause. When both positions are empty, the definiendum and the definiens rely on the

`notice` same agent and patient, as in *notice* 'animadverto' `know, see`.

20. FullClause → ComplexClause BinaryAtom ComplexClause

`polish` Finally, FullClauses have both the subject and the object slots filled. Example: *polish* `=agt cause_ surface[smooth, shine], =pat has surface`. 'agt polishing pat means that agt is causing the surface of pat to be smooth and to shine'.

Classroom experience shows that the system is learned relatively easily, with students providing remarkably similar, often identical, definitions after a few weeks. The exception is students of linguistics and philosophy, who really need to unlearn a lot, as they are professionally trained to have a fine ear for minute distinctions. The marriage of lexicography and encyclopedia-writing is never happy. Consider the definition of *potash* as given in Webster's 3rd:

> 1a: potassium carbonate, esp. that obtained in colored impure form by leaching wood ashes, evaporating the lye usu. in an iron pot, and calcinating the residue – compare pearl ash. b: potassium hydroxide. 2a : potassium oxide K_2O in combined form as determined by analysis (as of fertilizers) ⟨ soluble ∼ ⟩ b: potassium – not used systematically ⟨ ∼ salts ⟩ ⟨ sulfate of ∼ ⟩ 3: any of several potassium salts (as potassium chloride or potassium sulfate) often occurring naturally and used esp. in agriculture and industry ⟨ ∼ deposits ⟩ ⟨ ∼ fertilizers ⟩

What are we to make of this? The COBUILD project (Moon, 1987) and the resulting Collins-COBUILD dictionary, attempted to clarify matters by distinguishing three different senses:

1. another name for {potassium carbonate}, esp. the form obtained by leaching wood ash

2. another name for {potassium hydroxide}
3. potassium chemically combined in certain compounds

But is it now carbonate or hydroxide? Or, perhaps, both could be subsumed under 'certain compounds'? LDOCE (Procter, 1978) avoids chemistry altogether:

> any of various salts of potassium, used esp. in farming to feed the soil, and in making soap, strong glass, and various chemical compounds

In `4lang` we can accomodate the chemistry only by explicit reference to the encyclopedia *potash* `@potassium_carbonate` which resolves to the WP article which in turn offers a wealth of information on the subject, and similarly for potassium hydroxide. But what to do with all this artisanal knowledge about industrial processes, that leaching wood ash produces lye, that caustic soda is used in glassmaking, that farmers feed the soil with potassium salts, and so on? We use a much simpler style of definition whereby *potash* is simply `salt, contain potassium` and consider the pain of invoking scientific theories in the midst of dictionary building to be self-inflicted.

potash

The key takeaway from this section is that once lexicography is freed of this burden, it is possible to formalize definitions to such a degree that we can automatically convert them into equations, in this case `potash is_a salt` and `potash contain potassium`. How a symbolic equation `A is_a B` or `A contain B` get translated to more conventional vector equations will be discussed in 2.3. The overall strategy of converting definitions to equations is made more concrete in a step by step fashion throughout the book, with a summary provided in 9.5.

2

From morphology to syntax

Contents

Our goal is to develop a semantic theory that is equally suitable for the lexical material (words) and for the larger constructions (sentences) put together from these. In 2.1 we begin with the system of lexical categories that are in generative grammar routinely used as preterminals mediating between syntax and the lexicon. Morphology is discussed in 2.2, where subdirect composition is introduced. This notion is further developed in 2.3, where the geometric view is expanded from the standard word vectors and the voronoids introduced in Chapter 1 to include non-vectorial elements that express binary relations. These eigenspace techniques receive further use in 2.4, where some crucial relational devices of syntactic theory, *thematic relations, deep cases*, and *kārakas* are addressed. How much of syntax can be reconstructed with these is discussed in 2.5.

2.1 Lexical categories and subcategories

Whether a universal system of lexical categories exists is still a widely debated question. Bloomfield, 1933, and more recently Kaufman, 2009 argued that certain languages like Tagalog have only one category. But the notion that there are at least three major categories that are universal, nouns, verbs, and adjectives, has been broadly defended (Baker, 2003; Chung, 2012; Haspelmath, 2021). 4lang subdivides verbs into two categories: intransitive U and transitive V; retaining the standard N for noun; A for adjective; and also uses D for aDverb; and G for Grammatical formative.

While this rough categorization has proven useful for seeking bindings in the original 4 and in other languages, there is no theoretical claim associated to these categories, nei-

A. Kornai, *Vector Semantics*, Cognitive Technologies, https://doi.org/10.1007/978-981-19-5607-2_2

ther the universal claim that all languages would manifest these categories (or at least, or at most, these), nor the (four)language-particular claim that these categories are somehow necessary/sufficient for capturing the data. In fact, 4lang is a semantic system, and it says remarkably little about the system of lexical categories and subcategories, be they defined by morphological or syntactic cooccurrences. If anything, our findings lend support to the thesis of Wierzbicka, 2000 that cross-linguistic identification of lexical categories is to be achieved via prototypes rather than by abstract class meanings.

To the extent that none of the six lexical categories U,V,N,A,D,G is ever referred to by any definition or rule, 4lang holds fast to the autonomy of syntax thesis (by and large, one can think of the system as operating on categoryless roots). The categories are listed with each entry only to help the English-speaking user distinguish between e.g.

```
cook fo3l     coquitur gotowac1_sie1 822  U
     get heat
cook fo3z     coquo    gotowac1        825  V
     =agt make <food>, ins_ heat
cook szaka1cs coquus   kucharz         2152 N
     person, <profession>, make food
```

Here, and in most cases, the other three languages actually manifest the distinction either morphologically or in the choice of stem, but readers familiar with the largely (perhaps fully?) universal distinction between intransitive and transitive verbs and nouns can read off the distinction between the three senses of *cook* from the 7th column containing the categories U, V, and N.

This small example already displays some of the vexing problems of morphology that we need to consider here. First, whichever these three we take as basic, in English it would require phonologically null affixes to obtain the other two. Second, using these six categories creates a lot of ambiguity where there doesn't seem to be any, e.g. between verbal and nominal interpretation of English noun-verbs such as *divorce*, suggesting that six categories are too many. Third, there are obvious meaning distinctions e.g. between agent, action, and abstract nouns that share the category N: compare a *cook* 'the person who does the cooking' to a *shoot* which means 'preparing a segment of a film' rather than 'the person who does the shooting' or *addition* 'mathematical operation' to *addition* 'an extra room that is added to a building'. This widespread phenomenon would suggest that six categories are too few.

In regards to zero affixation, 4lang refrains from stating the categorial signature of elements even when it is obvious, e.g. that *-ize* (for which see 2.2) is N→V, producing a transitive verb from a nominal base (see Lieber, 1992 that in productive uses the resulting verb must be transitive). In informal contexts, where human readability is a concern, we go a step further and feel free to enhance English paraphrases by category-changing nonzero formatives such as *be, that, a/an, the, to, -ly* ... with the goal of making the English syntax come out right.

Pure category-changing, be it performed by zero or non-zero affixes, is modeled by change in the head (first, distinguished element) of a definition: consider

```
official hivalatalos  publicus    oficjalny 1065 A
         at authority
official tisztviselo3 officialis urzeldnik 2398 N
         person, has authority
```

When this process is regular, as with agentive *-er*, the suffix morpheme is given the same status in 4lang as any free-standing word would:

```
-er -o1 -tor/-trix -ac1/ic1 3627 G
    stem_-er is_a =agt, "__-er" mark_ stem_
```

which will in turn yield definitions such as

```
buyer vevo3  emptor   kupujac1y    3628 N
      =agt, buy, -er/3627
renter be1rlo3 conductor dzierz1awca 3632 N
      =agt, rent, -er/3627
seller elado1  venditor sprzedaja1cy 3629 N
      =agt sell, -er/3627
```

We started with affixation because this area (which we will discuss in a more systematic fashion in 2.2) offers laboratory-pure examples of change of syntactic category without change of meaning and, in the case of zero affixation, without change in form. We have seen that 4lang doesn't offer a full account of such phenomena. Needless to say, traditional lexicographic practice is no better off, with definitions often led by vague category phrases *used to, of or about, to be, someone who, relating to, done as, a way of, according to, to make, something that, a type of, the process of, . . .* which contribute very little beyond a hint for the syntactic type – we have already discussed a rather striking example, the verb *be*, in 1.4.

Traditionally, the idea of lexical categories, a notion that we would reconstruct in a purely syntactic means, see S19:4.2 and 6.3, come hand in hand with the idea of *class meanings*, 'the meaning common to all forms belonging to the same form class'. This is a highly contentious idea, of which Bloomfield, 1933 (Sec 16.2) has this to say:

> The school grammar tells us, for instance, that a noun is 'the name of a person, place, or thing'. This definition presupposes more philosophical and scientific knowledge than the human race can command, and implies, further, that the form-classes of a language agree with the classifications that would be made by a philosopher or scientist. Is fire, for instance, a thing? For over a century, physicists have believed it to be an action or process rather than a thing: under this view, the verb burn is more appropriate than the noun fire. Our language supplies the adjective hot, the noun heat, and the verb to heat, for what physicists believe to be a movement of particles in a body. (. . .) Class meanings, like all other meanings, elude the linguist's power of definition, and in general do not coincide with the meanings of strictly defined technical terms. To accept definitions of meaning, which at best are makeshifts, in place of an identification in formal terms, is to abandon scientific discourse.

On the other hand, there is something greatly appealing about the idea of conceptual parts of speech, and a type theory of one sort or another lies at the heart of many modern developments in philosophy, mathematics, and computer science. Here we compare 4lang to the ontological classification developed by Jackendoff, 1983, where eight major categories are distinguished: Thing, Event, State, Action, Place, Path, Property, and Amount.

Things In 1.4 we defined `object` by the cluster of properties that physical objects have: `thing, <has colour>, has shape, has weight, <has surface>, has position, <lack life>`. Jackendoff takes a more cognitively inspired view and singles out individuated entities within the visual field as the central aspect of his definition. However, it seems clear that he would assign *kidney, liver, lung, . . .* and other objects which rarely occur in the visual field (indeed, they are characteristically hidden, internal) as Things.

On the one hand, it is very clear that Bloomfield is right, we still don't have sufficient philosophical and scientific knowledge to formally define what a Thing must be. On the other hand, it is also clear that people will show remarkable inter-annotator agreement if we ask them whether some X is an object or not, very much including the fact that Jackendoff's 'Thing' and the `object` of 4lang designate pretty much the same class of entities.

We also grant full ontological status to abstract nominals like *happiness* or *equilateral triangle*. Since Jackendoff is committed to the same kind of theory that we are propounding, where meanings are concepts, ideas, things in the head (see 6.3), conferring first class citizenship on abstract nominals is not a problem for him, but in other settings the task is highly nontrivial (see Zalta, 1983 for a well worked out proposal, and Moltmann, 2013 for a different approach).

Events, Actions While we have much to say about events, actions, and event structure in 3.2 and 6.1, we don't have separate ontological categories for events or actions as such. Rather, we use *matters* as a convenient term that covers both Things, Events, and Actions in the Jackendovian sense, and we steer clear of the philosophical issues of whether events exist (occur) 'in the world'.

Contemporary philosophical theory is near unanimous in granting existence (ontological status) to Things, but far from unanimous in the treatment of events or actions. An important consequence of our parsimonious stance is that we see nominalization processes as purely morphosyntactic, with no corresponding change in meaning: we treat *breathe* and *breath, divorce* (V) and *divorce* (N) as alike.

States, Properties Again, philosophers are near-unanimous that sensory qualia exist, and we follow suit here in granting them ontological status. However, the distinction between these two classes is too subtle. Even the core examples, emotional states like *anger, fear, sorrow, joy, . . .* are trivial to view as properties, and core properties like *red, smelly, triangular, . . .* are in turn easily conceptualized as Things. In fact, States like feelings are generally treated as belonging in a well-circumscribed subclass of Things, *fluids*, which permeate the experiencer subject, and normally appear in constructions like *joy filled her heart* (Kornai, 2008).

Places, Paths That these require a separate ontological category is clearly seen from the fact that it requires some operation of *typecasting* or *coercion* to make sense of expression's like *Let's meet at Jim's*. It doesn't matter what kind of concrete object (e.g. Jim's house, restaurant, or office) is meant, it must be viewed as a place for it to function as the object of *at*. We discuss locations in 3.1, but note here that their treatment will involve conceptual schemas rather than a type theory with distinguished Place or Path types.

Amounts Again, no specific amount type is assumed. We consider 'being two' (in number) no different from 'being pink' (in color). See 4.5 for quantifiers, 3.4 for numerals and measure phrases, and 7.1 for adjectives.

Altogether, we have only three logical types: *matters*, used both for endurants and perdurants, *relations* (always binary), and *situations*, which are partial possible worlds, or, equivalently, equivalence classes of fully specified possible worlds. Events are constructed by linking a few (typically just one or two) participants to action verbs, but their conceptual classification is not any different from statives. In particular, we will not need a separate event variable (Davidson, 1980), and we treat the subclasses we see within matters, be they verbal (e.g. motion verbs `close`, `fall`, `fly`, `go`, `slide`, `turn`) or nominal (e.g. liquids `soup`, `water`) as part of the naive ontology, rather than as some deep-set cognitive ground.

2.2 Bound morphemes

The LDV contains a few dozen bound morphemes, the suffixes *-able -al -an -ance -ar -ate -ation -dom -ed -ee -en -ence -er -ery -ess -est -ful -hood -ible -ic -ical -ing -ion -ish -ist -ity -ive -ization -ize -less -like -ly -ment -ness -or -ous -ry -ship -th -ure -ward -wards -work -y* and prefixes *counter- dis- en- fore- im- in- ir- mid- mis- non- re- self- un- vice- well-* . These are tremendously useful both in reducing the size of the defining vocabulary, since *eat* and *eating* no longer need both be listed, and in making the definitions less complicated.

We cannot cover the entirety of English morphology as part of `4lang` here, but we do not consider the problems raised by bound forms to be qualitatively different from those raised by lexical semantics in general. As we shall see, most, if not all, of the problems raised by an effort to provide (non-compositional) semantics for morphology already arise in the course of analyzing this limited set of affixes. Obviously, languages are not uniform in where they draw the bound/free boundary: many concepts that are expressed by affixation in one are expressed by free forms in another, and dictionary definitions often contain these.

We will illustrate our methods on the suffix *-ize*, which means something like 'to cause to become', so *Americanize* 'cause to become American', *carbonize* 'cause to become carbon' and so forth. There are cases that do not fit this analysis (*agonize* doesn't mean 'cause to become agony' the same way *colonize* means 'cause to become colony') and there are other subregularities one may wish to consider, but the majority of the 200-300 English words ending in *-ize* fit this pattern well enough to consider it the leading candidate for a semantic definition. What we wish to state is a lexical rule roughly of the following form: for stem X, *stem+ize* means 'cause to become (like) X'. Using the notational conventions that were introduced in 1.3, we write this as *-ize*

-ize

```
cause_ {become <like/1701> stem_}, "_-ize" mark_ stem_
```

Here `cause_/3290` 'efficio' is a binary relation written with a trailing underscore to distinguish it from ordinary language `cause/1891` 'causa' because it is one of the few cases where we feel the technical sense is sufficiently different from the ordinary, naive sense to merit separate treatment (see 2.5). The curly braces denote a single hypergraph node (pictorially, all formulas will correspond to hypergraphs) and the angled brackets signify optionality, enclosing the default option. `mark_` is another technical notion, standing for the relation between signifier (a string, given in doublequotes) and the relevant element to be substituted, see 2.5. The node `stem_` is analogous to the variable X used above.

However, neither *like/1701* nor *become* are primitives (for the four-digit disambiguation number following the English binding see Section 1.3). `like/1701` 'sicut' is defined as `similar` (as opposed to `like/3382` 'amo') and *become* is defined as `after(=agt[=pat])` which for now we will paraphrase as 'afterwards, agent is_a patient' (thematic roles will be discussed in 2.4). For something like *John caramelized the sugar* this would be 'John caused the sugar to be <similar to> caramel afterwards'.

like
become

For the sake of readability, we will continue to make some concessions to English syntax, by adding agreement morphology, an article, a copula, and a preposition if needed, but eventually the reader will get familiar with the syntax of definitions that lacks all this niceties, and will fluently read `John cause_ after({sugar <similar> caramel})`. Since *similar* is not a primitive of the formal language of definitions, we can take this further by substituting its definition

```
=agt has property, =pat has property, "to" mark_ =pat
```

Since named nodes are unique in definitions, what this means is that in the construction X *(is) similar to* Y the agent will have the *same* property as the patient. As expected, the `mark_` relation is language-specific, for Hungarian we would want to say that the allative case *hoz/hez/höz* marks the patient. (`4lang` currently gives the mark_s only for English.)

We can omit the default (since it is a binary relation, this means substituting `is_a`) or we can expand it, to yield

-ize

```
cause_ {after({=pat has property, stem_ has property})},
     "_-ize" mark_ stem_
```

At this point, all our notions are sufficiently general, including not just the metalinguistic `stem_` but also the term `property_`, which is really underspecified as to what property it refers to. This fits well the definition of *similar* `2794 u A =agt has` `similar` `quality, =pat has quality, "to _" mark_ =pat`, which is underspecified exactly in this respect: compare *similar consequences* to *similar balloons*. The *has* relation is grammaticalized to different degree in different languages: here we opted for the ordinary meaning `has bi1r habeo miec1 288 p V =agt control` `has` `=pat, =agt has =pat` as opposed to a pure grammatical construct `poss` or `has_`, but its status as a primitive is indicated by the definiendum's presence in the definiens. The causative element in *-ize* is well known (Lieber, 1992; Plag, 1998), and the idea that we define certain verbs by their result state is standard. Temporal structure can refer to some state `before` or `after` the event, see 3.2. Comma-separated linear order, as in `=pat has property, stem_ has property` simply means conjunction (see Section 1.3), and as such it is independent of the order of the conjuncts.

In the fourth edition (Bullon, 2003) LDOCE defines *caramelize* as 'if sugar caramelizes, it becomes brown and hard when it is heated'. The first edition of LDOCE (Procter, 1978) does not define *caramelize* and has no self-recursion. The self-recursive definitions added to later editions may be a feature from the perspective of the human language learner, but they are definitely a bug from the definition substitution perspective. To parse this definition would lead us nowhere, since the definiendum is part of the definiens, and we don't have a theory for finding a minimal fixed point in *if sugar if sugar if sugar ...X, X becomes brown and hard when it is heated X becomes brown and hard when it is heated ....* What happens when it's not heated? Is it brown? Will it become brown? Is it hard when it's caramelized? Or will it become hard only when heated? How about caramelizing something other than sugar, say onions? This definition says nothing about the 'if not sugar' case, whereas the definition we derived above at least tells us that if onion is caramelized it will share some properties with caramel. `4lang` uses the appearance of the definiendum in the definiens to trigger a compiler warning in the handful of cases where we see no further reduction (see 2.5), and we see no reason to treat *caramel* or *caramelize* as a primitive.

Let us briefly discuss some of the challenges that arose in providing semantics for morphological operations such as *-ize*-suffixation. First, the rule is not uniform: there are notable subregularities and exceptions, such as *agonize, cannibalize, editorialize, ...* which do not at all fit the proposed semantics. Some of them lack an object, others may have a prepositional object (editorialize *about* something), yet others seem to be built on stems that are no longer actively part of the language *?extempore, ?proselyte, *tantal(os)*.

Stating a rule of *-ize* suffixation runs into problems of both underapplication and overapplication. There are seemingly excellent base candidates like *meat* from which we don't obtain **meatize* even though 'cause something to become (like) meat' would be a perfectly reasonable meaning, and the process actually exists (food producers are sometimes known to add wood pulp to meat products). This lack of productivity has led Chomsky, 1970 to proposing the Lexicalist Hypothesis that puts in the lexicon all processes that lack full generativity (see Bruening, 2018 for cogent criticism).

Conversely, no English speaker can extract a stem *ostrac for *ostracize*, and frankly, knowing that *ostrakon* means potsherd in Ancient Greek is not particularly helpful without a longer explanation. The problem is by no means specific to -*ize*, The same problem is seen with many suffixes like -*ify* (*modify, *ratify, ossify*) and even compounding (no *cran* for *cranberry*).

The rule-application approach also runs into subtle, and on occasion less subtle, problems at the string rewriting level. We use the standard underscore symbol '_' for rule focus, and doublequotes for string material, so that in an expression x mark_ y we can have some form (string) x and some meaning y. The expression of meanings, our main subject in this book, can be done by formulas as in 1.3, by vectors as in 1.4, or by graphs as in 1.5, we will see this in a great deal more detail as we go along.

For the expression of forms, our notation is at best indicative of intention, rather than a fully fleshed-out proposal for the appropriate morphophonological formalism. Creating such a formalism is a problem we don't take on board, since this would require importing a huge amount of technical machinery from phonology, and would not lead us closer to our goal of providing the semantics. (In S19 we argued that both morphophonology and semantics are well suited for finite state devices, but we leave this to the side here.) The intention of the shorthand "_-ize", while clear to the linguist, can only be made explicit in terms of machinery that we are not too keen on developing, as it goes far beyond what we could call 'naive'.

That said, in the metatheory we'll make use of several standard distinctions of morphology such as between roots, stems, and fully formed words, and between derivational and inflectional affixes. These distinctions, however, are not part of the naive theory of grammar, which begins and ends with words (see 2.5). We have already seen examples of roots in 1.2, where we discussed a pair of Hungarian suffixes -*it* and -*ul* which turn roots into transitive (resp. intransitive) verbs quite systematically (there are several hundred examples in the Hungarian vocabulary, more than for the average English derivational affix included in LDV). It is a standard assumption of morphology that roots themselves are categoryless, and obtain their lexical category only in the process of adding derivational affixes. In 2.1 we already discussed the 4lang system of lexical categories, and as we shall see, there are many elements in the 4lang lexicon that are better thought of as roots than as stems/lexemes or fully formed words. But given our focus on English, a ridiculously inadequate choice for the study of morphology, we will rarely encounter roots in this book. (For a more systematic study of (Latinate) English morphology, see Quirk et al., 1985 Appendix I; Plag, 2003; Hamawand, 2011; Schulte, 2015.)

What is stem_ in the definition of affixes? Ideally, it should be a string variable, which in a rule for *caramelize* will select the part before -*ize*, i.e. *caramel*. This clearly doesn't work well for most cases, as there is also some truncation going on: *deputy/deputize, colony/colonize* ... removes *y*; *economical/economize* removes *ical*, *feminize* involves removal of *ine* or *inity*; and so on. Here we take string to be no more than a call to associative memory, something that will be matched by *feminine, femininity, feminist, feminism* and perhaps even *effeminate*. The string associated to this element may be "femin". We use the maximal common substring, but treat this as a hack, and certainly don't want to elevate it to the status of a principle. Stems, in this sense, belong in the *naive* theory of grammar, see 2.5.

It is the meaning marked by this string, or more precisely, the meaning common to all words that are reached by the associative call, that is relevant for the use of the entry, as it is this meaning that must be used in linking the semantics of the derived word to the semantics of the stem or root. This linking, which we treat as a rule of lexical redundancy, rather than a generative rule, is imperfect, not so much because of the under- and -overapplication issues discussed above, but rather because the semantics computed along these lines is itself incomplete. Consider *editorialize*. In ordinary use, e.g *[T]his newspaper has editorialized about the disturbing achievement gaps between boys and girls* it is true that the meaning includes 'the newspaper made (the disturbing gaps) the subject of an editorial', but there is a lot more to this. On the grammatical side, the use of *about* is clearly necessary, something that will have to be made part of the full lexical entry of the word *editorialize* (see 2.4). But even more important, the meaning has shifted to 'use the editorial format as a means to publicly address an issue'.

This phenomenon, lexicalization, is the main driver of non-compositionality. Once a word enters the lexicon (permanent, community-wide repository of words), it can accrue meanings that go beyond the compositional meaning. This is evident for accrual of encyclopedic knowledge: *cook/2152* is not just a person who makes food in a professional manner, but also someone we picture as wearing a chef's uniform, complete with a *toque blanche*.

Paul Kiparsky (pc) calls attention to P. 3.2.135 *aā kves tacchīlataddharmatatsaādhu-kaārisu* which defines a class of agent nouns as denoting "habitual, professional, or skilled" actors. This would apply well for many agent nominals, not just those derived by zero-affixation as in *cook/2130 → cook/2152* or by the even more productive *-er/3627* (see 2.1), and the less productive, but semantically more transparent *-ist*. In 6.4 we will discuss how these can be compressed in a two-way disjunction between `person` and `professional`.

Since both grammatical and extragrammatical information can accrue, it should be no surprise that lexical information does too. This is particularly clear in the case of compound words: a *foursquare* building is not just one that has four squares, but rather one that 'has a solid appearance', a foursquare position is 'unyielding, firm'.

There is simply no way to derive the 'solid and strong' (Collins) or 'frank, forthright, blunt' (Webster) aspects either from *four* or from *square*. There may be some plausible story about how solid buildings gave rise to this meaning, but it is not even true that square buildings are more solid than cylindrical or hexagonal ones (as English can use *bastion* as a descriptor of strength/firmness just as well), and such stories are at best *post hoc* justifications for the lexical fact. In S19:5.2,6.4 we used the notion of *subdirect* composition to describe this phenomenon. Since this is the single most important element in any formulation of non-compositionality, be it formulaic (1.3), geometric (1.4), or algebraic (1.5), we will spend some time illustrating it from different standpoints. We will stay with binary operations, since, as argued in (Kornai, 2012) in detail, we never need to use ternary and higher arity operations.

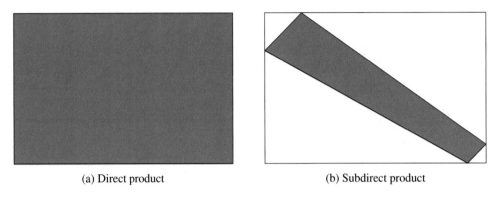

(a) Direct product   (b) Subdirect product

Fig. 2.1: Direct and subdirect products of the same two intervals $[0, 12]$ and $[0, 8]$

Given algebraic structures S_1 and S_2, we form their *direct product* $S = S_1 \times S_2$ by taking the base set to be the Cartesian product of the base sets of the S_i and performing operations coordinatewise. When S_1 is the interval $[0, 12]$ and S_2 is the interval $[0, 8]$, the result is the rectangle S shown in blue in panel (a) of Fig. 2.1. The *subdirect product* is a subalgebra S' of the direct product S that spans all coordinates, i.e. a subset $S' \subset S$ that is closed under the operations and satisfies $\pi_i(S') = S_i$ for $i = 1, 2$, where π_i is the i-th *projection* (a mapping that discards all coordinates other than the i-th). An example is shown in panel (b).

When the elements of the structures are formulas f and g, their direct product is simply their conjunction f, g. When they are polytopes in n-space, their direct product is their intersection. When they are hypergraphs, their direct product is their unification. Subdirect products are, by their very nature, underdefined: a typical subdirect product of formulas f and g may be f, g, h, where h expresses the additional non-compositional content accrued in the process of lexicalization.

What needs to be emphasized here is that 'compositional' versus 'non-compositional' is not a simple yes/no distinction, but rather a matter of degree: for the compositional case the contribution of h is negligible (either not there to begin with, or just irrelevant from the perspective of syntax, semantics, or both), whereas in the truly non-compositional cases h narrows the definition substantively. Recall Frege's definition of compositionality (S19:1.1):

The meaning of a complex expression is determined by its structure and the meanings of its constituents.

One way of abusing this definition is to use the 'structure' clause to sneak in some extra element h. This can be the order of function application: is the function to the left and the argument to the right, or is it the other way around? It could be the manner of application: is it to be performed elementwise, or on the whole set? Could it be sensitive to some hidden variable such as 'upward/downward entailing' that can only be picked out from the context? A well known case that requires some kind of ingenuity is the use of pronouns in quantified expressions, as in *Every farmer beats his donkey*, which we will discuss only after we have a theory of pronouns, see 3.3, and a theory of quantifiers, see 4.5.

Since we are explicit about performing non-compositional operations, we have less need for such tricks, but the issue remains relevant for cases where the operation depends on the signature of the entry. In 1.4 we distinguished ordinary content words, typically embedded as a distinguished vector in its own polytope, and relationals, which require change of scalar product. Next we refine these ideas in terms of a simple eigenspace calculus. This requires a large detour into the workings of neural nets, to which we now turn.

2.3 Relations

We will start with a simple (one layer, fully connected, recurrent) neural net, because it contains as special cases the more modern multilayer architectures we will discuss in 8.3. We follow (Hertz, Krogh, and Palmer, 1991) and use "neural"

> because much of the inspiration for such networks comes from neuroscience, *not* because we are concerned with networks of real neurons. Brain modeling is a different field and, though we sometimes describe biological analogies, our prime concern is with what the artificial networks can do, and why.

To this we add a similar disclaimer in regards to quantum mechanics: we will use standard bra and ket notation, because it is quite entrenched, broadly understood, it helps us easily distinguish inner and outer products (we use both a lot), and it is found in most of the mathematical work underpinning the models. This does not give us any license to wax philosophical about the quantum nature of grammar, and frankly, we will see precious little that would suggest that this would be a promising line of inquiry. Bras are row vectors (typically real valued in our case), kets are column vectors, and that's the end of the story. This is not to say that research into complex weights and connection strengths (Hirose, 2003; Guberman, 2016; Trabelsi et al., 2017) is a waste of time. To the contrary, this appears highly relevant in domains such as music or lidar where phase information is critical, but our problems are of a different nature.

We will use binary neurons numbered $1, \ldots, n$, mapping the two states on -1 and $+1$, and sigmoid squashing $\sigma_\beta(r) = \frac{1}{1+e^{-\beta r}}$. We have a n by n matrix V whose ij element describes the strength of the connection from j to i, so that incoming activations on i are simply $r = \Sigma_j V_{ij} \frac{s_j+1}{2}$. For now we assume a global threshold V_0, so the probability of neuron i firing at time t is given as

$$\sigma_\beta(\Sigma_j V_{ij} \frac{s_j + 1}{2} - V_0) \tag{2.1}$$

At time t, the entire network is described by a state vector (or, in the seductive terminology of Geoff Hinton, a 'thought vector') $\Psi(t) = |s_1, \ldots, s_n\rangle$. Without any input, external or internal, this follows a path on the n-dimensional hypercube determined by a 2^n by 2^n transition matrix P that defines the scalar product $\langle \Psi(t+1)|P|\Psi(t)\rangle$. (The notation suppresses the fact that each value of the thought vector is a complete configuration of $(s_1, \ldots, s_n)$, for a vector of 2^n coordinates altogether.) By external input we mean some neurons getting clamped to a value, as would be the case when we are presented with an image of a candle, something that may trigger recall of the lexical item *candle*. Our focus here will be with internal input, as part of the language understanding process, e.g. when we sense that the word *feminize* contains the string "femin".

We follow the seminal (Little, 1974) to see why we associate a permanent memory engram, such as a lexical item, with an eigenspace (in the limiting case, a single eigenvector) of P. There is a fixed (in the case of learning, adiabatically changing) connection matrix V that determines P by the product probability formula – as first approximation, we assume P is fixed and diagonalizable. (The latter may appear to be a strong assumption, but we note that a random matrix is diagonalizable with probability 1.) The evolution of the system after m steps is given by the mth power of P.

Every state vector Ψ can be expressed in the basis of unit length eigenvectors ϕ_r corresponding to eigenvalues λ_r (initially all assumed different) as $\psi(\Psi) = \sum_r \phi_r(\Psi)$, and since these are orthonormal, the scalar product simplifies to

$$\langle \Psi(t+1)|P|\Psi(t)\rangle = \Sigma_r \lambda_r \phi_r(\alpha(t+1))\phi_r(\alpha(t)) \tag{2.2}$$

Since every state is reachable (i.e. the transition graph is strongly connected), states cycle through $M = 2^n$ steps. Little, 1974 computed by standard methods the time average $\Gamma(\alpha)$ of the probability of the system being in state α as

$$\Gamma(\alpha) = \frac{\sum_r \lambda_r^M \phi_r^2(\alpha)}{\sum_r \lambda_r^M} \tag{2.3}$$

As long as there is a unique largest eigenvalue λ_1, for large M the contributions of all the other eigenvectors and eigenvalues will be negligible both in the numerator and the denominator of Eq.2.3, and we are left with

$$\Gamma(\alpha) = \phi_1^2(\alpha) \tag{2.4}$$

and we see that $\Gamma(\alpha, \beta) = \phi_1^2(\alpha)\phi_1^2(\beta) = \Gamma(\alpha)\Gamma(\beta)$ i.e. the long term probability distribution of β is totally uncorrelated to that of α after a large number of steps. Little interprets this as the system being incapable of having persistent states, and continues with the analysis of the case when there exist two or more largest eigenvalues λ_1 and λ_2, with corresponding eigenvectors ϕ_1 and ϕ_2, obtaining

$$\Gamma(\alpha, \beta) = \frac{\lambda_1^M \phi_1^2(\alpha) + \lambda_2^M \phi_2^2(\alpha)}{\lambda_1^M + \lambda_2^M} \tag{2.5}$$

It follows from the Perron-Frobenius theorem that the largest eigenvalue is isolated, but as Little notes, Eq. 2.5 is still valid if λ_1 and λ_2 are sufficiently close. In his words:

> we thus have the possibility of states occurring (...) which are correlated over arbitrarily long periods of time. It is worth noting too that the characteristics of the states which so persist are describable in terms of the eigenvectors associated only with the degenerate maximum eigenvalues. In this sense these persistent states are very much simpler to describe than an arbitrary state (...) for they involve only that small set of eigenvectors associated with the degenerate maximum eigenvalues, whereas other states (require) the full set of 2^n eigenvectors.

The above makes clear why we are constructing the geometrical model of semantics in terms of eigenvectors and eigenspaces: these, and only these, are naturally given to us as persistent building blocks. Since only a few will correspond to (near)maximal eigenvalues, the dimensionality of the problem is considerably reduced. Assuming a working lexicon of $10^4 - 10^6$ entries, the state space T of 2^n dimensions (an unimaginably large number, given that modern neural nets routinely contain $n = 10^4$ or even more units) is already reduced to below 10^6 dimensions, even if all vocabulary entries were independent.

Part of our goal here is to reduce this number even further, by taking the core elements, at most a few hundred, as our fundamental units, and showing that all lexical entries can all be characterized as sparse combinations of these. With $d = 10^2$ core lexemes, and definitions that are on the average 3 long, we can already capture 10^6 distinct words by three-hot vectors. For practical reasons, we will use softmax, where $|s_1, \ldots, s_d\rangle$, with $s_i = \pm 1$, are replaced by $|\sigma(s_1), \ldots, \sigma(s_d)\rangle$. Here the softmax function σ is given by

$$\sigma(s_i) = \frac{e^{s_i}}{\sum_{k=1}^{d} e^{s_k}}$$

Let us introduce some notation ands terminology. The entire *thought space T* is assumed to be very high dimensional (2^n), and is largely populated by non-persistent material. The *persistent linguistic subspace L* spanned by d persistent eigenvectors $p_1, \ldots, p_d$ corresponding to the core elements is much smaller, a few hundred dimensions at most. The *transient linguistic subspace* or *knowledge representation space* that we will denote by R will involve both L and copies of $L \times L$. One way of thinking of R is that it stores *contexts,* a knowledge representation of what was said so far, but as we will see, the situation is slightly more complex.

It is worth emphasizing that R has very little in way of material that needs to be learned. We need to acquire tens of thousands of words until we become fully functioning adults, but only a handful of these pertain to concepts like being, possession, causation, or purpose that cannot be satisfiably described as vectors. Typically, when we learn a new world, say *hinny*, we acquire a definition `animal, has father[donkey], has mother[horse]`. We often acquire partial definitions, such as for *liger* or *tigon*, defined for us by `animal, has parent[lion], has other(parent[tiger])` without necessarily knowing which is the female parent and which is the male.

We mention word learning only to say that we see it basically as acquiring new weights on existing lexical entries, i.e. as mapping out a new point in L. Exactly how many of the core elements will have to be innate, and how many can be learned constructively, based on the others, is a far-reaching question that we leave for Chapter 8. Our primary interest here is with the representation space R, and for this, L itself will be assumed fixed, changing only on the weeks/months/years timescale that can be neglected for sentence processing which takes place on the subsecond scale.

In the full analysis of linguistic behavior we would have to include states such as produced by particular activation patterns of the visual cortex that prompt us to recognize and name some object, and similarly for other sensory modalities, but we keep our focus on linguistic inputs, especially as we know very little about the full T. In principle we still have the operator P that describes the evolution of Ψ over the entire T, but for our goals it is sufficient to keep the first and second order terms describing points of L in terms for the eigenvectors $p_i \in L$ and scalar products in terms of $L \times L$. For a trainable neural net, our goal will be to learn the top left block of P_L that maps L on itself, and some mappings $L \to L$ which we roll into the segment of Ψ that corresponds to R. If $d = \dim L$ is just a few hundred, the entire R segment, composed of the rank 1 matrices over the basis $p_1, \ldots, p_d$ will have only d^2 (a few times 10^4) elements, so $r = \dim R$ still compares very favorably to $\dim T = 2^n$.

We will denote the projection of a thought vector Ψ of R by Φ, and the part of the transition matrix P that acts on R by P_R. We take the rectangular (landscape) block to the right of this, describing the weights coming from afferent neurons responsible for linguistic input, and the rectangular (portrait) block below it, describing the weights to efferent neurons responsible for linguistic output, as given. The second, much larger, diagonal block, devoted to non-linguistic perception, thought, and action, is outside the scope of our work. This simple partition of the transition matrix into these four major blocks constitutes the linear algebraic version of the broad research program implicit (often explicit) in much of cognitive linguistics, known as embodied cognition.

Besides restricting our attention to a thin slice, moving from T to R also involves a change in time scale: while vectors in T can diffuse in all kinds of ways over a few hundred milliseconds, within R the situation will be much more stable, and within L it will be reflected mostly in the shrinkage of the coefficients driven by the size of λ_1. Since this shrinkage is uniform (recall that for the persistent eigenvectors all eigenvalues are roughly as large as the first one), it is repaired by the softmax, and we can expect stability in R for a few centiseconds. Our new timescale is chosen to guarantee that ordinary linear algebraic operations such as vector addition or changing the scalar product can be performed on stable arguments by the neural units, which still operate on the millisecond scale. Within R, and for this new timescale, we can write actually write $P_R = \lambda I$, where λ is only very slightly smaller than 1.

After these preparations we are ready to take on the problem we started out with, the representation of relationals. What does our simplest (0th) relational, *is/is_a*, contribute to a sentence like *Bill is brave* or, to put it in stark terms, what is the meaning of *is* here? For now, we ignore the morphology that dictates the subject to be 3rd person singular and the tense to be present, and just consider the stem *be*. Eq. 1.1 makes clear, *is* coerces $\mathbf{v}(\text{Bill})$ into the $\mathbf{v}(\text{brave})$ subspace, by creating an R that makes $\langle \mathbf{v}(\text{Bill})|R|\mathbf{v}(\text{brave})\rangle$ positive. What *brave* actually means will be discussed in Chapter 7, here we can make the simplifying assumption that it is one of our eigenvectors p_j, and *Bill* is another eigenvector p_i. To obtain this modified geometry we need to replace $P_R = \lambda I$ by $\lambda I + s\pi_j^i$, where s is some positive scaling factor that we use for perturbing the state transition matrix λI, and π_j^i projects p_i on p_j (the softmax helps to make this non-zero).

What *is* accomplishes is to change the descriptor of system evolution from λI to $\lambda I + s\pi_j^i$. The way we formulated it so far would make this dependent on p_i and p_j, but we certainly don't want the meaning of *be* to depend on its arguments. Since π_j^i is just the outer product $|p_i\rangle\langle p_j|$, a d by d matrix of rank 1, we will write $R = \lambda I + s|\rangle\langle|$, as this removes the dependence on the subject $\mathbf{v}(\text{Bill})$ and on the predicate adjective $\mathbf{v}(\text{brave})$. The notation is a bit unusual, but the idea is clear: *be* is a two-variable functional scheme that will, for any transition matrix P_R produce a new transition matrix P_R':

$$P_R(t+1) = P_R(t) + s|\text{=agt}\rangle\langle\text{=pat}| \tag{2.6}$$

This meaning is reasonably abstract, and as we shall see, it covers remarkably well the ranges of concrete meanings that dictionaries such as LDOCE or Cambridge Online assign to *be*. Treating *Bill* and *brave* as unanalyzed (and really, in the current situation no further insight could be obtained by considering more detailed analysis of their respective meanings), the 4lang entry for *be* simply says:

```
be van sum bycl 2585 U
   =agt is_a =pat
```

and we see the *thematic* variables =agt, =pat in exactly their expected linking function here and in Eq. 2.6. We may want to make these, and equation 2.6 more symmetrical by adding in the other outer product, $|\text{=pat}\rangle\langle\text{=agt}|$ with the same scalar multiple s (which makes sense in terms of keeping the matrix describing the scalar product symmetrical) or with a different multiple t. Investigating such possibilities must be left for future research.

We will discuss the details of linking in 2.4, here a skeletal version is sufficient: first arguments (subjects) of relationals are by definition agents, second arguments (objects) are patients. Since three-place predicates are entirely eliminated (Kornai, 2012), the most complex entries we will have to deal with will be like *be*.

Simplifying English syntax considerably, the element to the left of the verb is the subject, the element to the right is the object, and for the sentence *Bill is brave* these are substituted in the graph, to yield `Bill is_a Brave`, and Eq. 2.6 to yield

$$P'_R = R + s|\mathbf{v}(\text{Bill})\rangle \langle \mathbf{v}(\text{brave})| \qquad (2.7)$$

Whatever situation obtained in R before the introduction of this sentence, afterwards we have a new scalar product that guarantees the extension of *Bill* to be included in that of *brave*.

This is precisely the sense put first by the Cambridge Online Dictionary: 'used to say something about a person, thing, or state, to show a permanent or temporary quality, state, job'. Given our abstract stance, we don't have to subtype the subject for 'person, thing, or state' or the object for 'permanent or temporary quality, state, job'. The semantics already covers, without any special effort, cases that fall between the cracks of these ad hoc categorizations. Consider *Countermanding an order from high command is insubordination*: is the countermanding of an order a person, a thing, or a state? Is insubordination a quality, state, or job? Or consider *Bacon is Shakespeare*: what is meant here is obviously the identity of singular terms, not just predicating some vaguely Shakespearean quality about Bacon.

Consistent with our approach, the modeling of the representation space R contains not just vectors from L, but also a matrix N that describes the static (lexically given) portion of the lexicon. The main component of N is the (transitive closure of) the `is_a` hierarchy, guaranteeing once and for all that *anger* is a *feeling* or *black* is a *color*. For those lexical entries that are given as conjuncts of other lexical entries (see 1.3) this much is

dust   sufficient: for example *dust* is given as `substance, fine, dry, particle, powder, <dirt>` so we add 1 to entries of N at the row corresponding to *dust* for all columns corresponding to *substance, fine, ...*

Given that *is* requires matrix operations to make sense of, and given the existence of three-argument predicate words like *sell*, the road seems open to higher tensor operations. This is the path taken in (Smolensky, 1990), a research program remarkable both for its analytical clarity and its lack of traction. Clearly, after thirty years of unabated Moore's Law growth, we still don't have what it takes to cope with superexponentiality, and this makes it imperative to throttle the dimension growth. In this book, we stop at two, and model all linguistic phenomena by using vectors for words but matrices for contexts. The recent work linking the tensorial program to BERT Moradshahi et al., 2020 will be discussed in 8.3.

So far, we have described the vectorial calculus for both contingent (dynamically changing) *is* and necessary (static) *is_a* by means of shifting the inner product. This is already sufficient for intransitive predication such as *John sleeps*. We define *sleep*

sleep   as `rest, lack conscious`, and we defer our dyadic negation primitive `lack` to

Chapter 4. $\mathbf{v}$(rest) is just one of our lexical eigenvectors p_i, and the mechanism simply adds $s|(\text{John})\rangle\langle p_i|$ to the previous transition matrix $P(t)$. How this actually takes place during sentence understanding is quite complex: the syntax must recognize that *sleeps* is a verb, and that *John* directly precedes it, therefore, in English, it is the *subject*. Ours in not a book about syntax, and we leave the can unopened, but we note that several computational systems (in particular, LFG and UD parsers) exist that perform this task quite well (Butt et al., 2002; Qi et al., 2020). The meaning of *sleep* is the operation

$$P(t+1) = P(t) + s|{=}\texttt{agt}\rangle\langle\texttt{sleep}| \tag{2.8}$$

Therefore, subjects (links of type '1') amount to replacing $=\texttt{agt}$ in Eq. 2.8 by the vector assigned to *John*, or whatever the subject might be. This would be straightforward β-reduction in λ-calculus, but for smoother handling of variadicity we will use a slightly different mechanism.

Grammarians have long noted that objects (links of type '2') differ quite a bit from subjects, and apparently these differences extend as far as differential loss in aphasia (Hanne, Burchert, and Vasishth, 2016). Here we will see that the semantic mechanism we associate to resolving $=\texttt{pat}$ links is very different from that used in the resolution of $=\texttt{agt}$. We again begin with a simple sentence, this time *John eats fish*. For now, we will entirely ignore the finer aspects of genericity, tense, and aspect, and concentrate on the basic meaning, that the relation of eating is holding between subject and object. This much (and only this) is expressed in the formula $\texttt{John eat fish}$. However, the method we used for intransitives cannot simply be repeated, for we cannot simply perturb the transition matrix $P(t)$ by some matrix E for $\texttt{eat}$, because there is no such matrix. Nor would we want to introduce a matrix for this, for this would bring in tow not just matrices (2-tensors) for every transitive verb, but 3-tensors for every ditransitive, and so on, perhaps up to 5-tensors for verbs like *promise* with separate modes for subject, object, recipient, issue date, and term date.

We take our cues from languages with object incorporation, found in many Iroquoian and Austronesian languages, and consider an analysis along the lines of $\texttt{John}$ $\texttt{fish-eat}$. The construction is clearly improductive in English (one can perhaps attempt back-formation *?to brewbake* or *to babysit* from synthetic compounds *brewbaker, babysitter*), but it is normal in many languages, and serves our goal here: once $\texttt{fish-eat}$ is admitted as a verb, we can deal with it in the manner of Eq. 2.8. For this, all we need is a vector $\mathbf{v}(\texttt{fish-eat})$ which we define as $\mathbf{v}(\texttt{fish}) + \mathbf{v}(\texttt{eat})$. Notice, that we automatically obtain $\langle\mathbf{v}(\texttt{fish-eat}),\mathbf{v}(\texttt{eat})\rangle > 0$ i.e. that *fish-eating is_a eating* even if $\mathbf{v}(\texttt{eat})$ and $\mathbf{v}(\texttt{fish})$ are orthogonal.

Perhaps more unexpectedly, we also obtain that *fish-eating is_a fish*. At first blush, this may seem counterintuitive, but in fact putting *fisheating* in conceptual space somewhere in the intersection of the *fish* and *eat* cells is reasonable. To be sure, the same can be said of the eating of a fish, i.e. eating performed by fish (cf. the classic *shooting of the hunters* examples) so the notion $\texttt{fish-eat}$ is somewhat underdifferentiated, but this is a small price to pay, especially as the proposed semantics will fit smoothly with commonsensical reasoning about grammatical objects, e.g. that if x is part of fish, eating fish will involve eating x or removing it prior to (or in the course of) the meal. The main point is that vector addition brings the fish in the scope of eating.

In terms of type signature, `fish-eat` is still a vector in the space L spanned by the core elements. Being derived on the spot by the semantic process means assigning semantics to the VP by means of replacing =pat in Eq. 2.9 by the vector that defines the object of the verb:

$$P(t + 1) = P(t) + s|\text{=agt}\rangle\langle\text{V + =pat}| \qquad (2.9)$$

While the object-incorporated `fish-eat` is helpful in deriving this equation, the result in no way depends on it: we have a declarative formula that links the subject, the verb, and the object in a compositional fashion. It is now only computational efficiency, or perhaps hacker esthetics, that determines how to implement this. Perhaps evidence could be gathered using quantified subjects and/or objects, but we leave this for later, noting here only that both classical-style S $\rightarrow$ NP VP analysis and more modern synchronous rewriting approach discussed in 1.6 are feasible. The key takeaway here is that we only need standard linear algebraic operations, namely vector addition, outer products, and linear transformations, to do all this. There are many key pieces still missing, but we will supply these as we go along without recourse to higher order multilinear devices.

Let us now consider a simple (though not entirely linguistic) example on four variables. People are quite adept at solving analogical puzzles like *Apple is to Steve Jobs as Microsoft is to X* or *Dollars are to the US as Y is to Great Britain*. This requires the discovery of some pivotal relation connecting the givens, something we could formalize as x[company] has y[CEO] or x[currency] used_by y[country]. How does this discovery proceed? In 4lang terms we have

```
Apple           company, @Apple
Microsoft       company, @Microsoft
US              country, @US
Great_Britain   country, @Great_Britain
Steve_Jobs      person, CEO, Apple has CEO
Bill_Gates      person, CEO, Microsoft has CEO
dollar      currency, US use, Australia use, Zimbabwe use
pound       currency, GB use
```

When we hear about Apple, we are not yet sure whether it's the fruit, Apple Records (the Beatles company), or the computer company, though we are likely primed toward the first and third options. But as soon as Microsoft appears in the same structural position, disambiguation is complete, we have reached the vector that has *company, corporation, computer, hi-tech, software* active, as well as the brand names *iphone, mac, ...* and some further encyclopedic knowledge such as *Steve Jobs, Steve Wozniak, Johnny Ive, ...* For Microsoft, we reach another set of products, brand names, people, headquarters, stock symbols, etc, but the *company, corporation, computer, hi-tech, software* coordinates of the Microsoft vector are equally hot. Once we hear about Steve Jobs, the nodes *Apple* and *person* are active, together with all sorts of encyclopedic knowledge about black turtlenecks, and maybe even a strong visual engram of his face. We know that he is the Apple person, so we search for a Microsoft person, and find it in Bill Gates.

In S19:5.7 we discussed how this search can be implemented by spreading activation, but here our concern is with word vectors. We know from (Mikolov, Yih, and Zweig, 2013) that $\mathbf{v}(Jobs) - \mathbf{v}(Apple) + \mathbf{v}(Microsoft)$ lands us in a polytope labeled by *Gates*. Since this works just as well with the sparse vectors we use here, the real question is how this formula comes about? The elementary building block used here is one that comes for free in any associative memory: reconstructing a whole from partial information. In general, this is the inverse of the projection, obtained for each p_i by considering the half-space it is in, i.e. all vectors q such that $\langle q, p_i \rangle > 0$. This half-space also includes elements of T not in L such as visual images and all sorts of encyclopedic knowledge.

As soon as *apple* is heard, the entire subspace $A \lhd T$ which falls in the positive half-space of *any* $\mathbf{v}(Apple)$ vector (including the one for the fruit) is activated. Since this much is common to the processing in any word w, we will denote by $T(w)$ the entire halfspace $\{x | \langle x, \mathbf{v}(w) \rangle > 0\}$. What the *x is to y as z is to w* construction triggers is a search for some scalar product R where both $\langle \mathbf{x}|R|\mathbf{y} \rangle > 0$ and $\langle \mathbf{z}|R|\mathbf{w} \rangle > 0$ are simultaneously satisfied. There will be many such Rs, but our interest will be with those that have $\mathbf{v}(Jobs)R\mathbf{v}(Apple)$. Since R is some binary relation such as 'founder of', 'CEO of', 'visionary leader of', using Eq. 2.9 we obtain $\langle \mathbf{v}(Jobs), \mathbf{r} + \mathbf{v}(Apple) \rangle > 0$, which will be best satisfied if $\mathbf{v}(Jobs) = \mathbf{r} + \mathbf{v}(Apple)$. In this situation, $\mathbf{j} = \mathbf{a} + \mathbf{r}$ we also have $\mathbf{j} - \mathbf{a} + \mathbf{m} = \mathbf{m} + \mathbf{r}$, the Mikolov parallelogram rule, and this will hold independent of the choice of $\mathbf{r}$ or R.

The main difference from (Mikolov, Yih, and Zweig, 2013) is not that the parallelogram was simply an observation there, whereas it was deduced here. (For a different form of reasoning that leads to the same conclusion see Gittens, Achlioptas, and Mahoney, 2017). However much we may care for theory and feel good about explaining a salient property of embeddings, clearly the two positions cannot be distinguished observationally. But, as our motto goes, there is nothing as practical as a good theory: here, for the first time, we can actually verify what a vectors like $\mathbf{v}(Jobs) - \mathbf{v}(Apple)$ or $\mathbf{v}(Gates) - \mathbf{v}(Microsoft)$ really mean. In our theory, these must correspond to the vectors for verbs like *direct* or *found*, while $\mathbf{v}(GB) - \mathbf{v}(pound)$ or $\mathbf{v}(Zimbabwe) - \mathbf{v}(dollar)$ must be the vector for *use* or *has*. Since we aim at compositionality with our vector calculus, we may even compare the differences to the vectors obtained from computing the semantics of *is the legendary CEO of* or *uses a monetary unit called the*. To compute such complex meanings requires more knowledge of the atoms like *of* and of the processes of syntactic combination that go beyond subjects and objects: these are the tasks we now turn to.

2.4 Linking

Our first task is to deal with the primitives we couldn't so far (as of Release 2.0) eliminate from 4lang. There are only a dozen or so such elements, listed here for convenience: =agt =pat before er_ for for_ gen has is_a lack other part_of wh. The two linkers =agt and =pat are different from the others. This is clearest from the hypergraph view where they are used for distinguishing edges, everything else is used for distinguishing (hyper)nodes. In the vectorial view, it is only these elements that trigger the creation of rank 1 perturbative matrixes as in Eqs. 2.6, 2.8, and as we shall see shortly, Eq. 2.12. This is not to say that a trainable system for static or dynamic vectors cannot be tricked into finding vectors corresponding to =agt, =pat, or even broader sets of linkers, but this takes quite a bit of effort, because the training data needs to enriched to graphs that show the links (Mohammadshahi and Henderson, 2020). While our main interest here will be with the linkers =agt and =pat, we need to first discuss the others, so as to get a better sense of how primitives in general are to be handled.

First, note that some of the remaining entries on the list are primitive only from the view of formulas or hypergraphs, but not from the vectorial point of view: for example gen is defined as the vector with equal components $1/d$ in each of the d dimensions, and is_a means set-theoretical containment of polytopes given the right scalar product. There are elements such as lack, wh, and other which make only sense in terms of a larger theory of negation (see Chapter 4), directed computation (see Chapter 7), and indexicals (see 3.3) and will be discussed in detail together with these theories.

Second, even internal to the formulaic or algebraic perspective, the primitivity of many entries is a matter of choice. For example, here we decided to leave part_of unanalyzed, but in fact the commonsensical *part* defined as part re1sz pars czelslcl 1997 u N in, connected could trivially be extended by "of _" mark_ =pat to yield a satisfactory analysis of this relation. Conversely, the definition of *in*, in -ban in w 2758 c G place, =agt at place, place contain =agt, "in _" mark_ place could be converted to one where *in* is left unanalyzed =agt in_ =pat. In fact, it is one of the strongest technical selling points of the algebraic approach that we don't need to make a hard and fast commitment to a fixed set of primitives. With the Appendix, we do make a specific choice, but only because there is no doubt that "the difficulty [of understanding] is compounded by the authors' apparent reluctance to be pinned down even temporarily to a particular explicit version of the theory" (Partee, 1985).

A longstanding observation, going back at least to Meillet, 1912, is that primitive or primitive-like vocabulary elements have a strong tendency toward grammaticalization. In 4lang this is evidenced by the fact that over half of our primitives p are classified as grammatical formatives G, and conversely, primitives, which make up 1.68% of the word senses considered, are over a quarter of the grammatical formatives.

In many cases, like the comparative, we have a clear tradeoff, either analyzing it as a grammatical primitive er_ or as a plain suffixal morpheme -er. In such cases, the underlying conceptual schema "there are two things, one $>$ the other" is fairly transparent, and analyzing the morpheme in terms of the schema makes sense, even if the schema itself must be treated as a primitive. The same can be said for `for/824` 'price', which is easily identified through the `exchange_` schema of 1.2. Here we actually succeed in doing away with primitivity, in that *exchange* lends itself to analysis `before(=pat at person),` `after(=pat at other(person))` and the 'commercial exchange' frame involves the conjunction of two such exchanges, that of the object and that of the money (see 3.3 for further discussion).

exchange

In general, the fact that we must resort to abstract conceptualizations requiring the use of Schankian scripts or Fillmorean frames is a strong indication that the word or morpheme in question is a (near) primitive. Consider another sense of *for*, `for_/2782` 'dative of purpose' as in *born for achievement*. We must recognize this in the grammatical sense, because the conceptual schema is somehow too diffuse to articulate even by a script or frame. Within some extremely broad pragmatic limits, anything can be for anything. JPEG uses the discrete cosine transform for compression. The men of Chou used the chestnut *(li)* for making the common people tremble *(li)*. As long as we cannot define the dative of purpose by a conceptual schema we must list it as a primitive.

Here we illustrate near-grammaticality on the prepositions *at*, (other prepositions with a clear spatial meaning are deferred to 3.1, where the `place_` schema is discussed in detail) and `for_/2782`. In LDOCE, *at* is defined as 'used to say exactly where something or someone is, or where something happens', in CED as 'used to show the position of a person or thing in space or time'. Since our analysis of *where* would be 'at wh' this would result in an analysis that has *at* on the rhs as well, making it a primitive. But even if it is a primitive, we can't afford to stay silent on what it means – to the extent we do we are left with trivial equations of the $x = x$ sort.

Clearly, *at* is a binary relation (we write these in SVO order) `=agt at =pat`, where `=pat` is strongly subtyped for location, be it spatial or temporal, so strongly that otherwise unspecified entities like *Jim's* have to be typcast to location if we are to make sense of expressions like *We meet at Jim's*. This selectional restriction on the second argument will be expressed by the clause `=pat[place]`. Using the same mechanism as in Eq. 2.6 we coerce the prepositional object to be a place by

$$P_R(t + 1) = P_R(t) + s|\text{=pat}\rangle\langle\text{place}| \qquad (2.10)$$

In contrast, the subject `=agt` is left untyped: it could be a physical object, a person, or even an event, what LDOCE describes as 'something happens'. What `at` means is that it is happening at the origin of the abstract `place_` where `=pat` is located:

$$P_R(t + 1) = P_R(t) + s|\text{=agt}\rangle\langle\text{origin}| \qquad (2.11)$$

In other words, while the ground (object) gets construed as a place, the figure (subject) gets coerced into the origin of the place coordinate system. Since *at* dictates to perform both of these operations, we have

$$P_R(t+1) = P_R(t) + s(|=\text{agt}\rangle\langle\text{origin}| + |=\text{pat}\rangle\langle\text{place}|) \tag{2.12}$$

and as usual, we leave it to the unification mechanism to guarantee that it is the origin of the ground (prepositional object) that the subject is coerced to, not the origin of some other coordinate system. Notice that the method of Eq. 2.12, adding two rank 1 matrices to produce a rank 2 matrix, could in principle be extended to the modeling of ditransitives by rank 3 matrices and so forth. But this would require the addition of further theta roles (variable-binding term operators) besides =agt and =pat, a step we will not take here.

What happens during the analysis of Bill at office? We must select the eigenspace for Bill. (This is not trivial, there may be several Bills around, and we need to do considerable work to choose the right one, see 3.3.) We also must select the right eigenspace for office, and most important, we must typecast Bill as figure and the office as ground, for this is what at means. For the ground, we have a complete abstract coordinate system, and as we shall see in 3.1, offices (and buildings in general) are trivially mapped to this. To conceptualize Bill as being at the office requires no more than applying to him the predicate *inside* that comes with this coordinate system.

At first blush, this may look as if we are just postponing the problem by reducing at to inside. But as we shall see, *inside* comes for free, as a prebuilt component of the coordinate system. The real work is in the typecasting, which creates a new instance of the standard coordinate system with the office at its origin, and maps many of the features of this system appropriately, in the kind of process described by Fauconnier, 1985. We call this process *coercion*, not because it is that different from what Fauconnier calls 'projection mapping', but rather because we wish to emphasize its forcible, Procrustean aspect. By understanding, mental reality is created. at forces Bill to be inside the office premises. We may entertain different notions, perhaps he is out shopping, but to understand the sentence is tantamount to having a concept of him in the office. We will return to the geometric interpretation of the coercion mechanism in 3.3.

Returning to the problem posed by at, we can reformulate this as computing a sequence of three thought vectors, the first one describing the state of the linguistic concept space after having heard (and recognized) *Bill*. This is simply $\Psi(1) = \mathbf{v}(Bill)$. The second one, $\Psi(2)$ is after having heard and processed *(is) at*, and the third one, $\Psi(3)$ characterizes the state of the mental space after having heard the entire expression *Bill (is) at (the) office*. We assume that *at* makes available the entire system of conceptual coordinates that we will describe in 3.1. The function of at described in Eq. 2.12 is twofold: it typecasts *Bill* as 'figure' and it also typecasts *office* as 'ground'.

Traditional constituency tests make it clear that we process the material in A(BC) rather than (AB)C order, and it is also evident from self-inspection that *Bill (is) at* is not a coherent thought, whereas *at (the) office* is, suggesting that $\Psi(2)$ will be hard to pin down beyond the obvious fact that it already contains $\mathbf{v}(Bill)$. The effect of combining at and *office* is to coerce the $\mathbf{v}(office)$ eigenspace (which would be an eigenvector if we assumed office to be unanalyzed) into a few dimensions of the ground construct, effectively equating the office location with the origin of the coordinate system, and its walls with the 'body' that we will describe in greater detail in 3.1.

We make no effort to describe the momentarily disconnect between the sequence of thought vectors $\Psi(1), \Psi(2), \Psi(3)$ that gets resolved only after subject and object are both substituted, though it would be fairly easy to bring the usual techniques of dynamic semantics to bear, and we leave it as an exercise to the reader to convince themselves that micro-parsing of *Bill is at the office*, with the addition of *is* and *the*, is still feasible in five steps. (Hint: assume Eq. 2.6 for *is* and assume that *the* contributes nothing. The only hard part is to make sure that the prepositional object *office* is combined with *at*, and it is this entire PP that is the object of *is*.) In 3.1 we will extend this treatment from `at` to a whole slew of locative prepositions (or postpositions, or case endings, depending on the language).

Let us turn to an explanation for more abstract, non-spatial binaries such as `for_`, `has`, `ins_`, `lack`, `mark_` and others using purposive `for_` as our example.

```
company valllalat   negotiatio firma      2549 N
        organization, for_ business
cutlery evo3eszko2z ferramentum sztuc1ce   3354 N
        knife is_a, fork is_a, spoon is_a, for_ eat
hand    ke1z        manus      re1ka       1264 N
        organ, part_of arm, human has arm, for_ [move gen],
        wrist part_of, palm part_of, five(finger) part_of,
        thumb part_of
handle  fogo1       manubrium  ra1czka     834  N
        part_of object, for_ hold(object in hand)
knife   ke1s        culter     no1z1       1256 N
        instrument, for_ cut, has blade<metal>, has handle
lens    lencse      lenticula  soczewica 3344 N
        shape, part_of camera, light/739 through,
        for_ clear(image), <glass>[curve],
        image has different(size), <look ins_>
money   pe1nz       pecunia    pienia1dze 1952 N
        artefact, for_ exchange, has value, official
norm    szaba1ly    regula     norma       3361 N
        good for_ society
useful  hasznos     utilis     przydatny 3134 A
        for_ gen
```

Just as we require a whole naive theory of space to make sense of locatives, we must invoke a whole naive theory of purpose to make sense of purpose clauses. The cardinal element of this is the premiss that artifacts are created for their utility. The naive defense of theism often relies on some form of this theory: since artifacts are created for a reason, there must be a creator. This is not to be confused with the Aristotelian notion that everything happens for a reason, which we interpret as a pure epistemological stance urging to find the causes, a matter we return to under `cause_` shortly. For now we state the premiss as our "Rule of facilitation":

$$gen\ use\ =agt,\ after(=pat[easy]) \tag{2.13}$$

This is another schema, one that we may consider the definition of `for_` (if we wish to go beyond the idea that the dative of purpose is an unanalyzed primitive) or even
`ins_`   the instrumental `ins_` , which we define as `=pat make =agt[easy]`. The operation of the facilitation schema can be illustrated on *John used a spoon for cutting the pizza*. Any parser will return something like `John use spoon, spoon for_ {cut pizza}` so 2.13 is invoked, we conclude that cutting the pizza was easier than to have done this by his bare hands. Clearly, this is not as good as cutting it with a pizzacutter or a knife, for which it is true that they make the pizzacutting task easy, and we use the dative of purpose precisely because we want to avoid the implication that spoons are tools in general use for cutting pizza. It is a means, but not the most effective means.

In S19:3.6 we discussed rules as being 'entirely outside the sphere of human (individual or social) ability to change, exceptionless, and strict'. This is not to say that naive rules like 2.13 are the final say in our understanding of the world. As we shall see in Chapter 5, we can, and do, have a better theory of probability than the naive theory thanks to Pascal, Laplace, and Kolmogorov; we have a better theory of space and time than the one articulated in Chapter 3 thanks to Euclid, Descartes, and Einstein; and so on. These theories are never hard-wired: they typically build on centuries of work by giants of intellect, they require considerable formal schooling to understand, and they rely on fields of knowledge, mathematics in particular, that have no support in natural language semantics. But the naive rules of how we perceive probabilities, space, time, cause, effect, and the like are built in, and it takes as much effort to unlearn them as to teach oneself to fly by means of controlling an airplane. Naive rules are exceptionless in the same way: once the conditions are met, we have no means to suspend their application. Once you learn that fish are animals that live in water, whales are fish, and it takes special effort to unlearn this implication.

What do we mean when we say that companies are organizations `for_` (doing) business, or that cutlery is `for_` eating? We mean that use of these devices makes the activity easier. This extends to 'activities' like society, which could easily be construed as nominals: norms make it easier to have, to govern, or just to live in, some kind of society; companies make it easier to do business, etc. What is common to all these definitions is that the object of `for_` refers to the matter made easier, and the subject of `for_` refers to the matter acting instrumentally. The resulting `after` state will be discussed in 3.2, but we note here that we treat this as a substantive part of the knowledge representation, one that may require different time-indexed copies of vectors already present.

Closely related to `for_` is `ins_`, which we use in `x has instrument y` rather than `x is instrument of y` order in all definitions (about 0.65% of the total) where it appears. For the most part, `ins_` is the inverse of `for_`:

```
bite  harap mordeo gryz1c1 1001 V
      cut, ins_ <tooth>
tooth fog   dens   za1b    827 N
      organ, animal has, hard, in jaw, bite/1001 ins_,
      chew ins_, attack ins_, defend ins_
```

where we could have just as well said

```
bite  harap mordeo gryz1c1 1001 V
      cut, <tooth> for_
tooth fog   dens   za1b    827 N
      organ, animal has, hard, in jaw, for_ bite/1001,
      for_ chew, for_ attack, for_ defend
```

Having clarified that primitive status is not an external given but rather a lack of ability to find a suitable definition, and that grammaticalization is neither necessary nor sufficient for primitivity, we can now turn to the most recalcitrant of our primitives, the linkers. Whether we keep `ins_` *(karaṇa)* as a primitive or accept the analysis given above, `4lang` covers the system of Pāṇinian kārakas reasonably well. Verbal 1-links point to subjects, which are for the most part Pāṇinian agents *(kartṛ)*, and we will even capture some of the definition as 'the independent one' (1.4.54). Note, however, that we also speak of subjects for prepositions, pure statives, and experiencer verbs, etc. that many grammatical theories prefer to handle by a variety of other means. `for_` is goal *(karman)*, locatives *(adhikaraṇa)* and ablative (source, *apādāna*) will be discussed further in 3.1, but their treatment is largely similar to that of `at`.

There is one notable sense in which our treatment is clearly inferior to the Aṣṭādhyā-yī, the preferential attachment of the kārakas. Pāṇini (1.4.42) uses the superlative *sādhakatamam* to define the instrument not just as the means, but as the *most effective* means to the goal, and similarly *īpsitatamam* as what is *primarily* desired by the agent (1.4.49). Needless to say, `4lang` will have the means to express superlatives – these will be derived using the comparative `er_ '>'` as `er_ all`. What it lacks is the kind of powerful metalanguage that the Aṣṭādhyāyī deploys in full. Our theory of naive grammar (see 2.5) simply doesn't have the means for comparing alternative derivations, even though such a facility would also be useful in phonology for implementing Optimality Theory. Regretfully, we must leave this for future work.

The one kāraka missing from our system is the recipient *(saṃpradāna)*. As discussed above, there is considerable computational pressure to avoid 3-tensors and higher multi-linear elements, and we will model ditransitives by decomposition:

```
give ad    do    dac1      113 V
    =agt cause_ {person has =pat}, dative_ mark_ person
buy  vesz emo   kupowac1 2609 V
    =agt receive =pat, =agt pay seller, "from _" mark_ seller
sell elad vendo sprzedac1 595 V
    =agt cause_ {buyer has =pat}, buyer cause_
    {=agt has money_}, dative_ mark_ buyer
```

Recall that in 2.2 we already derived the slot-fillers `buyer` and `seller` by the agentive suffix `-er/3627`. We treat *to* as a dative case marker, but we could just as well treat it as a genuine locative case: after all, the recipient will have the object in physical possession in the default case (see Hovav and Levin, 2008 for further discussion). For

cause_   `cause_` we adapt a *post hoc ergo propter hoc* analysis: we define `x cause_ y` by `x before y, after(y)`. This falls quite short of a proper analysis of single and multiple causes, and it encourages precisely the kind of errors that are rampant in the identification of cause-effect relations. But there is no reason to assume that sophisticated data analysis of the kind urged in (Pearl, 2009) can be replicated in natural language semantics, especially as the kind of statistics and probability theory that undergird the modern scientific understanding are not supported by natural language (see Chapter 5). We compare the commonsensical definition of causation to the counterfactual *sine qua non* definition in Chapter 6.

part_of   In the case of `part_of` the situation is different: there is no great conceptual gap between the naive theory of containment and set theory. Axiomatic set theory can of course approach a lot of problems that do not even arise in naive mereology, but we see no reason not to apply set theory here as well. Since we already have a containment primitive `in`, all we need is that `=agt` and `=pat` are `connected`. In spite of its reducibility, we keep `part_of` in the definitions, where it is used predominantly with body parts `nose part_of face` and parts of natural objects `fruit part_of plant`. This leaves one more relational to consider, `has`, which we use primarily in the notional sense of possession, as we handle inalienable possession by `part_of` already.

has   The Appendix reveals some ways we could further reduce our already small list of primitives. By defining `has` as `=agt control =pat, =agt has =pat`, we have identified only one defining aspect of ownership, control, but left `has` as primitive, since it occurs on the right hand side of the definition as well. Almost a third of our definitions contain `has`, but these could be often traded off for `part_of` as in *knife*

knife   `instrument, for_ cut, has blade<metal>, has handle`. As of Release 2.0 it is not yet clear how more abstract relationships, where control alone seems insufficient to explain what is going on, should be handled. Consider `way ult via`

way   `droga 2484 u N artefact, gen move at, has direction` or `black fekete niger czarny 761 e A colour, dark, night has colour,`

black   `coal has colour`. Perhaps we will want to say that colors are part of the object, or that the road controls its direction, but this is not evident, and for now `has` must be assigned a matrix to be computed on the entire set of definitions, see 9.5.

So far we connected, to the extent feasible, `4lang` to the Ashtādhyāyī. Sadly, we don't have a large body of machine-readable Sanskrit fully parsed for kārakas, and even if we did, the subtle interplay between tense, voice, and deep cases would fast overburden the skeletal grammatical mechanism provided here. We also explained how the mainstays of case/valency systems, such as datives, locatives, and instruments, can be reconstructed without assuming link types beyond '1' and '2', by taking these as relationals that typecast their arguments, the expressions that appear at the two ends of the named links.

In terms of the amount of fully analyzed text available, Universal Dependencies (UD) is the single most influential cross-linguistic framework of grammatical description (Nivre, Abrams, Agić, et al., 2018). While many other schools offer a broader variety of analyses, these, with the possible exception of tagmemics (Pike, 1982) and Relational Grammar (Perlmutter, 1980), rarely extend to a broad selection of languages. Also, the dominant style of linguistic analysis is the in-depth study of a restricted range of syntactic phenomena, ideally across many typologically diverse languages, rather than the in-breadth analysis of an entire language, which again makes it hard to link contemporary computational linguistics with linguistic theory. Here we assume the reader is familiar with UD, and compare 4lang to UD, pointing at other frameworks only in a few places. Generally, 4lang is on the sparse or 'lumping' side of the comparison, not just in relation to UD, but also in relation to other well-developed theories like LFG, HPSG, or MP.

Since UD distinguishes dependency links by the category of the head and the dependent, it naturally keeps notions like nsubj and csubj (nominal and clausal subjects) separate, and similarly for obj and ccomp. 4lang, with its roots in the theory of Knowledge Representation, where the proliferation of link types has emerged as a significant problem early on (Woods, 1975), admits only one other link type, '0' (is_a), which subsumes most of the other link types used in UD, such as amod, appos, nummod and advmod. In a strictly link-based system such as UD it is a practical necessity to have a separate link type for coordination: in 4lang we just use comma-separated concatenation.

Both UD, and other theories of valency (for a summary, see Somers, 1987) offer a broad variety of links, and our method of treating these as having their own subject and object remains applicable. A more radical step, one that is commonly taken in the study of thematic relations, is to assume that link types are acting as variable-binding term operators (VBTOs) so that we would have not just =agt and =pat, but also =goal, =source, =theme, =pos and perhaps several others. In Release 1.0 of 4lang Makrai, 2014 used several thematic role-like constructs, but this really stretched the ontological commitment (Quine, 1947) of the model beyond what is absolutely necessary, and by now only objects and subjects remain.

This is of course not to deny that there are such things as datives or locatives, only that we can handle the information content without recourse to additional VBTOs. In particular, we make do without the '3' or indirect object linker heavily used in Relational Grammar, which would call for a reanalysis for the broad variety of cases where this would come handy. On the theoretical side, we accept the arguments of Dowty, 1989 that =agt and =pat are sufficient – as a practical matter, these appeared in 178 (resp. 174) of the 1200 definitions whose headwords were listed in S19:4.8, while all others together appeared only 111 times. Consider the classic 'commercial exchange' schema we used in 1.4 to illustrate our use of voronoids as hypergraphs with nodes labeled by word vectors. This involves at least four participants: the seller, the buyer, the goods, and the money. Before the exchange, which can be conceptualized both in the *buy* and in the *sell* frame, the seller has the goods and the buyer has the money: afterwards the buyer has the goods and the seller has the money. This information can easily be captured using the formal language of 1.3: `before(seller has goods, buyer has money)`, `after(buyer has goods, seller has money)` and we will see in 3.2 how `before` and `after` can be treated geometrically.

As we already have agentive `-er` at our disposal, linking the verbs to this schema is effortless. Linking the nouns is more tricky, and it is not even obvious that *goods* 'things that are produced in order to be sold' (LDOCE); *product* (`4lang`); or perhaps a synthetic description *what seller sells* is the best way. In a spreading activation model, the LDOCE definition is reachable from *sell* in a single step (assuming, as we are, that *sold* is recognized as a form of *sell*), and similarly for `4lang`, where the definition of *product* is `artefact, for_ sell`. To synthesize a definition may also make sense, especially when the object of selling is not something that we would normally consider a product, as in *Mahema sold Sayuri's virginity to the Baron for 15,000 yen.*

Calling this nominal THEME offers no such advantage in reaching it, in fact it would negate the advantage of calling it `=pat`, which obviously facilitates link tracing. Note that the generally agreed definitions of themes, 'a participant which is characterized as changing its position or condition, or as being in a state or position' or 'an object in motion or in a steady state as the speakers perceives the state, or it is the topic of discussion' are so broad as to fit nearly all conceivable nominals including not just the money, but also the agent and action nominalizations. It is precisely because of the limited reachability from `goods` or `product` that we name this quadrant of the voronoid `goods_` or `product_`.

Today, the standard commercial exchange involves even more participants: the buyer has a credit card, or better yet, a cellphone that acts as one, the seller has a credit card terminal, the buyer and the seller both have bank accounts linked to these, and the exchange of money is effected by some protocol neither buyer nor seller are fully in control of. It would require an absurdly large array of thematic roles to reach all these participants from the actual keywords, yet the fact that they are available is evident from the fact that definite descriptions can be used without prior mention: *I wanted to buy a new pair of shoes. The card was rejected* (Kálmán, 1990).

2.5 Naive grammar

Here we begin to sketch, and discuss the limits of, a *naive* theory of grammar, offered in the spirit of the naive physics of Hayes, 1979, the naive psychology of Gordon and Hobbs, 2017, and the naive probability theory of Gyenis and Kornai (2019) (which will be discussed in Chapter 5). The fundamental elements of naive grammar are *words*. Many of our lexical entries do double duty as elements of the universal conceptual schema and as building blocks of the naive theory of grammar: we have already seen `stem_ szolto3 radix zlro1dl1o 3280 u N part_of word, stable` where the appended underscore serves to disambiguate from `stem`

`to3 stirps llodyga 2421 u N part_of plant, long, leaf on,`
`flower on, fruit on.`

There are other cases, such as `cause_` where the underscore signifies that we are interested in a substantive, if naive, theory of causation, something that must be available to support all kinds of decompositional analyses e.g. *kill* means 'cause to die'. Rather than criticizing the common decompositional style of lexical analysis that relies on supposed primitives like MOVE; BECOME; DO; and others, `4lang` simply uses ordinary lexical entries *move* `before(=agt at place/1026), after(=agt at` `other(place/1026))`; *become* `after(=agt[=pat])`; *do* `cause, =agt` `[animal], =pat[happen]`; and so on. Sometimes these are distinguished from the non-technical, everyday sense like `stem` versus `stem_` or *cause* `ok causa powold` `1891 u N reason` versus `cause_ okoz efficio spowodowac1 3290 u` `V before(=agt), after(=pat)`, but often the distinction between the 'grammatical' and the 'ordinary' use is so slight that we see no reason to even make the distinction: examples would be `part_of` (discussed in 2.4 above) and `is_a`.

The focal point of the the lexical semantics/naive grammar interface is the (primitive) relation `mark_`. As used in `4lang`, `mark_` is the relation connecting form and meaning. This corresponds well to the Saussurean notion of the sign, and will be sufficient for our purposes, even though a more sophisticated theory of signs (Kracht, 2011b) is available for the non-naive theory of grammar. Our main use of `mark_` is with function words (including bound grammatical formatives) as in

`-ing     stem_-ing is_a event, "_-ing" mark_ stem_`

Operationally, whatever precedes the formative `-ing` is considered a stem, and the whole form `stem+ing` is considered an event. There is clearly a great deal more that could be said about *-ing* suffixation, the notion of stems, the classification of junctures, or the conceptual classification of certain matters as events, but we make no apologies for not developing these notions as part of naive grammar, especially as `mark_` is used only in 0.5% of the vocabulary, and its treatment does not differ from those of other words. We don't actually develop a naive theory of morphology, and don't go anywhere near the issues of how a word is, or should be, defined in phonology, morphology, orthography, syntax, semantics, or lexicography (though we assume that the reader is somewhat familiar with the main proposals). For our purposes *word* is defined as `sign, speech,` and *sign* as `gen perceive, information, show, has meaning.`

With `4lang` we offer a kind of characteristica universalis, but fall conspicuously short of the lofty goal of a calculus ratiocinator in that we see no way to derive a sophisticated theory of grammar just on the basis of the everyday (naive) meanings of the terms it uses. This is not any different from other fields of inquiry: we may start by the method of analytic philosophy and consider the everyday usage of key terms, and in fact we will do so for e.g. the naive theory of probability (see Chapter 5), but this is not done with the assumption that the naive theory will somehow turn out to be superior. Consider *pain* `kiln cruciatus boll 1318 u N bad, sensation, injury` pain `cause_`. In S19:3.4 we wrote

> We are absolutely confident, based on primary sensory data, that boiling water will burn our skin. If the complete causal chain from heated nerve endings to the subjective sensation of burning pain could be exhibited, this would have far-reaching implications for example for the design of painkillers, implications that the naive theory lacks, so in this sense the detailed theory is superior to the un-analyzed statement.

This is a point worth reiterating: the naive theory is by no means the ultimate or the best theory.

Once we have words, the next issue to consider is part of speech. As discussed in 2.1, we do not consider the `4lang` system of lexical categories to be more than an expedient way to find bindings in many languages, and even for this very limited purpose, it is far from foolproof. An elaborate system of universal lexical categories is almost impossible to define and defend, but the distinction between nominals (in our system N and A) on the one hand, and verbals (in our system U and V) on the other, seems to be both defensible and desirable for any theory of grammar, naive or otherwise.

There are two fundamentally different ways to approach the nominal/verbal distinction. One is to consider the role of tense: typically, verbals are marked for tense and nominals are not. We defer discussion of this approach to 3.2, and focus on the other method, which links verbhood to the presence of arguments. Generally, nominals are viewed as free-standing, requiring no further information to access their meaning, while verbals are viewed as requiring a subject, and often an object as well. There are nominals that require implicit arguments, e.g. the nominals expressing family roles *brother, sister, father, mother* etc. are very hard to interpret without knowing whose brother, sister, etc. we are talking about. Similarly, there are verbal forms, imperatives in particular, that carry implicit arguments: when we say *run!* it is clear that the person who is supposed to do the running is the hearer, sometimes overtly expressed by the pronoun *you*, but more often not.

Also, there are notable corner cases, ranging from syntactically nominal elements like *danger* which tend to evoke a larger predicative frame *we are in danger* or *there is danger here*; all the way to syntactically verbal cases such as *run* in *the cable runs underground from the house to the pole* where there doesn't seem to be any action taking place (verbs of *fictive motion*, see 3.2). Another confounding factor is that when an array of criteria is used, it is rarely the case that the morphological, the syntactic, and the semantic criteria all yield the same classes. The idea of a more or less smooth transition between the major classes is also relevant for morphosyntactic processes such as nominalization that seem to preserve a great deal of the thematic structure of the stem.

Here we will outline a theory that keeps parts of speech maximally undifferentiated – wherever distinctions are made, these are tied to language-specific, rather than universal factors. The default case is for everything to be a nominal. Perhaps surprisingly for the semanticist brought up on the Montague Grammar tradition, this includes our generic quantifier gen, which we take to be a vector with equal nonzero components ($1/d$, where d is the dimension of L) at each coordinate.

As we discussed in 1.4, by default we assign a single vector to every element, and make very little distinction between the polytope and the distinguished point of it that is singled out by the labeling function. To the extent we can train embeddings, these use only one semantic type, vector. That this cannot be the entire story is well demonstrated by the analogical task discussed in 2.3. While people have an easy time with purely nominal tasks, or combination verbal-nominal tasks like *eat is to smoke as food is to __*, they are stymied by purely verbal tasks like *do is to have as stop is to __* even though logical solutions seem possible:

stopping is the end of doing, the same way as selling (or losing) is the end of having

In 2.3 we argued that the copula is better represented by a matrix, more precisely by a general equation that evaluates to the Gram matrix given by external multiplication of =agt and =pat, and that both subjects and objects also require treatment in terms of scalar product change, an operation that requires a matrix to keep track of. In our work on 4lang we also came to recognize a set of primitives, both verbs and prepositions, which require a matrix treatment. About half of these are spatiotemporal, and need a hidden intermediary element, the {place} schema (see 3.1), to be operational.

To analyze a simple statement such as Bill at office 'Bill is at the office' in a sequential model takes several steps, beginning with the recognition that Bill is =agt and office is =pat of at – we leave this task to syntax. It is a more semantics-flavored task to figure out *which* Bill is meant, we take this on in 3.3. Finally, it is a purely semantical task to figure out which sense of *office* is relevant here. Surely Hamlet is not complaining about the insolence of a building.

Since the meaning of at is to typecast its =agt as the _origin of the {place} that its =pat is typecast to, we get the disambiguation of *office* for free. There may be separate vectors for office$_1$ 'building' and office$_2$ 'officialdom', indeed we suppose there are, but it is only the former that lends itself to coercion to place. Geometrically (in the sense of linear algebra, not in the naive spatial model) a typical word sense will be a vector in a polytope, and polysemous words will have these polytopes close to one another, whereas homonymous words will have vectors in far regions of the same space. *Office* is something of an undifferentiated mixture between organizational positions in an institutional structure, the people who fill these positions, and the physical location of these people and the institution itself, but by projecting it to {place}, the non-spatial readings are mapped on zero. The coercion mechanism we posit goes some way toward explaining how bound morphemes like agentive *-er* (2.2) can attach to such undifferenti-ated nominal stems to yield *officer*, even though a verbal base *officate* would be available to yield **officater* or *officator*.

The critical element distinguishing *at* from *in* or *under* is that =agt must at the same time be mapped on the origin (as opposed to the inside, or the underside, see 3.1) of the place. In the algebraic view (1.5) we handle this by unification of graph nodes, which guarantees that the origin where we place =agt is the *same* origin that =pat, qua place, is endowed with. As discussed in 2.4, constituent structure analysis suggests that sequentially the coercion of the =pat to place happens first, a PP is formed, and coercing =agt to the origin of this place happens second.

Altogether, we have two kinds of transitives: those where the object is optional, e.g. *eat*, and those where it is obligatory, e.g. *betray*. Using the rudimentary 4lang system of lexical categories the first class alternates between U and V, while elements of the second are pure V. What makes the first class possible is not some kind of logical difference between the argument structure of *eat* and *betray*, for it is just as impossible to eat without eating something as is to betray without betraying something. Standardly, a distinction is made between adjuncts and complements:

> Adjuncts are always optional, whereas complements are frequently obligatory. The difference between them is that a complement is a phrase which is *selected* by the head, and therefore has an especially close relationship with the head; adjuncts, on the other hand, provide optional, extra information, and don't have a particularly close relationship with the head. (Tallerman, 2011)

(see also S19:4.2). Here we propose that the somewhat elusive idea of 'an especially close relationship' that obtains between betrayal and its object, but not between eating and its object, can be operationalized in terms of coercion. In one case, we have an object that is *food* only by default: a sentence like I can eat glass, it doesn't hurt me is perfectly grammatical, if unusual. We formalize this fact by marking the object as =pat[<food>] i.e. food by default only. Betrayal, on the other hand, invokes a frame where the object is something internal (a feeling, an internal state, a secret) and the object is a sign signaling externally the departure from the internal state. It will require building up a great deal of the formal apparatus before we can get to *naive psychology* in Chapter 6, here it is sufficient to say that people (and animate things in general) can have internal states or emotions that are observable only through indirect signs: *He tried to act angry, but his smile betrayed him.* This is not any different form the naive theory of signs/signals: we have overt signs, such as a spoken word, and a hidden element, the meaning, and we say that the former mark_s the latter. Examples include:

```
-er        -o1        -tor/-trix -ac1/ic1    3627 G
           stem-er is_a =agt, "_ -er" mark_ stem
buy        vesz       emo          kupowac1   2609 V
           =agt receive =pat, =agt pay seller,
           "from _" mark_ seller
command    parancs iussum      rozkaz       1941 N
           speak, has authority, cause_ person do =pat,
           dative_ mark_ person
conduct    vezet      transmitto przewodzic1 3353 V
           =agt cause_ {=pat at place}, "to" mark_ place,
           <energy[flow] in>
conform    megfelel aptus       sl1oda       3375 N
           =agt similar gen expect =pat,
           "to _" mark_ =pat
different ma1s       diversus   inny         1566 A
           =pat has quality, =agt lack quality, "from _" mark_ =pat
difficult nehe1z    difficilis trudny       1771 A
           act need large(effort), "to/3600 _" mark_ act
for        -e1rt      pro          dla         2824 G
           at exchange, "for _" mark_ price
use        haszna11 utor         uz1ywac1   1008 V
           =agt has purpose, =pat help purpose,
           "for _" mark_ purpose, "to _" mark_ purpose
```

This far, the naive theory is not any different from the theory of the Sausserean sign, which is precisely a binary relation between a form (for us, just a string in quotes, including an insertion locus __) and a meaning. A key further step, taken in (Bloomfield, 1926), is to distinguish between free and bound forms, define the morpheme as a minimum form, and the word as a minimum free form. We are quite content to use these technical developments without pretending that they are part of the naive theory, especially as these developments obviously lead to better formalization of the naive theory, preserving its main tenet, that words mean things.

As the examples show, the primary function of mark_ in this system is to help sort out which piece is which: stem_ is the part before -er, seller is the phrase marked by from and so on. Why *difficult* governs the infinitive, and *different* governs a PP[from] are things we consider to be historical accidents beyond the reach of explanatory theory. Perhaps we are giving up too soon, but a theory aiming at an actual explanation, as opposed to cataloging general tendencies, would need to take on board such an incredible array of facts from all the languages of the word as to put the task well beyond the author's scholarly powers.

The main takeaway here is that the semi-technical notion of a binary mark_ relation is not any different from the ordinary meaning of mark 'a sign which shows something'. That something, we claim here, is something that would be hidden without the marking, as in *X marks the spot*. The technical difficulties are not with keeping too little distinction between language and metalanguage (the opposite is true, our metalanguage (1.3) is quite distinct from natural language) but rather with describing free alternation between type U, a vector, and type V, a matrix, not just for somewhat 'metalinguistic' words like *mark*, but already for plain words like *eat*.

It is clear from the foregoing that the study of lexical semantics already relies on a non-eliminable grammatical core that goes beyond concatenation, encompassing at least some matters that require more than a simple vector addition calculus. Our lexical categories, skeletal as they are, offer a rich interface for connecting 4lang to issues that modern grammar (in this case, starting with Fillmore, 1968) has much to say on. For example, we will need a theory that connects intransitives to transitives, as in *The fire spread* and *The wind spread the fire*. In English, it is obvious that there is some relation between these two verbs, but in Latin there is no obvious reason to relate *distendo* and *sterno*. In Hungarian, the stems are derived from the same root by productive suffixation, so we have *ter-ül* and *ter-ít*. In the next Chapter, we will study this phenomenon with the aid of such roots that are truly neutral between U and V.

We have already expressly disavowed any idea of the naive grammar being the ultimate grammar, or even the metalanguage being the ultimate metalanguage. The system developed in this book is designed to support one thing, and one thing only, natural language semantics. There are many other semiotic systems from music to mathematics that would have very different semantics, and 4lang is simply not equipped to deal with these. Also, experience shows that naive theories are supplanted by more sophisticated ones for a reason, as the sophisticated theories are simply better. But they often rely on key components, such as arithmetic, or the analytic theory of continuous variables, that are out of scope for 4lang, expressly designed to deal with ordinary (as opposed to technical or scientific) language.

Since the issue is central to the development of generative grammar, we should make clear here that our position is not intended as an argument for, or against, the autonomy of syntax thesis. As a research strategy, we prefer a semantic formalism that is as autonomous as feasible, since this promotes modularity not just in the sense of Fodor, 1983, but also in the sense of enabling independent experimentation and research for both syntacticians and semanticists. We do not feel qualified to take sides in the debate, but if those who believe only in a limited autonomy of syntax are to mount arguments capable of convincing the opposing side, these arguments need to be cast in terms of the inadequacy of well-modularized systems, so even for those refusing to entertain full modularity the first order of business is to look at modular architectures.

strong   Let us briefly consider some irreducible *V* elements such as the predicate like has. We define *strong* as has force[great] and wish to derive *Bill has great force* from *Bill is strong* or, more precisely, we wish to derive the conjunct Bill has force, force is_a great, leaving it to the unification mechanism (see 1.5) to guarantee that the force in both clauses refers to one and the same instance. The key observation is that we don't wish to assert =agt has force from =agt is strong, for even a weakling has some force, what we want to assert is =agt has {force is_a great}. From the definition of strong we already know that this is a relation specifying that =agt possesses =pat, and that =pat, the thing being possessed by =agt, is itself a relation, force is_a great, which we handle by Eq. 2.6.

The effect of =pat is to automatically bring the object under the scope of the predicate. Since the object is already a complex state of affairs (and one that cannot be assumed to generally hold, since not every force is great), we must take the rank-1 matrix defined by 1 on the (force, great) coordinate and 0 elsewhere and bring it under the scope of has. On the standard view, has is a matrix H that has 1 on its ij coordinate if i has j. The lexical component of H is rather thin: buildings have walls, husbands have wives, but if Susan has a Ferrari this is a contingent fact.

This means we have to encode the projections themselves as vectors, and find a means of effecting a projection that fits into the natural evolution of the state. We do this by some small perturbation of V, using $V + W$ to compute P at least over the relevant subspace L. At least to first order, small changes in V will leave the eigenvectors largely unchanged (only the eigenvalues shift), so it will be convenient to proceed in the sparse eigenvector basis $p_1, \ldots, p_d$ we introduced in 2.3. All of semantics is played out in the linear space L spanned by these, the matrix space $R = L \times L$, and the higher tensorial spaces that we choose to ignore. Whatever construct we use must be expressible as vectors, matrices, and small, unordered sets of these.

For the majority of words, we need only one word vector to express its sense, or a handful to express its various senses. But there is a perceptible minority, about a quarter of the elements in 4lang, that require a matrix: these are the static elements of R. Their lexical content is rather sparse: even the most frequent one to appear in definitions, has, will have less than 0.1% of the full $d \times d$ matrix filled, often with mundane semi-encyclopedic content such as sheep has wool or cloth has thread. Other similar elements like cause_ use 0.01% or even less, and carry their own share of (naive) encyclopedic knowledge: alcohol liquid, <drink>, alcohol <cause_ person[drunk]>. As usual, we take everyday usage as basic, and ignore the fine encyclopedic details that people don't actually drink absolute alcohol, they drink it diluted, that modest amounts don't make one drunk, and so on. On the other hand, we consider it highly relevant wherever cross-linguistic evidence is available, e.g. that in Japanese *sake* is used in the general sense of 'alcoholic drink' even though *arukōru* is available.

The U/V (intransitive/transitive) distinction creates three categories: 'pure U' verbs like *sleep* or *run* that only have intransitive uses (formally, measure phrases appear in object position, as in *He sleeps all day*, but these are clearly not objects, as seen e.g. from the lack of passive *All day is slept by him*); 'pure V' verbs like *enclose, notice, realize, pierce* which demand a syntactic object; and 'mixed' verbs that alternate between intransitive and transitive usage, as in *divide* or *drop*. Here a word of caution is in order: when we say *The bacterium divided* it is the subject that undergoes the division, but when we say *John ate* it is not John being consumed, but rather we have a default object

<food> that is, similarly to *Mary is expecting (a baby)* etc. The only difference in the U/V alternations that have no default object is that in intransitives it is =agt, and in transitives it is =pat that undergoes the action indicated by the word. Even when there is a default object, intransitive readings are still possible, compare *For this contest, we will all cook* and *In this heatwave, with no airconditioning, we will all cook*. Conversely, when the default is intransitive, a causative transitive alternant often exists, consider *John flies, John flies the kite*. In such cases, 4lang rarely lists both alternants, since on the right hand side of definitions typically the intransitive version is used.

If the simple geometric picture presented by Eq. 2.9 is to be amended, it is by considerations of locality: normally, not only the subject is co-present with the event but also the object: subject is at, object is near (within arm's reach, see 3.1). Remarkably, a piece of naive physics, the prohibition on action at a distance follows from this consideration: whatever is the act performed on the object by the subject, the two must be near one another. (The prohibition on action at a distance, first stated by Aristotle, is actually preserved in modern physics, where such actions are mediated by force fields, now seen as having not just spatial extent but also energy and therefore mass as well.)

Our analysis of 'pure V' transitives extends smoothly to relational primitives like cause_. The *post hoc ergo propter hoc* analysis provided in 2.4 works well for the canonical cases like *Heart disease is a leading cause of death*, but we would want something more than mere temporal succession, we want to express the idea that the

cause_   cause actively contributes to the effect. The primitive closest to this in 4lang is *make* and a definition of cause_ as make happen is certainly possible. (As with 'cause to die', care needs to be taken to express the fact that the object of causation, just as the object of killing, must syntactically appear between the two terms, see Kornai, 2012.) At this level, 4lang lacks the resolving power to provide a deeper analysis of

make   cause_. To see this, let us trace the definitions of *make*, which is =agt cause_
happen   {=pat[exist]}; the definition of *happen*, which is simply change; and the defi-
change   nition of transitive *change*, which is after(=pat[different]) in the transitive, and after(=agt[different]) in the intransitive sense.

This gives us cause_ as make =pat change or, taking the substitution further, as make after(=pat[different]), but this is not particularly helpful since make makes reference to cause_. Altogether, we are not any closer to capturing the direct, active contribution of the cause to the effect – all this says that when we cause something, the causee either comes into being, as in *The faulty traffic light caused an accident* or changes *The remarks caused considerable consternation*. The moral of the exercise is clear: cause_ is a primitive, but still embedded in a web of definitions, rather than standing alone as an unanalyzable atom. We will offer a less circular analysis in 6.2, one that captures the essence of =agt cause_ =pat[change,exist] or =agt cause_ after(=pat[different, exist]), but this will come at the price of introducing a whole modal apparatus, though one justified by a multitude of phenomena besides causation.

Since our goals are remarkably close, and our motivations are nearly identical, readers of (Jackendoff and Audring, 2020) may find a direct comparison between the two formalisms helpful. We restrict ourselves to a single example, their $[_V A\text{-}en]$ *schema* (p 88):

Semantics: [BECOME (X, [<MORE> PROPer_TY]$_x$)]$_y$ (J&A 88:3)
Morphosyntax: $[_V A_x \text{ aff}_3]_y$
Phonology: $//\sigma$ [-son]$/_x$ ən$_6/_y$

which is to be compared to our considerably more impoverished

```
-en -ilt -o -cl 3594 G make stem, "_-en" mark_ stem (2.14)
```

The most obvious difference is that we have no Phonology section in our rule. Instead of string variables (which would bring a broad array of autosegmental complexities in tow) we are relying only on the hack/heuristics discussed in 2.2 that whatever precedes *-en* is considered the stem. We have nothing to offer for syllables σ, and also avoid the use of distinctive phonological features like [-son], as these are clearly beyond any naive theory. This is not to say that the care J&A lavish on phonology (and even phonetics, see especially their Ch. 6) is in any way misplaced. In fact, we believe their system of Relational Morphology (RM) to be largely 'upwards compatible' with the naive theory of mark_s that we sketched here, and key facets of RM, in particular their Relational Hypothesis (J&A Ch. 2.1) make the right call about generative use of rules/regularities.

Compared to the apparatus used in (J&A 88:3) and elsewhere, the morphosyntax embodied in Eq 2.14 is also very sketchy. If anything, *-en* is a laboratory pure example of a category-changing affix: the input is an adjective, and the output is a verb. But for a semantic theory that aspires to universality this is too much and too little at the same time. Too much, because the categories don't nearly line up perfectly across languages: Hungarian *-ít* has the exact same semantics as English *-en*, yet it applies not just to adjectives but also to (category-free) roots, and produces a transitive verb, in contrasts to Hungarian *-ul/ül*, which produces an intransitive (see 1.2). In fact, the purity of *-en* is questionable, given that it is capable of applying to nominal stems as well, as in *fright/frighten, heart/hearten, hight/highten, threat/threaten* etc. We resist the temptation to say that these are formed *because* the adjectival stem would be a mismatch for the phonology, as we see no reason to assume a secondary *-en*, with different phonology and morphosyntax but the exact same semantics. Even if we did, this would not stand in Elsewhere relation to the ordinary *-en*, so duplicating the lexical entry, always a dubious move, would still not result in a cause-effect explanation.

Without belaboring the obvious, there is nothing surprising in the fact that lexical entries in one language do not perfectly line up with those of another. For us, *-en* matters because it is one of the forty-odd suffixes permitted in the LDV. But Latin *-o*, which typically operates on the same root from which the adjective is formed, as in *laxus/laxo* or *rufus/rufo*, or uses an explicit make operation *facio*, would require building up a very substantive amount of morphology before the morphosyntactic pattern could even be stated, and our reductive effort would in no way be helped by building up all this machinery.

Turning to the semantic component, we avoid the complex system of linking and coindexation that J&A 4.13.1 rely on (the *equalizers* introduced in 3.3 will to some extent make up for this loss) because our primary goal is to concentrate on operations that make sense in linear algebra, in particular vector addition and matrix multiplication. We could easily translate the Semantics section of (J&A 88:3) to our own system, as long as care is taken to distinguish the intransitive and the transitive cases of the output. In most cases transitive and intransitive uses are equally felicitous: *The road widened/the county decided to widen the road.* In a few others, the output is primarily intransitive: *John reddened/??The hours under the blazing tropical sun reddened John.* The apparent lack of purely transitive uses can be chalked up to taking the intransitive form as basic, and deriving the transitives by causativization of some kind. This is the become analysis, the alternative make analysis takes the underspecified form as basic as in Eq. 2.14. Depending on language, and even depending on lexical entry, both may make sense.

While 4lang has the resources both for optionality (we use the same angled bracket notation, see 2.2) and for comparatives (see 2.4), we would argue against the optional <MORE> clause in (J&A 88:3), as it seems inconsistent with the canonical cases of *become*, as in *Mary became a doctor* – there is no implication that before this change she was already something of a doctor, just as there is no implication that the road become was already wide before widening. Our definition of *become* was vmvel_lesz fio stawacl_siel 2655 U after(=agt[=pat]). This simply says that in the result state =agt is_a =pat, which is simple enough that we don't even require *become* to be a primitive. Be it as it may, the differences in the semantic analyses (J&A 88:3) and 2.14 reflect not so much a difference in basic outlook as a different set of technical tools.

To a remarkable extent, the same can be said of the NSM approach (Goddard and Wierzbicka, 2014). We share their concern for 'soft' valences or, what is the same, permitting a loose set of valency frames for one and the same prime. Each of the frames may be more sharply delineated, e.g. "DO something" will have a 'minimal' frame (agent only); a 'patient' frame (both agent and patient); an 'instrument' frame (agent and instrument); and a 'comitative' frame (agent and co-agent). Taken together, these four frames permit a rather loose arrangement, especially as there is no prohibition on superimposing some or all of these as in *John hunted for deer with a bow and arrow*. We also share their view on how the primitives are expressed across languages:

1. A mere list is not sufficient, in itself, to identify the intended meanings, if only because many of these English exponents are polysemous (i.e. have several meanings), but only one sense of each is proposed as primitive. While it is claimed that the simplest sense of the exponent words can be matched across languages (i.e. that they are "lexical universals"), it is recognised that their secondary, polysemic meanings may differ widely from language to language. A fuller characterisation indicates for each

proposed prime a set of "canonical contexts" in which it can occur; that is, a set of sentences or sentence fragments exemplifying its allowable grammatical contexts.

2. When we say that a semantic prime ought to be a lexical universal, the term "lexical" is being used in a broad sense. An exponent of a semantic prime may be a phraseme or a bound morpheme, just so long as it expresses the requisite meaning. For example, in English the prime A LONG TIME is expressed by a phraseme, though in many languages the same meaning is conveyed by a single word. In many Australian languages the prime BECAUSE is expressed by a suffix.

3. Even when semantic primes take the form of single words, there is no need for them to be morphologically simple. For example, in English the words SOMEONE and INSIDE are morphologically complex, but their meanings are not composed from the meanings of the morphological "bits" in question. That is, the meaning SOMEONE does not equal "some + one"; the meaning INSIDE does not equal "in + side". In meaning terms, SOMEONE and INSIDE are indivisible.

4. Exponents of semantic primes can have language-specific variant forms (allolexes or allomorphs, indicated by in the table above). For example, in English the word 'else' is an alloex of OTHER; likewise, the word 'thing' functions as an alloex of SOMETHING when it is combined with a determiner or quantifier (i.e. this something = this thing, one something = one thing).

5. Exponents of semantic primes may have different morphosyntactic characteristics, and hence belong to different "parts of speech", in different languages, without this necessarily disturbing their essential combinatorial properties.

(quoted verbatim from https://intranet.secure.griffith.edu.au/schools-departments/natural-semantic-metalanguage/what-is-nsm/semantic-primes). These principles are sound – the main reason we don't adopt NSM in its entirety is that it lacks reductivity: there are likely words for *hunt, deer, bow, arrow* in many of the languages the NSM school has studied in detail, but it is quite unclear how definitions of these could be created by those outside the school, let alone by algorithms.

3

Time and space

Contents

We owe the recognition of a deep connection between time, space, and gravity to the 20th century, but people have used language to speak about spatial and temporal matters long before the development of Euclidean geometry, let alone general relativity. Throughout this book, we approach problems through language use, in search of a *naive theory* that can be reasonably assumed to underlie human linguistic competence.

Since such a theory predates all scientific advances, there is a great deal of temptation to endow it with some kind of deep mystical significance: if this is what humans are endowed with, this must be the 'true' theory of the domain. Here we not only resist this temptation (in fact we consider the whole idea of linguistics and cognitive science making a contribution e.g. to quantum gravity faintly ridiculous), but we will also steer clear of any attempt to bridge the gap between the naive and the scientific theory. The considerable difference between the two will no doubt have explanatory power when it comes to understanding, and dealing with, the difficulties that students routinely encounter when they try to learn the more sophisticated theories, but we leave this rich, if somewhat anecdotal, field for future study.

In 3.1 we begin with the naive theory of space, a crude version of 3D Euclidean geometry, and in 3.2 we deal with time. The two theories are connected by the use of similar proximities (near/far), similar ego-centered encoding (here/there, before/now/later), and similar use of anaphora (Partee, 1984), but there are no field equations connecting the two, not even in vacuum. The shared underpinnings, in particular the use of indexicals, are discussed in 3.3. Finally, the *naive* theory of numbers and measurement is discussed in 3.4.

A. Kornai, *Vector Semantics*, Cognitive Technologies, https://doi.org/10.1007/978-981-19-5607-2_3

3.1 Space

We conceptualize space from the perspective of the upright human, shown as a cylinder on Fig 3.1. A defining feature of naive space is the *gravity vertical* (Lipshits and McIntyre, 1999). Shown as the dot-dash axis on the figure, gravity is directly sensed by the inner ear, and as such, is a constitutive part of the body schema that we take to be fundamental to the perception of space. Relevant `4lang` definitions include:

```
up        fel            sursum       do_go1ry 763  A
          after(at position), vertical(position er_ gen)
down      le             deorsum      w_do1ll  1498 D
          vertical(gen er_)
vertical fu2ggo3leges verticalis pionowy  869  N
          direction, has top, has middle,  has bottom,
          earth pull in direction
fall      zuhan          cado         spadac1  2694 U
          move, after(down)
```

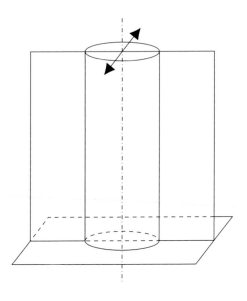

Fig. 3.1: Egocentric coordinates

Orthogonal to the gravity vertical we find a distinguished plane, the `ground`, defined as `surface`, `solid`, `at earth`. Actual ground may of course be sloping, but its default orientation is horizontal, as in Fig. 3.1. We use this to define `horizontal`:

```
horizontal vi1zszintes horizontalis horizontalny 3144 A
           direction, flat(ground) has, still(water) has
```

Here we must make the obvious distinction between cognitive *ground*, which we take to be the entire ego-centric model of space and denote by {place}, very much including the cylindrical figure at the center, and physical *ground*, which is the flat, horizontal component of this model. What we find under the ground plane is by definition under the figure and conversely, the ground plane is defined as the (top plane of the) underside of the schema. One way of saying "if and only if" is to assume that a part of the cognitive schema comes pre-labeled as the underside, and the general relation of under is given by the equation

$$P_R(t+1) = P_R(t) + s(|{=}\text{agt}\rangle\langle\text{underside}| + |{=}\text{pat}\rangle\langle\text{place}|) \qquad (3.1)$$

There is no circularity here. To define spatial notions, we do need an idea of space/place, and we assume this entire model, depicted in Fig. 3.1, to be primitive. (We see this as a prime example of *embodiment*, see S19: Ch.8, but we will not pursue this matter now.) The model has well-defined parts, and these parts are labeled by concepts such as *underside, ground, body,...* For primitives, and only for these, we must take the stance that the concepts are prior (inborn) and language learning consists in attaching names to these inborn concepts.

Next we turn to the cylinder, which we take to be a highly simplified, rotationally symmetric representation of the human body. In effect, the body is at the origin of the egocentric coordinate system: a {place} always comes with a body at the center, and never mind Cartesian geometry that requires the origin to be a single point with no extension in any direction. Further, the origin already has a definite orientation, it *stands* on the ground. Indeed, while the default of standing is on two feet, we consider it perfectly normal for objects with rough rotational symmetry, such as bottles or vases, to be described as standing on some flat surface.

Even for objects lacking symmetry, such pieces of furniture, it is normal to stand, as long as they have a well-defined top. Objects lacking this feature behave differently, for example it is strange to say that *??the soccer ball stands on the ground*. This is the motivation behind definitions such as

```
stand a1l1 sto stac1 74 U
       =agt[vertical], =agt on two(foot)
```

Similar generalizations are available for foot and top, which we do not at all consider metaphorical in expressions like *foot of the mountain* or *top of the hill*. Some further entries impacted by the egocentric organization of space include

```
base    alap       fundamentum podstawa    146  N
        part_of whole, at bottom, whole has bottom, cause_ whole[f
height magassalg altitudo     wysokoslcl 1583 N
        distance, vertical
root    gyo2kelr radix         korzen1     936  N
        under ground, part_of plant, support, at4 base/146
top     teto3      culmen      dach        2377 N
        part, at position, vertical(position er_ part[other])
```

in   The inside of the body is labeled `inside` and the outside is labeled `outside`. This
out gives our definition of *in* and *out* as

$$P_R(t+1) = P_R(t) + s(|{=}\mathtt{agt}\rangle\langle\mathtt{inside}| + |{=}\mathtt{pat}\rangle\langle\mathtt{place}|) \qquad (3.2)$$

$$P_R(t+1) = P_R(t) + s(|{=}\mathtt{agt}\rangle\langle\mathtt{outside}| + |{=}\mathtt{pat}\rangle\langle\mathtt{place}|) \qquad (3.3)$$

When our cognitive ground is a room, we effortlessly identify its 'skin' as the walls of the room, its 'top' as the ceiling, and its 'bottom' as the floor. This actually tells us what's inside the room and what's outside of it. We are not bothered by the fact that we can't identify the arms or the heart of the room, a partial mapping is sufficient for Eqs. (3.2-3) to work as intended.

The next notable feature of Fig. 3.1 is the plane bisecting the cylinder, which we take to be the frontal body plane, given by the maximum extension of the arms. Equally easy to define would be the sagittal plane as the locus of the mirror symmetry the human body enjoys. But we will have much use for arms shortly, whereas symmetries and higher notions of geometry would be hard to justify for the kind of minimalist schema we are developing here. If we permit symmetries, we may as well permit Bessel functions here.

The frontal plane defines the `front` and `back` halfspaces indicated by the two-headed arrow on top. The figure itself provides no clues in this regard, yet most readers will automatically assume that the body is depicted facing the reader, so it is the 7 o'clock arrow that points toward the front, and the 1 o'clock arrow that points toward the back of the body schema. This has to do with a phenomenon that we will discuss in more detail in 3.3: not only do we have our own body schema, one that moves with us as a matter of course, but we also assume that others will have theirs.

No matter how crude an image a cylinder provides for a human body, once we are told that it *is* the image of a body, we start making sense of it in a low-level, automatic fashion. Gordon and Hobbs, 2017 begin their discussion of naive theories with the classic Heider-Simmel test, which shows this phenomenon rather clearly. When we apply the body schema to the human body, it is clear that things `near` are those within arm's reach (something we could schematize by a larger cylinder around the body), and things `far` are those outside our reach. The space between the internal and the external cylinders can be labeled `about`, and it is only within this space that we can manipulate things (no

about action at a distance). Relational `about` is anchored to the *about* region of the body in a manner similar to Eqs. 3.1-3.3:

$$P_R(t+1) = P_R(t) + s(|{=}\mathtt{agt}\rangle\langle\mathtt{about}| + |{=}\mathtt{pat}\rangle\langle\mathtt{place}|) \qquad (3.4)$$

On the figure, it is clear to most readers which is the `left` and which is the `right` side of the body schema. Definitions affected by these notions include most body parts:

```
chin      a1l1    mentum      broda      73   N
          part_of face, at/2744 centre, under mouth
face      arc     vultus      twarz      177  N
          organ, surface, front, part_of head, forehead part_of,
          chin part_of, ear part_of, jaw part_of
forehead  homlok  frons       czol1o     1077 N
          part_of face, front, eye under, hair at, at temple/982
front     elej    pars_prior  przo1d     608  N
          part, first
nose      orr     nasus       nos        1912 N
          organ, part_of face, animal has face, front, at centre,
          smell, air[move] in, nostril part_of
left      bal     laevus      lewy       222  N
          side, has heart
right     jobb    dextra      prawy      1199 N
          side, lack heart
arm       kar     bracchium   ramie1     1231 N
          organ, long, human has body, body has, limb,
          hand at, wrist at, shoulder at
leg       la1b    pes         noga       1467 N
          limb, animal has, move ins_, support, low
limb      ve1gtag membrum     kon1czyna  3345 N
          part_of body, leg is_a, arm is_a
wrist     csuklo1 articulus   nadgarstek 438  N
          organ, joint, at hand, at end, arm has end
heart     szi1v   cor         serce      2210 N
          organ, cause_[blood[move]], love in/2758, centre
```

That the *heart* is not just the organ of blood circulation, but also the organ of emotions (6.3), love in particular, should come as no surprise: it would be virtually impossible to make sense of much human discourse about love without this assumption. Also, the heart (in Occidental metaphysics, in opposition to the brain, in Oriental, encompassing the brain) is somehow the most central, essential, ruling portion of the body, so that *the heart of the matter* is no more metaphorical than *the top of the hill*.

```
centre ko2ze1ppont centrum    centrum 1412 N
       middle
middle ko2ze1p      media_pars s1rodek 1410 N
       part, place/1026, near centre
side   oldal        latus       strona  1903 N
       part, <two>, centre far, oppose, object has
```

We defer full discussion of some dominantly temporal prepositions such as `follow`, `next`, `(un)til/to` and `through` to 3.2, but we note the strong association to their spatial senses here. By default, people are facing the future and have the past at their back. According to Núñez and Sweetser, 2006, Aymara is an exception, and we see a similar degree of accidentality in linking compass directions to the `place` schema: in Sanskrit `front` is East *pūrva-*, `right` is South *dakṣina*, etc. whereas in Finnish *etelä* 'South' is from *ete-* 'front' and *pohjoinen* 'North' is from *pohja-* 'bottom' (Paul Kiparsky, pc). This situation can be compared to the rule of the road: clearly it makes sense to drive on one side of the road, but it is a matter of convention whether a culture chooses this to be the left or the right side. Importantly, the conceptual schema for compass points may override the 'objective' arrangement of the cardinal directions as in Manhattan, where people will go *North* even when actually they go North-East (see Haviland (2000) for a summary of 'direction keeping' systems and Haugen (1957) for an even more elaborate example).

Of particular interest is the discrete view of space and time imposed by `next`. There are two things involved: a discrete sequence of matters, be they physical objects such as rooms or people standing in a line, abstract entities like numbers or events; and a notion of adjacency among these. When we say *x (is) next (to) y* this means both that x and y are near one another, and that there is no z between them. Typically, this means that x and y are touching, but this is not necessary: we can speak e.g. of *adjacent houses* irrespective of whether garden plots intervene. For an example where contiguity/touching is criterial

`on` consider *on*, which really means 'attached, touching' as in *horseshoes on hooves*. The most frequent (default) case is when the attachment is provided by gravity as in *the book on the table*. This is summarized in the definition `at, =agt touch =pat,` `<high(=agt er_ =pat)>`.

The notion of {place}, as developed in the foregoing, provides our second example of a conceptual schema of the kind words are constantly mapped on (recall {exchange} in Fig. 1.2). We could have called this schema *figure-ground complex* or *spatial model* just as easily, but English *place* is quite nebulous (dictionaries from Webster's 3rd to LDOCE list dozens and dozens of meanings) and our technical meaning covers most of these.

The geometry of this voronoid, given to us as as a collection of a few word vectors in L (a space of several hundred dimensions), has nothing to do with the approximate 3D geometry we model in 3.1. The means of guaranteeing that the body axis is aligned with the gravity vertical lie largely outside the domain of linguistic data, in the realm of embodied cognition (Gallese and Lakoff, 2005). When we use *inside* or *body* to label a

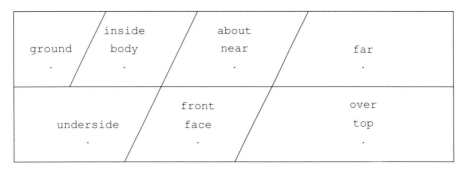

Fig. 3.2: {place}

polytope, this means that there is no inside without being inside something, the body_ segment of the voronoid is automatically invoked.

The relations that obtain between the polytopes are inherent in the schema. To get this effect, we need to go beyond the vectors (polytope regions) depicted in 3.2, and consider the matrices that model two-place relations. Whatever =agt under =pat means is derived from the fact that the matrix corresponding to under maps the underside polytope to the ground polytope. In other words, {place} is a conjunction of the vectors that make it up, and some canonical equations such as 3.1-3.3 that obtain among these vectors. In fact, we have expressions specifically devoted to signaling major mismatches between these canonical equations (the inherent content of the schema) and a particular situation, e.g. when something is turned *upside down* or *inside out*.

The body schema, sketchy as it is, already provides us with a mechanism to discuss the systematic differentiation that some languages make between intransitives and transitives, (Using the 4lang notation, we will often speak of U/V alternation.) English often leaves the distinction unmarked: especially in the core vocabulary we find a multitude of examples like *Mary changed* 'she became a different person' versus *Mary changed John* 'he became a different person'. Hungarian offers hundreds of roots that exemplify the intransitive/transitive alternation, and these can generally be translated with 'be(come) X' versus 'make X'. For example, *bús* means 'sad', *búsul* means 'be sad' and *búsít* means 'make someone sad, sadden someone'. Similarly, *but(a)* means 'stupid', *butul* means 'become stupid' and *butít* means 'make someone stupid'. Based on these two examples it may be tempting to think of the root as an adjective, but this is somewhat misleading, as the typical translation (at least to English, Latin, and Polish) is verbal.

In terms of the body schema, whenever the locus of the root X is inside the body, we treat the expression as intransitive, and whenever it is outside, but nearby (within arm's reach) we treat it as transitive. A clear example is *ford* 'turn', where *fordul* means =agt turn and *fordít* means =agt turn =pat, but the main class of what we called 'mixed U/V verbs' in 2.5 also belong here. In the transitive use *John dropped the keys* the locus of the dropping is the object, and in intransitive usage *John dropped* it is the subject, the body itself. We note here that since Fillmore, 1968, having a cognate object is often considered diagnostic of a verbal primitive, as it confers near-root status on the verb by virtue of being identical in intransitive and transitive contexts, even though the latter are limited to cognate objects. (See Höche, 2009 for a detailed discussion of cognate objects in English.)

The {place} schema is by no means the only conceptual schema we rely on in conceptualizing space around us, but to complete our discussion of the core spatial vocabulary we need to discuss only one other schema we will call {bound}. (Other notions grouped together by Buck (1949) under " Spatial relations: place, form, size" include *change/exchange*, see 3.3; *sign*, see 2.5, and *size*, see 3.4.) The {bound} scheme has two spatial participants, the distance, area, or volume that is being bound, which we will call volume_ irrespective of dimension, and a boundary_ which typically has one less dimension, e.g. a distance (line, one dimension) is bounded by points (zero dimensional). We could to some extent relate this to the {place} schema, comparing the volume_ to the body_, but really the 'skin' that bounds the body is derived from the boundary_ and not the other way around. Equally important, in a *bound* statement we don't particularly identify with the spatial viewpoint of the volume_ or the boundary_. Rather, the observer is floating somewhere, does not matter where.

A central instance of the schema is provided by *distance* 'the amount of space between two places or things'. In 4lang we define distance as space/2327 has size, space/2327 between, and either *distance* directly, or *between*, or perhaps both must make reference to the {bound} schema. In one dimension it is clear that the second argument of between is a collective noun, composed of two points. In two dimensions, more complex collective nouns are often seen: *[Ann Arbor's] Third Ward, bounded by Huron Parkway, Glacier Way, and US 23, [...].* Remarkably, the boundaries may be only implicitly given, as in *French Guiana is between Suriname, Brazil, and the Atlantic Ocean* rather than *??? between the Suriname-FG border, the Brazil-FG border, and the Atlantic seashore.* The choice between calling the schema {bound} or is_located_between is rather arbitrary – we were influenced by English *bind*, but note that e.g. in Hungarian the notions of being delimited and being bound are lexically unrelated, and the same is true for Latin *contineo/includo* on the one hand and *astringo* on the other.

3.2 Time

The simplest model of time is the one depicted in the right panel of Fig. 1.3: there are only now, and not-now, some other time. Sometimes it's light, sometimes dark, sometimes raining, sometimes not, sometimes we hunger, sometimes we are full. In principle, we could consider an even simpler model, depicted in the left panel, where there is only one time, which we can call *now* just as well as we could call it *eternity*. But this doesn't quite amount to a model of time, because we can't have landmarks, we are always in *now* or, what is the same, always in eternity. Ecclesiastes, as good an exposition of the eternal mode of thinking as any, actually relies on a two-state model:

to every thing there is a season ... a time to be born, and a time to die; a time to plant, and a time to pluck up that which is planted.

The standard mathematical model for now/other is the cyclic group of two elements, C_2, and for n states, C_n, Many temporally marked utterances already make perfect sense in C_2, and not just the ones marked for present tense. What does it mean that *it has rained?* According to Ecc. 1.9 this is no different from *it will rain,* since "the thing that has been, it is that which shall be". Languages like Chinese make no tense distinctions as such in the grammar (this is not at all the same as not making temporal distinctions conceptually), and several languages stop at C_2. For some of these, like Arabic and Japanese, the "other" state is past, so that "now" is lumped together with the future. For others, such as Quechua or Kalaallisut, it is the future, so that the present and the past are lumped together. We will deal here mostly with the C_3 case, with the standard division into past, present, and future, but it is well known that more complex systems exist, by subdividing the past in two (remote and recent) or even three (historic, remote, recent) and similarly for the future.

While cyclic groups are eminently suitable for seconds, minutes (C_{60}), hours (C_{24}), days (C_7), weeks, and in general for calendar devices, we will not spend a great deal of effort on exploring this connection, since calendars are culture- and language-dependent, whereas our focal interest is with universal semantics. For C_2, the conceptual relation between modulo 2 counting and grammatical conceptualization is evident in frequentative forms, which enforce some cyclicity in the way we conceptualize time, but this no longer works for C_3 and higher moduli.

Aside from the particular world-view presented in Ecclesiastes, we consider the past gone, the future unwritten, and the idea of "nothing new under the Sun" that connects these two is uncharacteristic of everyday thought, where no amount of going forward in the future can take us back to the past. One area that makes the weakness of the connection with C_2 evident is iterating *other,* as seen e.g. in the treatment of redoing, which we will briefly inspect in a somewhat underused corner of the lexicon. One meaning of the English prefix *ana-* that we see in Latinate words such as *anabaptism* 're-baptising'; *anabiosis* 'return to life'; *anaclasis* 'reflect, turn back'; *anacrusis* 'pushing back'; *anadiplosis* 'repetition of a prominent word'; *anaphoric* 'repetition of a word'; *anaphylaxis* 'severe reaction to second or later administration of a substance'; *anatexis* 'melting again'; *anatocism* 'the taking of compound interest' is precisely this redoing. (The most frequent use of *ana-* is in a different sense, 'up', but clearly none of our examples involve this sense.)

Doubling back, returning, does not take us to the original concept, it takes us to the concept *again,* with some temporal marker or counter updated. Getting rebaptized is not at all the same as getting baptized. Extreme reaction to encountering some allergen the second time is sufficiently different from getting this reaction on the first encounter to have its own name. The rocks formed by re-melting and subsequent re-hardening can be distinguished from those formed by hardening alone. Interest on interest adds up, rather than taking us back to the capital. The temporal marker must be (for all cases other than

Ecclesiastes) more complex than C_2, but cannot be as complex as the integers, let alone the reals (see 3.4). Perhaps unsurprisingly, we will reach very similar conclusions in relation to negation in 4.4, where we will conclude that double negation is not the same as affirmation.

before   Relative to now, past events are `before`, and future events are `after`. Of these two, a reductionist theory such as `4lang` can treat one as primitive, and define the other based on the observation that

$$x \; before \; y \Leftrightarrow y \; after \; x \tag{3.5}$$

We will take `before` to be a primitive, irreducible notion, even though there is a tenuous connection with spatial *before*, as in *The knight kneeled before the king*, and we will not take full advantage of Eq. 3.5 because we will have both `before` and `after`
after   anchored in `now`, but in slightly different ways. *after* is defined by regular succession, `follow, in order/2739`, and will have many uses in the core vocabulary
burn   to capture result states, as in *burn* `fire, <=agt wood, after(ash)>` or *stop*
stop   `after(=agt lack move)`.

In contrast, `before` appears mostly in definitions that also have an `after` compo-
move   nent, such as *move* `before(=agt at place/1026), after(=agt at`
exchange   `other(place/1026))`, or *exchange* `before(=pat at person), after`
`(=pat at other(person))`. By lumping 'before' and 'now' together these could be reduced to pure resultatives, but we will not follow this path here, especially as there
sudden   are other lexical items whose definition refers to pure preconditions: *sudden* is given as
win   `lack warn, before(lack (gen know))` or *win* as `best, succeed/2718,`
`before(compete), before(effort), get/1223 <prize>`.

What the foregoing suggests is that to make sense of temporal effects in terms of world vectors we need not just one world model V, but three: V_b, V_n, V_a 'the world *before* the event, *now*, and *after*'. Aside from the time indexing, we assume these are given to us in the same basis. (In C_2 we would use only two, V_n and V_o.) The key to understanding temporality is that these worlds themselves are timeless, and the time spent between them is underspecified. Silently, automatically, V_n becomes V_b, and V_a becomes V_n, and we need predicates only when objects and their properties do not persist. We only need to list the changes to obtain a full picture of the next time instance based on the present one, and conversely, we generally only find changes to be worthy
former   of reporting. A good example is *former*, which simply means that the object is in V_b at the relevant coordinates, but not in V_n. When exactly that former state obtained (a few minutes, days, or many years ago) is something that pure tense marking, such as a PAST morpheme, leaves unspecified, especially in systems where it is not contrasted to historical/remote past marking.

In 1.4 we sketched the `exchange_` frame we invoke in analyzing commercial exchange (buying and selling), as distinct from the word `exchange` which has no commercial aspect, cf. *they exchanged knowing glances*. Now we can extend this to a fully temporal analysis. We have four participants: Buyer, Seller, Money, and Goods, and we assume the exchange itself is taking place now. Therefore, in V_b we have `B has M`, `S has G` and in V_a we have `B has G`, `S has M`. This is invariant of the choice of *buy* or *sell* perspective, and more important, it is the most exhaustive statement we can make about the temporal ordering without introducing spurious ambiguities. In reality, the exchange of money may precede, follow, or be synchronous with the exchange of the goods: we don't know, and we conceptualize the whole an act of buying/selling independent of this detail. The frame actually carries no hidden presupposition that normally the goods are handed over first, or the other way around. It's not that the various orders cannot be expressed, they can, but it requires special effort to disambiguate between them.

There are significant differences between the naive view of time presented here and the scientific model. In fact, the differences between the classical Newtonian and the modern relativistic view, significant as they are, pale into insignificance when comparing either of these to the naive view. Crucially, the naive view is based on discrete time instances, whereas the scientific view relies on continuous variables and differentiability. In 6.1 we will return to the question of how much dynamics can be stated commonsensically – here we offer only a few pertinent observations.

First, there is no guarantee that the left difference Δ_{bn} and the right difference Δ_{na} are in any way similar, in fact there is good evidence to the contrary. Consider *pause*: under any theory it will mean something like `lack action, before(action)`, `after(action)`. The usual epistemic limitations, which we will turn to in 6.3, apply to any use of the word: when we say *Hearing the extraordinary noise, John paused typing* there is no guarantee that he will resume typing, the noise may just be the building collapsing on him, but our normal expectation is that he will. We defer a more detailed (nonstandard) analysis of pausing to 6.2, but note here that the speed with which the activity is paused, on the order of human reaction time, say 200 milliseconds, can be very different from the timescale on which the activity is resumed, say after a few seconds of contemplation.

Second, there are key cases where we can't even estimate Δ_{bn} and Δ_{na}, only Δ_{ba} is available: consider *move* defined above as `before(=agt at place/1026)`, `after(=agt at other (place/1026))`. Clearly, the naive theory is too weak to support Aristotelian dynamics (where speed, as opposed to acceleration, is proportional to force) let alone Newtonian, for this would require second derivatives where we don't even have first ones. What little dynamics there is follows neither Aristotle's law that things return to their natural 'rest state' nor Newton's law of inertia that things will keep moving as long as counteracting forces are absent. The best we will be able to provide in 6.1 is Buridan's theory of *impetus*, including the scientifically false, but commonsensical idea that planetary orbits are to be explained by circular impetus. This is consonant with every child's expectation that things on a circular path will continue their circular motion.

From a cognitive perspective, this lack of dynamics is as it should be, especially as *move* is applicable in a great number of cases where motion does not involve physical motion at all: consider *the lecture moved from theory to practical issues* or *they were*

pause

moved to tears. The same phenomenon of using motion verbs where there is no physical, or even emotional, motion, is seen in verbs of fictive motion: *the pipe runs underground, the fence zigzags from here to the house, the mountains surround the village, ...* (Talmy, 1983). There are a number of theories addressing these: at one extreme we find Jackendoff, 1983, who denies that motion is taking place, at the other we find Langacker, 1987, who is basing his theory on the motion of the scanning focus of the observer. (While our sympathies are with this latter view, we cannot possibly adjudicate the matter here, and refer the reader to (Waliński, 2018) for further discussion.)

The paucity of testable predictions in regards to dynamics can be contrasted to the richness of grammatical evidence about perspective. As before (S19:6.4), we consider a Reichenbachian view, distinguishing four different notions of time: (i) *speech time*, when the utterance is spoken, (ii) *perspective time*, the vantage point of temporal deixis, (iii) *reference time*, the time that adverbs refer to, and (iv) *event time*, the time during which the named event unfolds. So far, we spoke only about event time, which we can fairly identify with V_n. Temporal adverbials, such as *quick* defined as `act in short(time)` refer to the size of the temporal interval between V_b and V_a. Speech time and perspective time rarely coincide. Even in blow-by-blow descriptions given in the present tense *so I'm walking down the street, minding my own business, when this guy starts shouting in my face and ...* we automatically assume a perspective time prior to speech time.

quick

Within the confines of this volume we cannot pursue the issue of *Aktionsart* in any detail, but a few remarks are in order. Obviously, the use of `before` and `after` is closely related to lexical aspect, but on our view semelfactives (Comrie, 1976) like *blink* have `before` and `after` clauses. Analogous to our analysis of `pause`, *blink* would be defined as `before(eye[open])`, `eye[shut]`, `after(eye[open])`. In contrast, statives like *know* and possessive *have* are defined without reference to an `after` state, and their well-known durativity (once you know something, you keep on knowing it, and once you have something, you own it forever) is due to a general law of default continuation (see 6.4). Telic words exhibit a contrast between their their `before` and `after` (goal) states: for example *release* `before(keep)`, `after(free)` or *drown* `before(breathe)`, `after(dead)`.

We maintain temporal deixis by means very similar to the ones used in maintaining spatial deixis, chiefly by *indexical* expressions, to which we now turn.

3.3 Indexicals, coercion

Perhaps the conceptually simplest way to specify *when* and *where* is by means of absolute coordinates: *the party will start at 2PM on July 29 2020 at (47.55625, 18.80125)*. This powerful combination of simplicity and precision is achieved by reliance on highly complex notions, such as the real line, or spherical coordinates, that are relatively recent developments. Natural language has supported locating matters in space and time for many millennia without absolute coordinates. As in most domains, the central method of transferring information is by relative coordinates, tying the new information to something assumed mutually known, as in *The battle took place in the 32nd year of King Darius' reign.*

In between the relative and the absolute mode there lie centuries of standardization efforts gradually moving us from highly subjective measures like *a few hundred steps* or *two day's journey* to the contemporary metric system of units, made ever more precise by metrology (Mohr, Newell, and Taylor, 2016). Most of the units relevant to natural language semantics are, by contemporary standards, highly imprecise: to keep good track of *years* already requires astronomical observations, *seasons* depend of the vagaries of the weather, *days and nights* are not of equal length, what is seven days' journey for one party may only take six days for another, and so on.

Here we follow in the footsteps on Meinong (see in particular Parsons, 1974, Parsons, 1980) and consider words to be capable of denoting objects about which we only have partial information, partial even to the extent their very existence and identity are uncertain. These denotations are greatly similar to the *pegs* of Landman, 1986, especially as we already have a naturally defined partial order on our hands, containment of polytopes in Euclidean space. Since containment is affected by choice of scalar product, things are a bit more complicated than in the data semantic view proposed by Landman, but on the whole we see no need to introduce new, special entities for indexicals.

There are, broadly speaking, two schools of thought: under the dominant view indexicals are variables that obtain their value in reference to external objects present in the real-world context or elsewhere in the discourse. Under the minority view that we follow here, indexicals are just words, not particularly different from other nouns, common or proper, in the degree to which they are underdefined. We can liken them to *bobbers:* much as the float fisher's bobber keeps the bait at a certain fixed depth, bobbers are partially defined individuals already tied to some properties that can be effortlessly computed from regions of the thought vector that can lie outside the linguistic subspace L. When the water level rises, the bobber rises with it, and so does the bait linked to it by a fixed length of string.

The string has zero length with indexicals like *now* – as absolute time moves on, so does *now*. We don't have a full understanding of circadian clocks (the 2017 Nobel prize in medicine was awarded for discoveries of molecular mechanisms controlling the circadian rhythm in fruit flies) but by definition the state of the suprachiasmatic nucleus, and indeed the state of the entire of hypothalamus, is included in Ψ, and we need no special mechanism for *now* to key off of Ψ. With words like *today* the string is longer, and an absolute value cannot be specified without reference to the current time, but a definition `day, now` is sufficient. For *yesterday* we have `day, after(today)`. (This is not a typo: `after` refers to the result state of what happens after the definiendum. After yesterday, today happens.) In terms of the geometric view (1.4) indexicals are simply polytopes whose distinguished point is obtained by projecting the whole thought vector Ψ in the linguistic subspace L discussed in 2.3.

today

yesterday

In the spatial domain, the zero length case is *I*, computed effortlessly from the real world speech situation based on `person, speak`. As we discussed in 3.1 *here* is egocentrically attached to the origin of the coordinate system of the speaker at `I`, unless accompanied by a pointing gesture as in *we should plant the tree here.* 2nd person singulars are again automatically resolved to the hearer, but 3rd person requires either deixis or some circumlocution, as does *there, then*. In terms of simplicity, we consider the direct deictic reading of indexicals to be prototypical, but there is often an indirect reading, tied e.g. to perspective time rather than speech time, the coordinate system of the protagonist rather than the ego. Consider *Roxanne hasn't seen such enthusiasm for years* – clearly, *such* refers to the enthusiasm she sees at event time.

Interrogatives (on our analysis, the morpheme *wh*) are simply requests for a resolution. That they can often be satisfied by a pointing gesture goes to show that the answer is typically obtained by a mechanism outside *L* proper, engaging those parts of the thought vector that are clamped to visual input. This mechanism of going outside one's own linguistic state vector is also responsible for direct manipulation of the listener's thought vector in rhetorical questions, and in the case of informational questions, by reliance on the knowledge state of the listener.

With a rough understanding of indexicals in place, let us now turn to the general mechanism of *coercion*, what (Fauconnier, 1985) calls 'projection mapping'. It is this, as opposed to the more widely used variable binding mechanism, that we make responsible not just for the interpretation of pronouns, but also for most conceptual analysis. We begin with a simple example we already touched on in 1.4, the *commercial transaction* or `exchange_` schema.

There are four participants: the buyer, the seller, the goods, and the money. Of these, the two agentive forms are compositionally named (see 2.1 where the suffix *-er/3627* is discussed), meaning that *buyer* is agent, and so is seller. As we already noted, the name *money* is somewhat imperfect for the 'thing of value' that is used in the exchange, and *goods* is a very imperfect name. Nevertheless, whatever was the patient of the exchange is forced or *coerced* into this role by a rule of English grammar that the NP following the verb is the patient (see Fig. 1.2). Even more remarkably, whatever appears in the fourth slot is now a 'thing of value', even if it's just a bowl of lentil stew.

If this happened in the interpretation of a single sentence we could claim the effect is due to the preposition *for*, but as the story is told (Gen 25:29), Esau is asking for food, and Jacob asks Esau to *sell* his birthright. Esau was only asking for food, and it is Jacob who invokes the exchange schema, with the slots `seller` filled by Esau; `buyer` by Jacob; and `goods` by the birthright. Subsequently the schema is ratified by Esau swearing to it, and fulfilled by his eating the bowl of lentils. That this food is the 'thing of value' is unquestionable, but how does the vectorial semantics reflect this?

The four vectors $\{\mathbf{v}(buyer), \mathbf{v}(seller), \mathbf{v}(goods), \mathbf{v}(money)\}$ are the defining elements of the exchange schema as a set (we use curly braces to emphasize that their order is immaterial). Together, they define a polytope, the intersection of the positive half-spaces. The other 4 vectors in our example, $\mathbf{v}(Jacob), \mathbf{v}(Esau), \mathbf{v}(birthright)$

I

here

and $\mathbf{v}(food)$ are just points (or small polytopes) in L. What we are looking for is an equaliser Q such that after applying Q to the representation space R that reflects the state of affairs before Jacob making the offer, we obtain R', where not just $\mathbf{v}(buyer) = \mathbf{v}(Jacob)$, $\mathbf{v}(seller) = \mathbf{v}(Esau)$ and $\mathbf{v}(goods) = \mathbf{v}(birthright)$ but also $\mathbf{v}(thing\_of\_value) = \mathbf{v}(bowl\_of\_lentils)$ holds.

These equations are created by several different mechanisms. The first two equations come from resolving pronouns corresponding to speaker and hearer: the sentence *Sell me (this day) your birthright* is addressed to Esau, making him the seller, and spoken by Jacob, making him the buyer. Since the birthright appears in the patient slot of this sentence, we obtain the third equation by the same syntax mechanism we discussed in S19:5.3. The last equation is supported by the mechanism of pragmatic inference discussed in S19:5.7: we know from earlier sentences that the *food* and the *bowl of lentils* are the same, we know Esau is faint, and he himself acknowledges that at this point the food is more important to him than his birthright: *Behold, I am at the point to die: and what profit shall this birthright do to me?* This establishes, from the seller's perspective, that the thing of value to be received for the goods is indeed the food.

By change of scalar product, the vectors corresponding to the discourse entities can be easily moved to the respective positive half-spaces, as discussed in 2.3. But here we want to express not just containment. `Esau is_a seller`, but rather equality, that Esau is (uniquely) tied to the seller slot in this particular instance of the exchange schema, hence the need for equalizers. It adds to the challenge that `exchange_` is not a word, something we could describe by a single vector: we need four vectors to make sense of it, and we know that a great deal of additional knowledge is implicated, such as the reversal of ownership of both money and goods on completion of the schema.

In general, none of the relationals we discussed in 2.4 has a clear and unambiguous word we could use to name it, with the possible exception of `er_`, which has a good, albeit morphological rather than word-level, expression in the English comparative suffix `-er`/14. These relations (the list includes not just spatials but also `cause_`, `for_`, `has`, `ins_`, `lack`, `mark_`, and `part_of`) have in common the obvious requirement of using at least two vectors to characterize a single instance, but otherwise they are rather different. More detailed analyses are provided for `has` in 2.2; `for_`, `ins_` and `part_of` in 2.4; `mark_` in 2.5; `lack` in Chapter 4; and `cause_` in 6.2.

3.4 Measure

Counting and measuring things is central to civilization. Buck (1949) lists "Quantity and number" as one of the semantic fields he uses to organize the IndoEuropean material, containing entries not just for the cardinals *one, two, . . .* and the ordinals *first, second, . . .* and fractions *half, third, . . .* but also for less specific notions of quantity such as *much, many, more, little, few* and for broad measure phrases like *full, empty, whole, enough, every, all, ...* As Buck (13.31) notes, "no class of words, not even those denoting family relationship, has been so persistent as the numerals in retaining the inherited words". Given the semantic coherence of the class, and the difficulty of subtle shifts in meaning, it is not surprising that this phenomenon is not limited to IndoEuropean – similar coherence is seen e.g. in Bantu, now tentatively extending to Niger-Congo (Pozdniakov, 2018).

From the mathematical perspective, the first thing to note about the system is that there is no system. It is only in hindsight, from the vantage point of the modern system of natural numbers $\mathbb{N}$, that we see the elements of counting, the cardinals, as being
last   useful as ordinals as well. But certain notions like *last*, 'part_of sequence, at end' which make eminent sense among ordinals, have no counterpart among the car-
first   dinals, while others, like *first* 'lack before, second/1569 follow', do. Key elements, *one* in particular, are used not just for counting and ordering, but also for signifying uniqueness 'unus, unicus' and separateness, standing alone.

The idea of using functions from objects to $\mathbb{R}$ to gain traction of measure phrases such as *three liters of milk* is common in mainstream logical semantics (Landman, 2004; Borschev and Partee, 2014) but, as will be discussed in greater detail in 4.5, we view this approach as highly problematic both in terms of empirical coverage and in terms of bringing in an extra computational stratum. 4lang has no problems handling vague
many   measures of quantity, like *many* 'quantity, er_ gen' or *few* 'amount (gen
few   er_)', though these present the modern, more precise, theory with significant difficulties. However, it does have problems with the modern quantificational readings of *all*
all   and *every*, since it defines the former as 'gen, whole' and the latter as gen. As we
every have noted elsewhere (Kornai, 2010b), actual English usage (in newspaper text) is characterized by generic readings, and the episodic readings are actually restricted to highly technical prose of the kind found in calculus textbooks.

Thanks to the foundational work of the late 19th and early 20th centuries we now have a simple, elegant method for extending $\mathbb{N}$ to the rationals $\mathbb{Q}$. These, or even finite precision decimals, would arguably be sufficient for covering much of everyday experience, especially ordinary measure phrases like *This screen is 70" wide*. Since the Message Understanding Conferences (Grishman and Sundheim, 1996) special attention is paid to the extraction of *numerical expressions* (NUMEX) such as monetary sums and dates. The notion of calendar dates has been extended to cover more complex time

expressions (TIMEX), and for most of these, there is a standard Semantic Web representation scheme, ISO TimeML associated to instances, intervals, etc. which grew out of earlier work on providing semantics for time expressions (Pustejovsky et al., 2003; Hobbs and Pan, 2004). Extracting this information from (English) text is difficult (Chang and Manning, 2012) and the parsing and normalization of time/date expressions is still an active research area (Laparra et al., 2018).

These representation schemas, both for direct time and space measurements, and for the more abstract quantities like monetary sums, implicitly rely on the standard theory of the real line $\mathbb{R}$. Tellingly, all work on the subject has an important caveat (Hobbs and Pan, 2004):

In natural language, a very important class of temporal expressions are inherently vague. Included in this category are such terms as *soon, recently, late,* and *a little while*. These require an underlying theory of vagueness, and in any case are probably not immediately critical for the Semantic Web. (This area will be postponed for a little while).

Here we turn this around, and treat expressions like *soon* 'a short time after <now>' or *late* 'after the time that was expected, agreed, or arranged' as normal, and vague only from the vantage point of the arbitrary precision semantics imposed by using real numbers. From this vantage point, every term we use in ordinary language is vague: for example *water* does not precisely demarcate how many milligrams/liter mineral content it may have. From the vantage point of ordinary language, it is not just the real numbers $\mathbb{R}$ that require special semantics, the problem is already present for natural numbers $\mathbb{N}$: iterative application of the Peano Axioms (or even the axioms of the weaker system known as Robinson's Q) is not feasible given the simple principle of non-counting that we have argued for in Kornai (2010b):

> For any natural language N, if $\alpha p^n \beta \in N$ for $n > 4$, $\alpha p^{n+1} \beta \in N$ and has the same meaning

Since we simply can't make a distinction between *great-great–great-great-great-great-great-grandfather* and *great-great-great-great-great-grandfather* unless we start counting on our fingers, we conclude that the only feasible approach is to do the work outside `4lang` by means of a separate equation solver. This is the approach taken in modern systems aiming at word problems such as Kushman et al. (2014), which derives the equations from text using standard NLP methods, and solves them by Maxima.

Unlike ordinary language understanding, solving word problems, or even setting up the equations, is a skill that Kahneman (2011) would consider 'slow thinking'. Whereas ordinary semantic capabilities are 'fast thinking', deployed in real time, and acquired in everyday contexts by all cognitively unimpaired people early on, solving word problems is a task that many fail to master even after years of formal schooling.

Once we permit an external solver, there is no need to restrict the system to (finite precision) rationals, and sophisticated methods using reals $\mathbb{R}$ and even complex numbers $\mathbb{C}$ are also within scope. What we need is a system to extract the equations from the running text. This is effectively a template filling task, originally considered over a fixed predetermined range of templates by Kushman et al. (2014), and more recently extended to arbitrary expression trees by Roy and Roth (2016). This is a very active area of research, and we single out Mitra and Baral (2016) and Matsuzaki et al. (2017) as particularly relevant for the linguistic issues of assigning variables to the phrases used in the question and in the body of the word problem.

Altogether, the proto-arithmetic that is discernible in systems of numerical symbols, e.g. Chinese 一, 二, 三 or Roman I, II, III or from reconstructed proto-forms that give 7 as '5 + 2' or 8 as '4 · 2' is haphazard, weak, and both theoretically and practically inadequate. This is evident not just from comparing the axiomatic foundations of arithmetic to that of 4lang but also from evolutionary considerations, as the modern system of Arabic numerals has displaced all earlier ones such as the Babylonian, Chinese, Roman and Maya numeral systems.

It doesn't follow that every semantic field will require a specific, highly tailored system of Knowledge Representation and Reasoning to get closer to human performance, but certainly 'slow thinking' fields will. Such systems actually have great intrinsic interest: for example Roy and Roth (2017) offer a domain-specific version of type theory (better known in physics as dimensional analysis) to increase performance, a deep domain model on its own right. But our interest here is with precisely the kind of 'fast thinking' that does not require deep domain models. We return to the matter in Chapter 8, where we will discuss a central case, trivia questions, which we can capture without custom-built inferencing.

size
dimension

To elucidate the 'fast thinking' theory of quantity further, we consider the notion of *size*, which 4lang defines simply as me1ret magnitudo rozmiar 1605 c N dimension. In turn, *dimension* is dimenzio1 dimensio wymiar 3355 c N quantity, size, place/2326 has. We again see a near-mutual defining relation, but with the added information that dimension, and by implication, size is a quantity, one that place/2326 has. Tracking this further, *place/2326*

place/2326
place/1026

is given as te1r spatium przestrzen1 2326 c N thing in, related to the {bound} schema we discussed in 3.1, as opposed to *place/1026* hely locus miejsce 1026 c N point, gen at, which is related the the {place} schema. It may be possible to unify these two schemas e.g. by assuming that the body used in place/1026 is also a place/2326 has, but we see no compelling reason to do so, especially as this would bring in the human size scale as default to both, a step of dubious utility.

Our treatment of measure is geared toward raw measurements, as in *John is tall* or *It was a huge success*, as opposed to measure phrases like *John is six feet two inches tall* or *The earthquake measured 7.1 on the Richter scale*. Raw measurements are treated

big
small
large
little
-est
all

as comparisons with averages: *big* is defined as nagy magnus duzly 1744 e A er_ gen, and *small* as kis parvus mally 1356 u A gen er_. (*large* is defined as big, and *little* as small.) This yields a three-pont scale: big/large – medium – small/little, which can be extended to five points by adding superlatives, typically by means of the suffix *-est*, defined as leg-bb -issimus naj-szy 1513 e G er_ all. Here all is not some new quantifier, but simply another noun, mind omnis wszyscy 1695 u N gen, whole. We defer a fuller discussion of quantifiers to 4.5, but note here that 4lang treats them as more related to pronouns than to VBTOs.

In Chapter 5 we will use an even finer, seven-point scale to describe the naive theory of probability, but this should not obscure the plain fact that no n-point scale, for however large n, can capture modern usage, which relies on real-valued (continuous) measure phrases, for which we must rely on equation solvers we see as entirely external to natural language semantics. To deal with *six feet two inches tall* we would need some mechanism that shows this to be equivalent to *188 centimeters tall*. This requires not just the foot/inch and inch/cm conversion, but also the capability to recognize that for practical purposes the unrounded value of 187.96 must be rounded. We can't measure people's height to a millimeter, but if we are talking about a uranium rod in a nuclear power plant, we may well insist on this, if only to guarantee that it will fit some precision-manufactured container.

This is not to say that we cannot write a grammar capable of recognizing the measure phrase. To the contrary, building such a grammar is near trivial when the measurement unit is explicit in the text (but can lead to expensive mistakes when it is not), and standard rule-to-rule compositional semantics, specified in terms of ordinary arithmetic operations, can be used to compute values to arbitrary precision. But doing so is irrelevant to our main goal, which is to characterize human semantic competence, rather than the competence of ALUs.

4

Negation

Contents

Our goal in this chapter is to provide a formal theory of negation *in ordinary language*, as opposed to the formal theory of negation in logic and mathematics. In order to provide for a linguistically and cognitively sound theory of negation, we argue for the introduction of a dyadic negation predicate `lack` and a force dynamic account of affirmation and negation in general. We take the linguistic horn of the dilemma first articulated by Benacerraf, 1973:

> (...) accounts of truth that treat mathematical and nonmathematical discourse in relevantly similar ways do so at the cost of leaving it unintelligible how we can have any mathematical knowledge whatsoever; whereas those which attribute to mathematical propositions the kinds of truth conditions we can clearly know to obtain, do so at the expense of failing to connect these conditions with any analysis of the sentences which shows how the assigned conditions are conditions of their *truth*.

The linguistic background is sketched in 4.1. We are equally interested in lexical semantics and the semantics of larger constructions recursively (compositionally) built from lexical elements. In 4.2, we provide a systematic survey of the negative lexical elements in 4lang. We turn to compositional constructions in 4.3, again aiming at exhaustiveness, including many forms that involve negation only in an indirect fashion. We offer a simple, finite state formalization that embodies a more nuanced understanding of affirmation and negation, seeing these as opposing forces in the force dynamic setting (Talmy, 1988). Once the frequent cases are treated, we turn to less frequent cases that

A. Kornai, *Vector Semantics*, Cognitive Technologies, https://doi.org/10.1007/978-981-19-5607-2_4

are nevertheless often seen as diagnostic, such as double negation, discussed in 4.4, quantification and scope ambiguities in 4.5, and disjunction in 4.6.

4.1 Background

Boole, 1854, building upon thousands of years of work in the Scholastic tradition, reformulated parts of, and in important ways extended, Aristotle's logic. The structures that today bear his name, Boolean Algebras (BAs), have several features that make little sense from a linguistic standpoint, such as the commutativity of conjunction (really, *I had dinner and went home* is quite different from *I went home and had dinner*), and the basic 'Boolean' duality that stems from treating negation as a unary operation that is involutionary: $\neg\neg = id$. It is important to emphasize at the outset that what follows is a formalization of the cognitive structures underlying negation, not a critique of the standard (Boolean) negation we rely on in logic and mathematics. As we shall see, the two are very different: the economy, elegance, and tremendous usefulness of BAs came at the price of significant loss of linguistic and cognitive realism. To quote Horn, 1989:

> (...) the form and function of negative statements in ordinary language are far from simple and transparent. In particular, the absolute symmetry definable between affirmative and negative propositions in logic is not reflected by a comparable symmetry in language structure and language use. Much of the speculative, theoretical, and empirical work on negation over the last twenty-three centuries has focused on the relatively marked or complex nature of the negative statement vis-a-vis its affirmative counterpart.

In many adjectival oppositions, normally handled by some version of scalar semantics, it is very easy to pinpoint the asymmetry that Horn talks about, and assign negative value to one side of the scale unambiguously – for a summary of standard marked/unmarked diagnostic tests see Lehrer, 1985. For example, *invisible* carries overt negative marking relative to *visible*, so we conclude that conceptually it is invisible things that have no visibility, rather than visible things that lack invisibility. Yet other oppositions, such as between *full* and *empty*, offer no overt morphological cues, but are nevertheless trivial to classify, because their definition hinges on words (in this case *presence* v. *absence* of filling material) one of which is broadly synonymous to overt negatives: in this case, *absence* to *lack* or *want* (Merriam-Webster).

We will make use of the information-theoretic insight that positives, the unmarked case, are not just more frequent but, as befits a communication system, have less information content (require fewer bits). While there is no strict quantitative correspondence between frequency and the size of the code of the kind we find in artificially constructed codes (Huffman, 1952), the tendency is unmistakable in natural language and has been noted as early as Zipf, 1949.

Syntactically, the key distinction is between positive and negative polarity items (abbreviated PPI and NPI respectively), and the classification of contexts in which they

appear as positive or negative (Giannakidou, 1997). Typical NPIs are like English *ever,*
any, either which occur naturally in negative contexts but not in positive ones: compare
He hasn't seen one ever to *He has seen one ever; He hasn't seen any* to *He has seen*
any; or *He hasn't seen it either* to *He has seen it either*. In other words, polarity items
in English behave much like gendered items in languages like German or Polish which
can appear only in the appropriately gendered context. We will not attempt to sort out
the syntactic properties of polarity items here (especially as these, much like gender
systems, show significant variability across languages), but will discuss their semantic
import in 6.1.

4.2 Negation in the lexicon

About 12% of the 1,200 word defining vocabulary of Release V1 (S19 Appendix 4.8),
144 items altogether, involve some form of negation: *accept accident acid arrive atom*
bad bar behind bend black block building burn calm catch chance child clean close coal
continue continuous cover curve dark dead destroy different dry eager easy elephant end
fail finish firm first flat free full gas gradual green hang hard hide ill instead jump laugh
leave light limit long lose mean middle must narrow natural necessary need negative
new night no nothing object off offensive one only open opinion oppose out park per-
manent plant police practice preserve prison private protect public quiet reach remove
rest right romantic rough rubber rude sad safe same send separate serious sharp short
simple sincere single sleep slope smoke smooth soft solid sometimes special steady steal
stiff stop straight strange stupid success sudden sure surprise take tent thick thin tie tight
together twist unless waste water weak without wrong. This list is actually a bit shorter
(139 elements), because in the 144 we count with multiplicity elements that are homo-
phonic in English, such as *thin* as in *thin paint* `hilg liquidus rzadki 1038`   `thin/1038`
`flow(er_ gen)` and *thin* as in *thin reed* `velkony tenuis chudy 2598 gen`   `thin/2598`
`er_ {distance between surface}`. (Since we avoid spurious duplication of
entries for metaphorical senses, treating e.g. *acid* in *vinegar is an acid* and in *an unnec-*
essarily acid remark by one and the same lexical item, disambiguation is rarely called
for.) In Release V2 (see the Appendix) the proportion of negatives is even lower, 8%.

The list has many elements such as *water* which seem to lack any negative aspect.   `water`
But a closer look at the definition `vilz aqua woda 2622 u N liquid, lack`
`colour, lack taste, lack smell, life need` shows how negative statements enter the picture. (Recall from 1.5 that in `4lang` dyadic predicates are given in
infix notation (SVO order), so `life need` means that the subject of need is life, and
the object is the definiendum, whereas `lack taste` means that the object of lack is
taste, and the subject is the definiendum.) The example already shows how our central
innovation, the dyadic negation predicate `lack` works, and the definition `hialnyzik`   `lack`
`desum brak 3306 p V =agt lack =pat` show it to be irreducible (primitive).
As we shall see, `lack` is sufficient: we will not require other primitives, not even an
unanalyzed unary `no`.

In many cases like *dirty* or *blind* the lexical entry carries a negative (prejudicial) sentiment, but not all of these are amenable to an analysis that contains a negative. Every analysis of blindness invokes a logical negative: 'sightless' (Merriam-Webster) 'unable to see' (Longman), etc. Within the bounds of our defining vocabulary, we can write this as `lack sight`. The critical observation here is that lack signifies the absence of a default: people (generic individuals) are sighted, which is the unmarked (default) case, but *blind* contains lexical prespecification overriding this default. Returning to *dirty*, which at first sight is defined as 'not clean'; and to *clean*, definable as 'not dirty', in terms of lack it is obviously *clean* that needs to override the default of things, in their natural state, being somewhat dirty, whereas *dirty* is definable in terms of dirt, mud, dust, soil, etc. just as *sight* is definable without recourse to negation as a form of perception that relies on eyes.

The same treatment can be effortlessly extended to many antonym pairs, e.g. defining *good* as `gen want`, i.e. the object of `want`, where `want` is given as `=agt feel {=agt need =pat}`. Given a positive definition of `good`, we can define *bad* as `lack good`. Antonyms such as *left/right* make clear that `lack` is in some sense the dual of `has`: *left* is `side, has heart` and *right* 'dextra' is `side, lack heart`. Similarly, *same* can be defined as `lack different` but *different* need not be defined as 'lack same', we have a better definition relying on Leibniz' Principle The Identity of Indiscernibles `=pat has quality, =agt lack quality, "from _" mark_ =pat`. We also rely on this principle in the definition of the pronoun *self* which we take to be `=pat[=agt], =agt[=pat]` without any recourse to negation. In all such cases, it is really a matter of lexicographic taste whether we choose to mark antonymy on both members or just one: *invisible* means lack of visibility, and we could redundantly mark *visible* as lacking in invisibility, but we see no compelling reason to do so. Indeed, by omitting these antonymy clauses from the unmarked members of the antonymic pairs, the list we started with can be reduced considerably, a process we carried through for Release V2. Remarkably, we don't have a single example of irreducible antonymy, where both definitions would have to refer to the opposing element.

There is of course an entire class of lexical items whose primary function is to negate: the words *no, not,* the clitic *n't,* the prefixes *un-, im-, de-, non-, anti-* and the like. Ideally, we wish to represent these by a unary negation operator, provisionally written as `no`. This brings into sharp focus the issue of double negation, a matter we will discuss more fully in 4.4, but illustrate here on a contender for the title of longest English word.

Establishmentarianism is the 'movement or ideology advocating the principle of an established Church with special rights, status, and support granted by the state', an issue most people never heard of and most likely stand neutral on. *Disestablishmentarianism* is the directly opposed 'movement or ideology advocating the withdrawal of special rights, status, and support granted an established church by a state', and *antidisestablishmentarianism* is of course the movement or ideology directly opposed to this. People who prefer the status quo will likely be antidisestablishmentarian, but not establishmen-

good
want
bad
left
right
same

different
self

tarian, since neither of these movements/ideologies would be content to leave things as they are.

A shorter and more common, but conceptually not any easier, case is provided by *open* versus *close* (shut). Unlike in topology, where close/open have such specialized meanings that sets can satisfy both predicates at the same time, in ordinary language no ordinary object can be clopen. Yet a third state of affairs, where the status of an object is not known, exists, just as in topology, where a set need be neither closed nor open. Tertium datur. We will denote this third state by ⊙, and use ⊕ and ⊖ to denote the positive and the negative states.

If we don't insist on lexical semantics, compositional cases, which we will treat in more detail in Section 4.3, offer much simpler examples of double negation failure. Consider *up* and *down*. Let's say we are at a construction site, perhaps standing on a ladder, and receive the instruction *move up!* which we want to defy. This can be achieved not just by moving down, but also by moving sideways, or by not moving at all. All three of these acts will conform to the negated command *don't move up. Don't move* or *rest* are contrary to *move*, and *move down* is contrary to *move up*, but these simply don't exhaust the entire space of possibilities, which also contains moving sideways, an action contrary to *rest, move up*, and *move down* alike. In the terms of philosophical logic, natural language negation creates *contraries* rather than *contradictionaries*. Thus, the classical Boole/De Morgan picture where negation satisfies the involution law is simply not tenable for natural language – we present our own solution in Section 4.3, and return to double negation in Section 4.4.

4.3 Negation in compositional constructions

From our perspective, the traditional Square of Opposition (Parsons, 2017) is inhomogeneous. "A" statements of the form *every s is p* are simply written `p(s)` or `s is_a p` (the two styles of writing are just syntactic variants). But a word of caution is in order: these formulas are not aimed at the logical sense of *every* (∀), but rather at the everyday sense, which admits exceptions (Moltmann, 1995; Lappin, 1996). Also, such formulas typically appear in the translation of restrictive modifier clauses, where they have existential, rather than universal import.

For example, when we say in naive physics (Hayes, 1978) that *atoms* are small particles that have nuclear energy (never mind how well this definition fits modern physics, our target is ordinary language), the definiens is formulated as `small, particle, atom has nuclear(energy)`, and here `nuclear(energy)` doesn't embody the claim, not even in naive physics, that all energy is nuclear. Only the much narrower claim, that the energy that atoms have is nuclear, is part of the definition. In this respect, generic is_a is closer to "I" statements of the form *some s is p*.

Of particular interest here is the style of default inference supported: if energy is provided by atoms, that energy is nuclear, if a cane is owned by a blind person, that cane is white, and so forth. This is indeed in opposition to "E" statements *no s is p* whose central goal is to block similar inferences: persons have organs, these organs are typically functioning, so persons can walk, talk, see, etc. – this all goes without saying. The inferences are highly automatic/preconscious, yet we rely on such inferences in the process of making sense of natural language utterances all the time.

`blind`   Clearly, the raison d'être of the word *blind* is to guarantee that some of these inferences are blocked, hence our definition `lack sight`. Further, this prohibition on the inference is absolute, we treat a blind person with a black cane as unusual, exceptional, out of the ordinary, but reality overrides the default, whereas we treat a blind person that can see as paradoxical, impossible, and our best interpretation strategy upon encountering a situation like this is to say that the person was not really blind, that this has something to do with some technical definition 'legally blind' rather than the everyday meaning of blindness.

Finally, "O" statements, *some s is not p* mean lack of implication from *s* to *p*, a view equally compatible with Aristotle's original formulation *not every s is p*, which need not carry the existential implicature that many take for granted in the analysis of *some*. This becomes a bit clearer if we take into account the Aristotelian view that the predicate inheres in the subject: there is no difference, other than surface form, between *Joe is fat* and *Joe has fatness* or *Joe fat(ten)ed*. Whether the predicate is expressed adjectivally, nominally, or verbally has no bearing on its relation to the subject, which is one of subsumption. On this view, O forms are simply `s no p` which leaves it ambiguous between `s isa no p` (adjectival/nominal form using the copula) and `s (no p)` (overtly negated verb). To make the type theory work out, in 2.1 we assumed a broad type of *matters*, which are neutral between things (ordinary nominals), action nominals, events, actions (verbal elements) and properties (adjectival elements). English verb-nouns such as *divorce* furnish a rich class of surface examples.

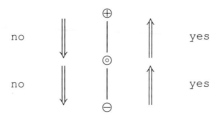

Fig. 4.1: Forces in negation and affirmation

The outstanding issue is explaining why unary `no` is absolute while binary `lack` is generic. `lack` signifies that the predicate in question does not inhere in the subject. What does `no` signify? It is at this point that the information-theoretic view comes to the fore. By the logic of compressibility, `no` must be adding some extra information, but this is not simply negating the statement, as the Boolean solution would have it, but rather *applying a force to make it negative*. As in naive physics (Hayes, 1979) we assume that *matters* have three basic states, positive, zero (default, resting state), and negative: we will depict this in a three-state finite automaton arranged top to bottom as in Figure 4.1.

A word of caution is in order: while finite state automata of the sort depicted here are capable of limited counting, e.g. `no yes yes no no` would move the current state from the initial ⊚ to ⊖, this really doesn't correspond to anything in natural language. Motion, both ordinary physical motion of objects and more general 'movements' or 'processes' provide another example of the same tripartite characterization that we have seen in Fig. 4.1, this time with *start, steady,* and *stop* states.

To see how the state transitions actually work, and to refine the picture to include not just negation but also affirmation, we analyze some ordinary language expressions here. We start with imperatives, both because these are a major source of negatives and because they justify some of the key features of our model. Consider the negatives *Don't smoke!* or *No smoking* and their paired affirmatives *Smoke!* or *#Smoking*.

Normally, locations are unspecified for smoking/nonsmoking, though there are many places where the default is nonsmoking and some where the default is still smoking. A sign that simply says *No smoking* has the same force as one with an overt deontic operator *Smoking prohibited*. The opposite of this is a sign *smoking (permitted)*, and not *#smoking mandatory* which would carry a much stronger affirmation of smoking. This is not because we don't find obligatory rules, there are many from *seatbelts mandatory* to *you must agree to our privacy policy first*, but rather because we find smoking increasingly restricted to special settings like dedicated smoking rooms at airports.

Returning for a moment to our starting example, it is clear, even if we don't take overt morphological marking into consideration, that the normal (default) state of things is to be visible, and invisibility, to the extent it exists, is the marked case. The primary goal of prohibitions is to designate their object as abnormal. Consider *You shall not kill.* Biblical Hebrew (and English at the time of King James) made no distinction between imperative and future negative, the normative effect (of an ideally kept command) is that in the future there is simply no killing (*retzach*). In `4lang` we can write this as `after(gen lack kill)`.

As we have briefly discussed in 3.4, we frequently encounter antonyms that fit well with the tripartite picture of Fig. 4.1: *heavy* means `weight(er_ gen)` and `light`   heavy
means `weight(gen er_ )`. Since the generic will unify with the subject, the effect   light/1381
that (Parsons, 1970) illustrates with the example of *enormous flea*, that such a flea is still rather small, is easily explained: such a flea has size much larger than `gen`, but this automatically refers to a generic flea, not any generic object.

Returning to our theory of *You shall not kill*, `gen` is the same proquant that we use elsewhere to denote a non-specific entity. After the utterance of the command who does no killing? Somebody. Everybody. People. Recipients of the command. It is precisely the generic nature of the subject that guarantees the universal import of the prohibition. This gives an answer to the question we raised above: we will not need a unary negation operator *no* since it can be defined as `gen lack`. From the logic point of view `gen`   no
may appear mysterious, but from the vector semantic view it is simply the row vector   gen

(bra) $\langle 1/n, \ldots 1/n|$ which has the same nonnegative constant $1/n$ at each coordinate, where n is the dimension of the linguistic subspace L (see 2.3 for the notation). gen may be a primitive in the 4lang dictionary, not reducible to other words, but it is not a primitive in vector semantics, in fact it is one of the elements most easily defined.

The picture of negation that emerges from these considerations is very nontraditional. From the mathematical side we have already seen that it requires an intuitionistic proto-logic of some sort, since it admits at least one extra value in addition to true and false. And from the linguistic side, instead of the standard, unary negation operation no that is analogous to Boolean $\neg$, we have a dyadic operation lack that signifies that its first argument does not have some defaults normally associated to it, with the second argument determining which default gets overridden. For example, persons are assumed to have fully functioning organs (in fact, this assumption is held for all living beings, and is inherited to persons via animals) so person, lack sight defeases an entire chain of inferences whereby eye is_a organ and living_being has organ(working) that would normally lead us to believe that persons have working eyes i.e. they are sighted. Compositional no is derived as gen lack, the unary nega-tion operator is formed by quantifying over the first argument of the dyadic lack. Since gen is a fixed vector, and lack is a fixed matrix, the unary no operator is simply the vector obtained from postmultiplying gen by lack. In particular, this is not a matrix like $-I$ (with -1 at each diagonal element and zeros elsewhere), it's just a vector.

How the (primitive) dyadic negation operator lack and the (derived) unary no inter-act with auxiliaries, main verbs, adjectives, and adverbials is a complex matter. We can't possibly do justice to the syntax of negation in this book, especially as this changes from language to language. But the semantics is constant, and is simple enough to derive some major conclusions that appear to have syntactic import as well. In particular, note that lack is a sparse matrix with only a few dozen prespecified elements in the uroboros

`night` set given by equations such as *night* $\overset{d}{=}$ period, follow sunset, sunrise

`opinion` follow, dark, lack sun, <sleep at> or *opinion* $\overset{d}{=}$ thought, person has, person[confident], person lack proof. The simplest of these con-

`public` tain just one clause: *public* $\overset{d}{=}$ lack owner or *lose* $\overset{d}{=}$ after(lack).

`lose` We call attention to the fact that lack is not a pure negation operator (scalar product of a vector with -1) but rather the subtraction of something that is normally there. A person normally has intact bodily functions, including sight, so *blind man* is perfectly normal, while *#blind stone* is markedly infelicitous, just as *#sighted stone* would be. In geometric terms, negation is better explained as partial complementation, i.e. a subset of the complement, not the entire complement. We return to this matter in 7.2, where we discuss scalar semantics.

4.4 Double negation

In general, double negation is out. Negative imperatives are easy (in English, they require *do*-support, but this is exceptional), from *go!* it is easy to form *don't go!* with the intended meaning *stay!*. But double negatives *???don't don't go* are hard to produce, people tend to express the intended meaning by *don't stay*. A British National Corpus (BNC) search reveals 40 examples of *don't don't*, all in live conversation (as opposed to writing), and all with the meaning 'emphatically don't' as in *Charlotte please don't don't go noisy* or *Don't don't you think that there's a conflict of interest there*. This is from a total of 92,334 *don't*s in the corpus. The asymmetry is not restricted to imperatives: consider a grocery store with a sign *no bananas (today)*. Once the shipment arrives, they will not advertise *???no no bananas*. To quote De Mey, 1972:

> 'Natural' negation only involves objects or elements a speaker or listener is attending to . . . It makes no sense to instruct a listener to suppress a thought he is not considering or an idea he is not having.

The only standard case of double negation is when the first negative is syntactic and the second morphological: *a not unhappy person, a not unfriendly letter,* . . . (see Horn, 1989 5.1.3). What is remarkable about such cases is that they are no longer about the negation of some default: there is no assumption that people are generically happy or letters are friendly. It is the unhappiness of a person that is being negated here, an idea that we couldn't reasonably assume to have already been present in the listener's mind as a default assumption. Rather, it is the compositional meaning `person is_a unhappy` that gets negated in its entirety. We conclude that `no`, as a syntactic operator, negates the main predicate, so from aRb we obtain $a(\neg R)b$ by the corresponding compositional semantic rule. We assume, without argumentation, a rule-to-rule hypothesis (Montague, 1970; Bach, 1977; Gazdar et al., 1985) between rules of compositional syntax and semantics.

In this case, the negation of the predicate is easy: both `¬is_a` and `¬has` can simply be taken as `lack`, so we obtain `person lack unhappy`. To negate *John ate fish* we need to invoke some form of *do*-support on the syntactic side to obtain *No, John didn't eat fish*. Note that the main predicate *John ¬eat fish* is coordinated with *No*: to obtain the desired result that this is a singly negated statement about eating we take $\neg X$ to be headed by $\neg$ rather than by X. Since our meaning representations can't have nodes with multiplicity (without the use of the `other` operator), the sentence-initial `no` is unified with the `no` of `no eat`, and we obtain `John no eat fish`. Returning to `person lack unhappy`, we can accept this as is, or proceed syntactically from *not (unhappy person)* or from *(not unhappy) person*. We investigate both possibilities.

Since standard tests of constituency (Wells, 1947) support the second analysis, we start with *not unhappy* and substitute, salva veritate, the definition of unhappy, to obtain `no (gen lack happy)`. As we have seen, the syntactic negation operator affects the main predicate, in this case `lack`. A suitable candidate for `¬lack` will be `has`, which means 'doesn't lack' after all. This way, we obtain `gen has happy` which, when applied to `person`, will yield the desired `person has happi(ness)`.

In the other analysis, we start with *unhappy person* with the semantics `person is_a unhappy`. Again substituting salva veritate, we obtain `person is_a gen lack happy`. Here `person` can unify with `gen` and to yield the more specific `person`, and similarly `is_a` can unify with `lack` to yield `lack`, so altogether we have `person lack happy`, a very reasonable semantic representation of *unhappy person*. Negating this by the syntactic `no` again amounts to negating the main predicate, so we obtain `person has happy` as before, irrespective of the constituent structure we started with.

When both `no`s in a double negation are compositional, the above analysis would yield `gen lack gen lack` which, without special pleading, will simply reduce to `gen lack` i.e. to single negation, a result we are not unhappy with, given the absence of real-life examples suggesting otherwise. For the better attested *Don't you ever NOT clean up after yourself!* we can invoke extra rules, e.g. that the contrastive stress actually keeps the second negation distinct from the first, and indeed, such sentences sound natural only with contrastive stress/intonation.

4.5 Quantifiers

Following Frege, 1879 and Russell, 1905 the treatment of a restricted class of lexical elements, quantifiers, has become virtually inseparable from the treatment of negation. In this regard, our treatment is a considered return from Montague, 1973 and subsequent work to the earlier tradition, whose last significant exponent was Peirce (Böttner, 2001). While Montague Grammar eventually treated nominals as generalized quantifiers (Gärdenfors, 2007; Badia, 2009), we move in the other direction, and treat quantifiers as nominals whose compositional behavior is largely dictated by their semantic content, rather than as special term-binding operators. In doing this, we make purposely very little distinction between an individual fox, the species *Vulpes vulpes,* the set of foxes in the world, or the class of potential foxes in all possible worlds.

That some kind of quantificational ur-element is needed is already clear from a closer look of our definition of *good* as the object of `want`. To write out the definiens in infix (SVO) order, it is not enough to write `want good`, for this would be interpreted as the definiendum filling the subject slot, saying in effect *(the) good wants (the) good,* or worse yet, *(the) good wants itself.* Since the intended meaning is that *good* is what people want (a consensus theory of value), who is the subject, one person, an exemplary and perhaps even God-like person, or just anybody? We will use the same generic `gen` that we used in 4.3 to fill the subject slot, but caution the reader that this element doesn't have universal import, it's just a placeholder that 'plugs up' the valence. The closest overt element in English with roughly the same meaning and distribution is *one* used generically, as in *One should take an umbrella if the sky is cloudy,* but we use `gen` so as to avoid confusion with numerical `one`. Unlike `one` whose semantics clearly involves the singular, gen, being at the top of the subsumption hierarchy, will unify with any x. Whereas one, book means a single book, gen, book is simply book, and we leave it open whether this means an arbitrary book, the set (or class) of all (actual of potential) books, or some abstract notion of 'bookness' as in *the book of nature*.

Lexicalized quantifiers either in their base form *some, any, no, …* or in a subtyped form *someone, somebody, something, somewhere, somehow, anyone, anybody, anything, anywhere, anyhow, noone/no-one, nobody, nothing, nowhere, …* will be treated on a par with pronouns, including interrogatives, as members of a new lexical category *proquant*, whose crosslinguistic coherence (but not the name proquant) is argued for by Szabolcsi, 2015. Many, if not most, of the proquants are either lexical primitives, or have a compositional analysis that directly relies on abstract primitives such as the wh morpheme responsible for interrogatives. Here our focus is on overtly negated elements such as *nobody*, and the main question is whether these require a unary negation operator no.

One area where the standard theory appears vastly superior to the one presented here is assigning semantics to obviously compositional quantifier structures such as *at most seven, no more than ten*. But this is accomplished at the price of sweeping under the rug the fundamental problem we started out with, assigning semantics to the atomic units. What is the semantics of *seven*? The dictionary suggests 'the number 7', but this is not exactly helpful, since '7' is left undefined.

Could we actually use here the standard mathematical semantics that rests on the Peano axioms? The requisite formulas $\leqslant 7, \neg(> 10)$ seem to capture the intended meaning quite nicely, and the task of assembling them in a rule-to-rule fashion appears feasible. Yet the same approach is notoriously problematic for common 'fuzzy' cases like *at least a few, some, many/much* …. A more subtle problem is posed by overgeneration: the standard semantics smoothly extends to zero and negative integers, yet expressions like *at most minus one* are hard to interpret by ordinary speakers, and the more math we apply the clumsier the corresponding natural language expressions become. Do we have to translate *greater than i* as denoting the complex plane with the unit disk removed? If so, why don't we assign this as the meaning for *greater than 1* as well? If not, how do we account for expressions like *greater than z*, with z any complex number, which are perfectly common and ordinary in complex function theory?

Altogether, the standard logical approach is inappropriate for handling what little overlap there is between the semantics of logical and natural language expressions. It offers spurious precision, not just in the handling of 'fuzzy' quantifiers but also for any number above the magical number 7 ± 2 (Miller, 1956). Since the standard theory was developed in order to overcome the well-known limits of human numerosity (Dehaene, 1997), it is incapable, by design, of accounting for these limits. A fuller discussion would go beyond the scope of this book, but a step in the right direction is already taken in Gordon and Hobbs, 2017, who restrict Peano arithmetic to the metatheory, and concentrate on the cognitively relevant structures like 'half orders of magnitude'.

Using this notion, we can assign meaning to lexically complex quantifiers such as *somewhat* in constructions such as *It will be somewhat warm(er)* which we take to mean 'it will be perceptibly warm(er)' where *perceptibly* means 'by half order of magnitude'. Since this is arguably an adverbial meaning, we will concentrate here more on the pro-quants, where *some-* has a pure existential import. Deriving the lexical meaning of quantifiers is made easier by the fact that in most languages they share a sortal type with pronouns, so we will have interrogatives *who, what, where, when, ...* and follow the same typing *everyone/anyone/someone/noone, everything/anything/something/nothing, everywhere/anywhere/somewhere/nowhere, everytime/anytime/sometime/never.*

The sortal types are quite transparent: *who* requires a `person`, normally spelled out in English as *one*; *what* requires a `thing`; *where* requires a `place`, spelled in these proquants as *where* but historically *ere* (also seen in *here, there*); *when* requires a `time`; and *how* requires a proadverbial, spelled variously as *how (anyhow, somehow)* or as *way (anyway, someway, no way/nohow).* Another suppletive form is *never*, with *no+ever* used interchangeably with *no+time.*

who
what
where
when
how
manner

As standard (Katz and Postal, 1964; Langacker, 2001), we analyze *who* as `person, wh`; *what* as `thing, wh`; *where* as `at, wh`; *when* as `time, wh`; and *how* as `manner, wh`, where `manner` is `quality, do has`. By taking *some-* to mean `exist`, arguably a primitive, we obtain for *someone* the definition `exist, person` and similarly for *something, somewhere, sometime, somehow*. We take *every-* to be synonymous with `gen`, and again use the conjunctive combinations `gen, place` to define *everywhere*; `gen, manner/1706` to define *everyway*, etc.

In systems of Knowledge Representation (KR) such as Cyc (Lenat and Guha, 1990) it is common to distinguish individuals, e.g. some particular poet, say Allen Ginsberg, from the class Poet, of which Ginsberg is an InstanceOf. The semantics of *any-*, however conceived, will have to express the choosing of one particular instance from a class, the central element of the meaning being that it doesn't matter which instance (Kadmon and Landman (1993) call this the 'free choice' reading of *any*). Here we take advantage of the mechanism that we have at our disposal independent of negation and quantification, thematic roles (Dowty, 1986) and the fact that we already have a fundamental `is_a` relation in the system. With this, we can define *any* as `<one>, =agt is_a`. We note that optionality (the use of defaults, see 6.4) is another feature of the system that has broad justification already on the quantifier-free fragment (Reiter and Criscuolo, 1983). When we say *any poet* this will mean any (one) x such that x *is_a poet*, and it is the same semantics that we apply to *anyone, anything, anywhere, ...*

any

With the other proquantal roots out of the way, we can turn to our central subject matter here, the semantics of *noone, nothing, nowhere, ...* This requires no special effort, in that *no-* is already defined as `gen lack` and the sortal types just unify with `gen`, leading to `person lack` for *noone*; `thing lack` for *nothing*; etc. Thus *noone slept* is simply `person lack sleep`, and the key scope effect, that this really means 'nobody *among the people relevant in this context* slept' is obtained by reading `person` in this manner. Unlike the Generative Semantics tradition, where this scope restriction is obtained via tracing the scope of (typically covert) high-level speech act operators that act indexically (Lakoff, 1970; Kaplan, 1978), here we take the genericity as basic and find, to the very limited extent one can (Kornai, 2010b), episodic readings by special effort. In this regard, our system is closer to the database logics that rely on a locally closed world assumption (Doherty, Lukaszewicz, and Szalas, 2000) than to classic Montague Grammar.

Compare *Everyone on Cormorant Island speaks two languages* to *Two languages are spoken by everyone on Cormorant Island*. There is a sense that the active sentence does not require these to be the same two languages for everyone, whereas the passive sentence does. But how strong is this sense? Early generative theory (Katz and Postal, 1964) assumed that both readings are available for both sentences. This left explaining which reading is preferred in which context to factors that go beyond syntax and semantics such as communicative dynamism (Firbas, 1971), as there is a similarly strong sense that the active sentence is about the inhabitants of Cormorant Island while the passive is about two languages. Also, it is worth keeping in mind that the entire phenomenon is somewhat marginal. The ratio of passives to actives is somewhere between 4% and 18% depending on genre (Givón, 1979), e.g. the BNC has 662 instances of *killed by* compared to 4407 instances of *kill*. Quantifier phrases (nearly 70k examples in the BNC) will appear in the *by-* phrase only in about 1.5% of the cases.

In 4lang the active sentence means `person in Cormorant, person speak language(two)` (recall that the two instances of `person` that appear in the linearly rendered formula are automatically unified). The passive sentence means `language(two) is_spoken_by person in Cormorant Island`. It is unclear whether these become the exact same thing as soon as we acknowledge a lexical redundancy rule (Bresnan, 1982) that relates active *V* to passive *is V-ed by*: there are surprisingly many design choices even within LFG where the idea that the active/passive relation is to be captured in the lexicon is taken for granted (Genabith and Crouch, 1999).

Here we consider, very briefly, the other proquants. *Anyone on Cormorant Island speaks two languages* versus *Two languages are spoken by anyone on Cormorant Island* has the same level of uncertainty in regards to judgments of grammaticality and readings as the *everyone* examples we started out with. To avoid bracketing, we will write `Cormorant_Islander` or just `C_I` for `person in Cormorant Island`. With this abbreviation the active sentence can be paraphrased as `C_I speak language(two)` and `lg(two) is_spoken_by C_I` and again the outcome depends on the status of the redundancy rule (or in other generative treatments, the transformation) that relates actives to passives. *Someone* does not bring in the same ambiguity problem, since `exist C_I speak language(two)` is implicationally equivalent to `lg(two) is_spoken_by C_I, exist C_I`, no matter how we handle active/passive.

Finally, let us consider the examples most relevant to our subject matter, negated universals or "E" statements. Clearly, *Noone on Cormorant Island speaks two languages* means `C_I lack speak language(two)` and this is subject to the downward entailment issues that smart alecs often play on: ... *but Joe here speaks seven!* More important, we see `lack` as negating a non-default proposition, as in the double negation cases discussed in 4.4, indicating that the mechanism we proposed there is available for these cases as well.

As for "E" passives, we get `lg(two) is_spoken_by lack C_I` which says, in a somewhat clumsy fashion 'among the people who speak two languages we don't find Cormorant Islanders'. This offers the same episodic reading as the active, and is subject to the same downward entailment problem. Note, however, that the phenomenon is even more marginal: *by noone/nobody* phrases are just 0.1% of the total occurrences of *noone/nobody* in the BNC, for a total of 8 sentences among over ten million. One would really have to be superbly confident about having already captured 99.9999% of English grammar before seeing these as a descriptive challenge.

4.6 Disjunction

In BAs, De Morgan's Laws connect conjunction to disjunction in a perfectly symmetrical fashion. But in natural language semantics conjunction is the default operation: unless some other particle is present we interpret phrases and clauses conjunctively. In case of proper nouns, we treat the conjunct as a collective (Scha, 1981). Given that negation is a marked operation, there is no way to follow the BA technique and reduce disjunction to conjunction. In fact, `no (A and B)` ends up negating the head predicate, so we get `A ¬and B`. This is tantamount to the well-known deontic paradox: *No food and drink* is actually obeyed by a person who only brings food but no drink. The obverse of this, Ross's Paradox (Ross, 1941) brings in the same concerns.

It is fair to say, then, that our interest is with a positive, rather than a De Morgan-style definition of disjunction. While we take the rather unsurprising route that `or` is a primitive, not at all reducible to `and` and `no`, let alone to `and` and `lack`, there is more to disjunction than 'well, it's a primitive'. The cognitive import of `or` is clearly to keep both disjuncts open, whereas in conjunction a higher (collective) node is formed and or the conjuncts themselves are no longer active. We define *or* by `"_ or _" mark_ choose`, but note that it is unconnected to the broader system: since not one `4lang` definition contains `or`, it is eliminable from the uroboros set.

A systematic study of `or` in the larger lexical domain must await later releases, but it is worth noting that almost 40% of LDOCE (Bullon, 2003) definitions uses this word. Most of these seem trivial from the `4lang` perspective: for example *abandon* (V) is defined as 'to stop having a particular idea, belief, or attitude'. Here we could simply add much-needed generality by saying 'stop having =pat', for this would add back more literal senses of *abandon* like 'to leave someone, especially someone you are responsible for'; 'to go away from a place, vehicle etc permanently, especially because the situation makes it impossible for you to stay'; and 'to stop doing something because there are too many problems and it is impossible to continue' as in *The soldiers abandoned the battlefield/their weapons.*

Manually checking over 32,800 cases (the 1st edition of LDOCE (Procter, 1978), is far more frugal in this regard) is beyond our powers, but the number of cases seems large enough to justify a deeper study. What seems clear is that eschewing *or* leads to a less precise description of synonyms and of selectional restrictions, as in in *zonked* 'very tired or suffering from the effects of drugs or alcohol, so that you do not want to do anything'. The vectorial perspective may make this considerably easier: to the extent we define conjunction by intersection of polytopes, we may be able to define disjunction by their union, but we must leave this at the conjectural level for now. This may work well even if negation does not correspond to complementation, as we have argued above.

At sentence level, *or* signifies either a future choice to be made, or a past, unknown, choice. This makes `or` more closely related to exclusive or (xor) than to standard Boolean $\vee$, though this is often hard to discern since the alternatives are disjoint to begin with. Further, while natural language `and` must involve incrementing the time index on successive verbal conjuncts (cf. the example we started out with, *I went home and had dinner*), `or` has no temporal update associated to it, which again highlights the lack of duality between these two. Another diagnostic pointing at the same conclusion is the clear ability of *or* to introduce alternatives that are counterfactual: *It can wait, or they would have called us by now.*

There is no question that the proposal made here sacrifices quite a bit on the mathematics side: conjunction is not commutative, Boolean duality is gone, and there are many ripple effects through the entire system we haven't even discussed, e.g. that existential quantification no longer amounts to infinite disjunction. But the gains on the linguistic side are considerable: we have a formal theory of word meaning whereby we can assign semantics to morphological operations in a manner that smoothly extends to compositional semantics.

`4lang` captures well the key observation that negation is not an involution, and in general offers translations whose processing difficulty correlates inversely with their frequency. Clearly, the theory is a better fit with the classical Knowledge Representation tradition (Brachman and Levesque, 1985; Brachman and Levesque, 2004) and with database logic than with the first- and higher-order (intensional) calculi familiar from MG and related theories. We do not see this as a loss, especially not from the learnability perspective, a matter we shall return to in 5.3.

We started this chapter with Benacerraf's observation that sentences in natural language and in mathematics are different enough to merit separate semantic frameworks. Were this not so, it would actually be hard to explain why Boolean Algebra, and modern logical calculi in general, took so long to develop from Aristotle's logic. The approach presented here, in many ways a considered return to a more Aristotelian perspective, is not an attempt to 'reform' standard mathematical logic, which we consider to be the correct theory of the domain. Rather, our goal is to build, with the same care, a formal theory of natural language semantics, even at the price of finding this theory insufficient in the mathematical domain the same way as e.g. in the measurement domain discussed in 3.4.

5

Valuations and learnability

Contents

In this chapter we describe a rational, but low resolution, model of probability. We do this for two reasons: first, to show how a naive theory, using only discrete categories, can still explain how people think about uncertainty, and second, as a model for fitting discrete theories of valuation (which arise in many other contexts from moral judgments to household finance) into the overall 4lang framework. In 5.1 we introduce *likeliness*, which we take to be a valuation of propositions on a discrete (seven-point) scale. In 5.2 we turn to the inference mechanism supported by the naive theory, akin to Jeffreys-style probability updates, and argue that valuations are, for the most part, computed rather than learned. After these preparations, in 5.3 we address what we take to be the central concern for any cognitively inspired theory, learnability. We divide the problem in three parts, learning of (hyper)nodes, learning of edges, and learning of valuations. We argue for a system powered by embodied cognition, with all three parts operating simultaneously.

5.1 The likeliness scale

Historically, the theory of probability emerged from the efforts of Pascal and Fermat in the 1650s to solve problems posed by a gambler, Chevalier de Méré (Rényi, 1972; Devlin, 2008), and reached its current form in Kolmogorov, 1933. Remarkably, not even highly experienced gamblers can extract high precision probability estimates from observed data: one of de Méré's questions concerned comparing the probabilities of getting at least one 6 in four rolls of one die ($p = 0.5177$) and getting at least one double-6 in 24 throws of a pair of dice ($p = 0.4914$). Four decades later, Samuel Pepys is asking Newton to discern the difference between at least two 6s when 12 dice are rolled ($p = 0.6187$) and at least 3 6s when 18 dice are rolled ($p = 0.5973$).

© The Author(s) 2023
A. Kornai, *Vector Semantics*, Cognitive Technologies, https://doi.org/10.1007/978-981-19-5607-2_5

Here we make this phenomenon, the very limited ability of people to deal with probabilities, the focal point of our inquiry. These limitations, we will argue, go beyond the well understood limits of numerosity (Dehaene, 1997), and touch upon areas such as cognitive limits of deduction (Kracht, 2011a) and default inheritance (Etherington, 1987). We introduce *likeliness*, which we take to be a valuation of propositions on a discrete (seven-point) scale. We defer the issue of "computing" with these values, the inference mechanism supported by the naive theory, to 5.2. For the case at hand (low resolution probabilities, this will be much like Jeffreys-style probability updates, but the same mechanism is available for other, non-probabilistic updates as well.

We use the term 'likeliness' for a valuation on a 7-point scale $0,\dots,6$ which only roughly corresponds to a discretized notion of probability (we avoid the more natural-sounding 'likelihood' as this already has a well-established technical sense). 0 is assigned to *impossible* events, $l(e) = 0$, and 6 to *necessary* ones. Note that in this regard l corresponds better to everyday usage in that zero probability events ($p(e) = 0$) do occur, and $p(e) = 1$ guarantees only that the event e has measure zero exceptions of occurring. $l(e) = 2$ means *unlikely*: an example would be traffic accidents. $l(e) = 1$ means *conceivable*, events that are unlikely in the extreme, but not forbidden by physical law. An example would be being struck by a meteorite.

There is a duality between x and $6 - x$ as in Łukasiewicz L_7, so $l(e) = 4$ is assigned to *likely* events such as traveling without an accident and $l(e) = 5$ to *typical* or *expected* ones. Almost all lexical knowledge falls in this last category: chairs are by definition furniture that support a seated person, and if a particular instance collapses under ordinary weight we say it failed (whereas we don't conclude that my car failed when I get in a traffic accident – alternative hypotheses such as driver error are readily entertained). Events that are neither likely nor unlikely are assigned the value 3.

Clearly, using exactly 7 degrees is somewhat arbitrary, but it is evident that using only 3 (say impossible, unknown, possible) would be a gross oversimplification of how people deal with probability, and using a very fine scale would create illusory precision that goes beyond people's actual abilities. With 7, we stick to a relatively small but descriptive enough scale. Even if one could argue that, say on cognitive grounds, 5 or 9 degrees would be better, the overall methodology would be the exact same, and everything below could be easily modified and worked out with that scale. Altogether, our choice of having a 7-degree scale is more of an illustration than a commitment, albeit one well supported by practical experience with semantic differentials (Osgood, Suci, and Tannenbaum, 1957).

The commonsensical valuation, which is our object of study here, differs from probabilities in several respects. The most important from our perspective is *lack of additivity*. At this point, it is worth emphasizing that the theory of likeliness valuation is not intended as a replacement of the standard (Kolmogorov) notion of probability, which we take to be the correct theory of the phenomena studied under this heading, but rather as an explanatory theory of how the *naive* worldview accounts for these phenomena. The fact that as a computational device the standard theory is superior to the naive theory is no more a reason to abandon study of the naive theory than the superiority of eukaryotes is reason to abandon study of prokaryotes.

By lack of additivity we don't just mean lack of σ-additivity, but something that is already visible on finite sums. Consider the Law of Total Probability, that $p(A)$ can be computed as $\sum_n p(A|B_i)p(B_i)$ where the B_i provide a (typically finite) partition of the event space. The equivalent formulation with likeliness normed to 1 would be

$$l(A) = \bigoplus_i l(B_i) \otimes l(B_i \to A) \tag{5.1}$$

Here we retain the assumption that likeliness is a valuation in a semiring where addition $\oplus$ and multiplication $\otimes$ are defined, but instead of conditional probability we will speak about relevant implication $\to$ having a valuation of its own. The semiring of greatest interest is the one familiar from n-valued logic, where $\otimes$ is min, and $\oplus$ is max. In this simplified model, we allow two types of propositions only: standalone sentences A and sentences in the form of an implication $A \to B$ (see 5.2).

To put lack of additivity in sharp relief, consider the following commonsensical example: *all men are mortal*. If we take A to be eventual death, we have $l(A) = 6$. If we ask people to elicit causes of death B_i, they will produce a handful of causes such as cancer or heart attack that they consider likely ($l = 4$); some like accidents or tropical diseases they consider neither very likely nor very unlikely ($l = 3$); some like autoimmune diseases or freezing to death they consider less likely ($l = 2$); and some they consider conceivable but extremely unlikely such as murder/suicide or terrorism ($l = 1$). Needless to say, such valuations are not precisely uniform across people, but they do have high intrasubjective consistency (as measured e.g. by κ statistics). Since $l(B_i \to A)$ is by definition 6, we are left with an enumeration of causes:

$$l(A) = \bigoplus l(B_i) = \oplus_{i=0}^{6} \oplus_{l(B_j)=i} i \tag{5.2}$$

The problem here is that no amount of heaping on more of less likely causes will increase the $\oplus$ above the valuation of its highest term. The phenomenon is already perceptible at the low end: if we collect all conceivable causes of death from lightning strike to shark attack, we have 'death by (barely) conceivable causes' which itself is unlikely, not just conceivable.

In actual mortality tables, this phenomenon is reflected in the proliferation of categories like 'unknown', 'unspecified', and 'other', which take up the slack. Depending on the depth of tabulation, the catchall category typically takes up between .5% and 5% of the total data, which corresponds well to the lack of sensitivity below 1% observed in the de Méré and Pepys examples we started with.

Another obvious difference between the standard and the naive theory is the way extremely low or extremely high probability events are treated. When we want to draw the line between impossible and conceivable events, we don't rely on a single numerical cutoff. But if we take the proverbial 'one in a billion chance' as marking, in some fuzzy sense, the impossible/conceivable boundary, and use log odds scale, as argued by

Jaynes, 2003, the next natural order of magnitude (Gordon and Hobbs, 2017) brings us to $p = 0.0014$, which we can take to mark the conceivable/unlikely boundary, and the one beyond that to $p = 0.1118$, which marks the unlikely/neutral boundary.

In this reckoning everything between $p = 0.1118$ and $p = 0.8882$ is considered $l = 3$, neither particularly likely nor particularly unlikely. Likely events are between $p = 0.8882$ and $p = 0.9986$, while typical events are above that limit though still with a one in a billion chance of failure. As at the low end, the naive theory lacks the resolution to distinguish such failure rates from necessity (total absence of failure).

We should emphasize here that it is the overall logic of the scheme that we are vested in, not the particular numbers. For example, if we assume an initial threshold of one in a million instead of one in a billion, the limits will be at 0.0125 and 0.2008 (and by symmetry at 0.7992 and 0.9875), but the major characteristics of the system, such as the 'neither likely nor unlikely' category takes up the bulk of the cases, or that $l = 2$ cases are noticeable, whereas $l = 1$ cases are barely detectable, remain unchanged. Further, we should emphasize that such limits, however we set them, are not intended as a crisp characterization of human classification ability, the decision boundaries are fuzzy. Returning to lack of additivity, there may well be several likely causes of death beyond cancer and heart attack, but no closed list of such is sufficient for accounting for the fact that eventual death is typical (as assumed by Christian doctrine that posits Jesus as an exception), let alone necessary, as assumed by the irreligious. For this, we need a slack variable that lifts the $\oplus$ of the likely $l = 4$ causes to $l = 5$ or $l = 6$, which we find in B_n 'death by other causes'. We note that historically old age was seen as a legitimate cause of death, and only very recently (since the 1980s) do US coroner's reports and obituaries find it necessary to list the failure of a specific organ or subsystem as the cause of death, and a catchall category, *geriatric malady*, remains available in many countries.

Finally, in contradistinction to the standard theory, $\oplus$ can extend only to a handful of terms, especially as the terms are implicitly assumed independent. By the above reckoning, it takes less than 80 unlikely causes to make one neutral, and less than 8 neutral to make a likely one. The geometry of the likeliness space is *tropical* (Maclagan and Sturmfels, 2015), with the naive theory approximating the log odds (max) semiring.

5.2 Naive inference (likeliness update)

We have two types of propositions: stand alone sentences A and sentences in the form of an implication $A \to B$. A context is a (finite) collection of propositions, which can be represented by a directed graph: nodes of the graph denote propositions A and edges of the graph denote implications $A \to B$. The likeliness function is an evaluation acting on the graph: both vertices and edges can have numeric values between 0 and 6, 0 representing impossibility, 6 representing necessity.

Values $l(A \to B)$ belong to the inner model (adult competence, see 5.3) therefore they are hardly subject to change. Take the following example as an illustration. Snowbird is a ski resort in Utah. Say, for a typical European, Snowbird is related to traveling, skiing, and snowing with the likeliness

$$l(\text{Snowbird} \to \text{traveling}) = 5$$
$$l(\text{Snowbird} \to \text{skiing}) = 5$$
$$l(\text{Snowbird} \to \text{snowing}) = 5$$

Such likelinesses express *typicality* of these relations. Skiing is related to some extent, say, to ski-accident, and ski-accident to death. Take the example below (for the sake of example we differentiate between ski-accidents and accidents; the latter excludes accidents occurring while skiing).

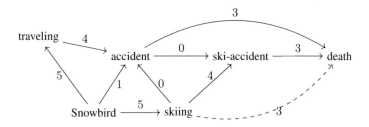

Fig. 5.1: Skiing, accidents, Snowbird

In a typical scenario one does not have any likeliness of the implication Snowbird→death inside the inner model. However, naive inference works: Snowbird *typically* implies skiing; skiing is *likely* to imply ski-accident; finally, it is neither likely nor unlikely that ski-accident results in death. Therefore, one may say [visiting] Snowbird is neither likely nor unlikely to result in death, i.e.

$$l(\text{Snowbird} \to \text{death}) = 3$$

In a similar manner, one could obtain the likeliness $l(\text{skiing} \to \text{death}) = 3$ by saying that skiing is *likely* to ensure a ski-accident, while it is neither likely nor unlikely that ski-accident results in death.

In virtue of the examples above we give a formal model. Let assume we have a finite directed graph $G = (V, E)$ and an evaluation $l : E \to \{0, \dots, 6\}$. We would like to evaluate edges of the complete graph on V that are not in E. Pick two vertices $a, b \in V$, $a \neq b$ and suppose $(a, b) \notin E$. Let $p = (v_1, \dots, v_n)$ be a path in G from $a = v_1$ to $b = v_n$. We write

$$l(p) = \min \big\{ l(v_i \to v_{i+1}) : i = 1 \dots n - 1 \big\} \tag{5.3}$$

The value $l(p)$ expresses how likely the inference $a \to b$ is in case we are relying on the chain of already evaluated implications belonging to the path p. Then the value $l(a \to b)$ is obtained by

$$l(a \to b) = \min \{l(p) : \ p \text{ is a path in } G \text{ from } a \text{ to } b\} \qquad (5.4)$$

In the example above vertices of the graph did not have likelinesses. Suppose we get new information about John: he is *likely* to be in Snowbird, i.e. $l(\text{Snowbird}) = 4$. What consequences can we draw? Being a typical European, if John is in Snowbird, then he must be traveling and it is really typical that people travel to Snowbird to ski. The information that $l(\text{Snowbird}) = 4$ propagates via the edges of the graph: the likeliness of those propositions that are related to Snowbird (that is, they are connected by an edge in the graph to Snowbird) will be updated given new information: $l(\text{traveling})$ and $l(\text{skiing})$ become 4. In the formal model, given the value $l(a)$ and a path $p = (v_1, \dots, v_n)$ from $a = v_1$ to $b = v_n$, using the definition of $l(p)$ in equation (5.3) we can update the likeliness of b writing

$$l(b) = \max \{l(a), l(p) : \ p \text{ is a path in } G \text{ from } a \text{ to } b\} \qquad (5.5)$$

This process of updating iterates: neighbors of just updated vertices get updates in the next round, etc. Supposing the graph is connected, all vertices are assigned with likeliness:

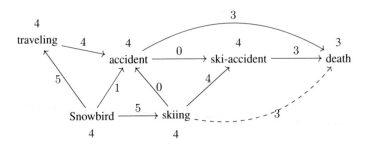

Fig. 5.2: John in Snowbird

Let us now suppose that we learn that John died abroad. The first column of Table 1 describes the default likelinesses we assign to various causes of death, with subsequent columns showing the updates based on whether we learn ($l = 6$) that the death took place in Reykjavík, Istanbul, or on a tourist trip, destination unspecified. Some rows are easy to explain: for example death at home in bed is considered likely, but if we know that John was on a tourist trip the implication is that he is not at home, and the likeliness is demoted to 1. Not 0, because there are extremely unlikely but not inconceivable scenarios whereby he fell in love with the place, bought a home, and resettled there, cf. Jaynes, 2003 5.2.2. This is a scenario that is, perhaps, worth considering if we know only that John went to Reykyavík or Istanbul and tourism was merely an inferred, rather than explicitly stated, goal of the trip, but if we *know* it was a tourist trip and nothing more (last column) this is logically incompatible with being at home.

Cause of death	Default	Reykjavík	Istanbul	trip
in hospital	4	4	5	4
by accident (non-ski)	4	4	4	5
at home in bed	4	1	1	0
in war	1	0	0	1
by homicide	1	1	1	1
by suicide	2	2	2	1
by forces of nature	1	4	1	2
by ski accident	1	2	1	1

Table 5.1: Likeliness of cause of death

The same logic is operative in the next row (war): since we know there is no war in Reykjavík or Istanbul the likeliness is demoted to 0, but for a generic trip it is not, since we do know that there are war zones on the globe and John may have visited one of these.

We obtain that death by ski accident is less likely in Reykjavík (2) than in Snowbird (3) not because skiing is inherently more safe in Iceland, but simply because one can travel to Reykjavík for many reasons, and the likeliness that one goes skiing there is 3, perhaps 4, whereas to ski in Snowbird is typical (5). In connection of Reykjavík we are much more likely to think of death by forces of nature, as there are many natural dangers nearby, from volcanoes to geysers and sneaker waves, indeed this class rises to the top category (4).

This line also illustrates the nonmonotonic nature of the calculus: in general we consider death by forces of nature conceivable but unlikely in the extreme (1), knowing that John went to a tourist trip increases this to 2, but further learning that he went to Istanbul, not particularly known as a natural danger zone, demotes this back to 1.

With this, our rational reconstruction of the naive theory of probability is complete. This theory is not as powerful computational device as the standard theory, and generally only leads to rough estimates of likeliness. However, it is better suited for studying human cognitive behavior, as it requires very little data, and extends to a broad range of cases where the statistical data undergirding the standard theory is unavailable.

Importantly, the idea of updates extends well beyond probability/likeliness. Equation 5.2 and the update formulas remain meaningful for all other kinds of valuations. Following Osgood, May, and Miron, 1975 in essence, but not in terminology, we take the first principal component of their analysis as our primary example. Unluckily for the present discussion, they called this factor EVALUATION, but here we use a more specific name, GOOD/BAD, since it is just one (though clearly the most important) of many valuations. We take the evolutionary roots of this valuation to be hardwired. In S19:3.4 we wrote:

> The pain/pleasure valuation is largely fixed. A human being may have the power to acquire new tastes, and make similar small modifications around the edges, but key values, such as the fact that harming or destroying sensors and effectors is painful, can not be changed.

In terms of the thought vector analysis we sketched in 2.3, a word like *good* is strongly anchored as the density center of the projection of pleasurable mind states on the linguistic subspace, and a word like *bad* is similarly anchored as the density center of painful mind states. These two may not be orthogonal (clearly a mind state, including external and proprioceptive sensor states, can be both pleasurable and painful at the same time), but their existence is sufficient for setting up an initial valuation (say, on a scale of -3 to $+3$) that applies to novel mind states as well. It requires further acquisition work to generalize this from current sensory state to anticipation of future events, `bad`   which is what our definition of *bad* as `cause_ hurt` assumes. This requires nothing far-fetched, given that primary linguistic data, parents' utterances of *bad*, will typically refer to events and behaviors which, upon continuation, would indeed lead to bodily harm. Other valuations, such as Osgood et al.'s POTENCY or ACTIVITY are also richly embedded in sensory data, making them quite learnable early on. As the likeliness case shows, we don't actually need for every word (fixed linguistic data) or set of propositions (transient linguistic data) to have a stored valuation: it is sufficient for there to be deductive methods for dynamically computing such valuations on stored data.

5.3 Learning

How the adult system of mental representation, called the 'inner model' above, is formed in regards to probabilities? Before we turn to the theory of learning, a word of caution is in order: it is not our goal to supplant the existing theories such as (Tomasello, 2003), especially not when it comes to descriptive detail of child language acquisition. Rather, our goal is to provide *explanatory adequacy* in the sense of Chomsky (1973) who states this quite clearly:

> [t]he fundamental empirical problem of linguistics is to explain how a person can acquire knowledge of language.

To paraphrase (Hertz, Krogh, and Palmer, 1991) whom we quoted in 2.3, we will speak of children, because much of the inspiration for learning comes from developmental psychology, *not* because we are concerned with actual persons as opposed to algorithms. "Brain modeling is a different field and, though we sometimes describe biological analogies, our prime concern is with what the artificial networks can do, and why." When we say we consider something 'innate', what this means is that we assume a learning algorithm that has its search space restricted ab initio. This way, explanatory adequacy becomes the issue of how an algorithm, as opposed to a person, can acquire human language, and all we can promise is to keep appeals to 'innate' material within the same bounds for algorithms as are routinely assumed for humans.

Recall that the elementary building blocks of 4lang, the vertices of a graph, correspond to words or morphemes. There is a considerable number, about 10^5, of these, and we add an empty (unlabeled) node ·. These are connected by three types of directed edges: '0' (is, isa); '1' (subject); and '2' (object). Our theory of types is rather skeletal, especially when compared to what is standard in cognitive linguistics (Jackendoff, 1983) or situation theory (Barwise and Perry, 1983; Devlin, 1991), theories we share with a great deal of motivation, especially in regards to common-sense reasoning about real world situations. When we say that a node is (defeasably) typed as Location or Person, this simply means that a 0-edge runs from the node in question to the place/1026 or man/659 node (see 2.1). This applies not just to nodes assumed present already, but also to hypernodes (set of nodes with internal structure, see Definition 5 in 1.5) created during text understanding.

There can be various relations obtaining between objects but, importantly, relations can also hold between things *construed as* objects, such as geometrical points with no atomic content. Consider *the corner of the room is next to the window* – there is no actual physical object 'the corner of the room'. Relational arguments may also include complex motion predicates, as in *flood caused the breaking of the dam*, and so on. To allow for this type-theoretical looseness, arguments of relations will be called *matters*, without any implication that they are material. We use edges of type 1 and 2 to indirectly anchor such higher relations, so the subject of causing will have a 1-edge running from the vertex cause/3290 to the vertex flood/85, and the object, the bursting of the dam, will have a 2-edge running from the cause/3290 node to the head of the construction where *dam* (not in 4lang) is subject of burst/2709. For ditransitive and higher arity relations, which are tangential to our main topic here, we use decomposition (see 2.4).

In general, we define *valuations* as partial mappings from graphs (both from vertices and from edges) to some small linear order L of scores. There is no analogous 'truth assignment' because in the inner models that are central to the theory, everything is true by virtue of being present. On occasion we may be able to reason based on missing signifiers, the dog that didn't bark, but this is atypical and left for later study. Learning, therefore, requires three kinds of processes: the learning of nodes, the learning of edges, and the learning of valuations. We discuss each in turn.

Learning new vertices We assume a small, inborn set of nodes roughly corresponding to cardinal points of the body schema (Head and Holmes, 1911) and cardinal aspects of the outside world such as the gravity vertical (Campos, Langer, and Krowitz, 1970), to which further nodes are incrementally adjoined (see 3.1). This adjunction typically happens in one shot, a single exposure to a new object like a boot/413 is sufficient to set up a permanent association between the word and the object, likely including sensory snapshots from smell to texture and a prototypical image (Rosch, 1975). The association is thus between a phonologically marked point, something that, by virtue of being so marked, is obtained by projecting the entire thought vector on the persistent linguistic subspace L (see 2.3).

As the child is repeatedly exposed to new instances of the category, or even pre-existing instances but seen from a different perspective, against a different background, etc. they gradually obtain a whole set of vectors in L, together forming a point cloud that is generally (but not always, see *radial categories* below) describable by a probability distribution with a single peak, the prototype. This model is very well suited for the Probably Approximately Correct (PAC) theory of learning (Valiant, 1984), and is commonly approximated in machine learning by Gaussian density models. This is not to say that Gaussians are the only plausible model – density estimation offers a rich variety, and remarkably, many of the approaches are directly implementable on artificial neural networks.

On rare occasions, children may learn abstract nodes, such as `color/2207`, based on explicit enumerations 'red isa color, blue isa color,. . .', but on the whole we don't have much use for post hoc taxonomic categories like *footwear*. Many of these taxonomies are language- and culture-dependent, for example Hungarian has a category *nyílászáró szerkezet* 'closure device' which in English is overtly conjunctive: *doors and windows*. In this particular case, the conjuncts are explicitly nameable, but cognitive semantics considers many other cases that Lakoff (1987) calls *radial* categories, where no single prototype can be identified. Here we illustrate the phenomenon based on (Hanks, 2000), where the homonymy/polysemy distinction is considered from the perspective of the lexicographer, using a standard example:

bank/227	bank/1945
is an institution	is land
is a large building	is sloping
for storage	is long
for safekeeping	is elevated
of finance/money	situated beside water
carries out transactions	
consists of a staff of people	

Hanks, much as Lakoff and Wittgenstein before him, pays close attention to the fact that radial categories may be explained in terms of a variety of conditions that may `bank/227` or may not be sufficient. The actual `4lang` definitions are more sparse `bank bank` `bank/1945` `argentaria bank 227 u N institution, money in` versus `bank part` `ripa brzeg 1945 u N land, slope, at river`. We could use defaults to extend this latter definition with `<long>` or perhaps even *elevated*, though we do not at all see how to derive this latter condition for submerged banks such as the famous Dogger Bank.

More important than the details of this particular definition are the fact of common 'metaphorical usage' (snow bank, fog bank, cloud bank) which, many would argue, are present in `bank/1945` as well, with metonymic usage of the institution for the building, or perhaps conversely, using the building metonymically for the institution, as in *The*

Pentagon decided not to deploy more troops. One way or another, this is the key issue for radial categories: surely a large building does not consist of a staff of people.

In a plain intersective theory of word meaning we would simply have a contradiction: as long as `bank/227` is defined as the intersection of the *building* set and the *carries out transactions* set, we obtain as a result the empty set, since buildings don't carry out transactions. We will illustrate how `4lang` solves the problem on the definition of *institution* `intelzmelny institutio instytucja 3372 e N organize at, institution work at, has purpose, system, society/2285 has, has long(past), building, people in, conform norm`. This is a lot to unpack, but we concentrate on the seeming contradiction between `system` and `building`. Our understanding of real-life institutions is assumed to be encoded in very high-dimensional thought vectors, and the word *institution* is only the projection of these vectors on the permanent (stable) linguistic subspace L given to us as the eigenspace of the largest eigenvectors (see our discussion of Little (1974) in 2.3). Within L there is a whole subspace S, spanned by vectors (words) related to systems such as *machine, automatism, process, behavior, period, attractor, stability, evolution,* and so on. There is also a subspace B devoted to buildings, spanned by words such as *wall, roof, room, cellar, corridor, brick, mortar, concrete, window, door* and so on and so forth. By accident, there may be some highly abstract words such as *component* that are applicable in both S and B, but we may as well assume that the two subspaces are disjoint. However, thought vectors can have non-zero projection on both of these subspaces at the same time, and our claim is that this is exactly what is going on with `institution`. Since by definition `bank/227 is_a institution`, the word sense `bank/227` just inherits this split without any special provision.

This has nothing to do with the homonymy between `bank/227` and `bank/1945`: we have two disjoint polytopes for these, rather than one polytope with a rich set of projections. There is no notion of 'bank' of which `bank/227` and `bank/1945` could be obtained by projection as there is a single sense of *institution* of which both the building and the system are projections. Importantly, humans perform contextual disambiguation effortlessly: Hanks (2000) makes this point using real life examples

> people without bank accounts; his bank balance; bank charges; gives written notice to the bank; in the event of a bank ceasing to conduct business; high levels of bank deposits; the bank's solvency; a bank's internal audit department; a bank loan; a bank manager; commercial banks; High-Street banks; European and Japanese banks; a granny who tried to rob a bank

on the one hand, and

> the grassy river bank; the northern bank of the Glen water; olive groves and sponge gardens on either bank; generations of farmers built flood banks to create arable land; many people were stranded as the river burst its banks; she slipped down the bank to the water's edge; the high banks towered on either side of us, covered in wild flowers

on the other. Compare this to the case *the bank refused to cash the check*. The victim is typically quite unable to say whether it was the system that is to blame or the staff, actually acting against the system in an arbitrary and capricious manner. There may be some resolution based on a deeper study of financial regulations and the bank's bylaws, but this takes 'slow thinking', what Kahneman (2011) calls 'System 2', as opposed to the 'fast thinking' (System 1) evident in the `227/1945` disambiguation process, and requires access to a great deal of non-linguistic (encyclopedic) knowledge.

In fact, the learning of nouns corresponding to the core case of concrete objects is now solved remarkably well by systems such as YOLO9000 (Redmon et al., 2016) and subsequent work in this direction, lending credence to the insight of Jackendoff (1983) taking "individuated entities within the visual field" as the canonical case for these. Outside this core, the recognition of abstract nouns like *treason* or attitudes like *scornful* are still in their infancy, though sentiment analysis is making remarkable progress.

Learning new edges Again, we assume a small, inborn set of edges (0,1,2), and an inborn mechanism of spreading activation. The canonical edge types are learned by a direct mechanism. Let us return to `boot/413` for a moment and assume a climate/cultural background where the child has already learned `shoe/377` first. Now, seeing the boot on a foot, and having already acquired the notion of `shoe/377`, the child simply adds a '0' edge 'boot isa shoe' i.e. a '0' edge to the graph view of their inner model. In vector semantics, it is the task-specific version of Eq. 2.6 that is added to the system of equations that characterizes the inner model:

$$P_R(t+1) = P_R(t) + s|\text{boot}\rangle\langle\text{shoe}| \tag{5.6}$$

The case of '1' edges, the separation of subject from predicate, is a bit more complex, especially as two-word utterances are initially used in a variety of functions that the adult grammar will treat by separate construction types such as possessives *dada chair* 'daddy's chair'; spatials *ricky floor* 'Ricky is on the floor'; imperatives *papa pix* 'daddy, fix (this)' and so on. Subjects/subjecthood may not fully emerge until the system of pronouns is firmed up, but our central point here is that what Tomasello (1992) calls "second-order symbols" (for him including not just nominative and accusative linkers but all case markers) are learnable incrementally, on top of the system of what he calls first order symbols (typically, nouns). What is learned by learning verbs is not just some actions, but an entire Fillmorean frame with roles, and markers for these roles. Remarkably, machine learning systems such as Karpathy and Li (2014) are now capable of

recognizing and correctly captioning action shots with verbs like *play, eat, jump, throw, hold, sit, stand,...*, see Karpathy's old webpage for some examples.

In 3.1 we speculated that subjects and objects are initially undifferentiated, and it is the same action that we see either performed by the body *John turns* or on something within arms reach *John turns the wheel* for a large class of motion verbs showing intransitive/transitive alternation. But the same incrementality applies to all adpositions/case markers that act as second order entities, e.g. that the dative would indicate the recipient. Let us consider the situations `boot on foot`, which has direct visual support, and `boot for_ excursion`, which also has strong contextual support, but outside the visual realm.

If the parents are skinheads, the association 'boot for excursion' may never get formed, since the parents wear the boots on all occasions. But if the boots are only worn for excursions (or construction work, or any other specific occasion already identified as such by the child) we will see the *boot* and the *excursion* or *construction work* nodes jointly activated, which will prompt the creation of a new purposive link between the two, just as a joint visual input would trigger the appropriate locative linker.

Again we emphasize that the gradual addition of links described here is not intended as a replacement for actual child language acquisition work such as (Jones, Gobet, and Pine, 2000), but rather as an indication of how such a mechanism, relying on training data of the same sort, can proceed. We note that the ab initio learning of semantic frames (Baker, Ellsworth, and Erk, 2007) is still very hard, but the less ambitious task of semantic role labeling is by now solved remarkably well (Park, 2019).

Learning valuations Probabilities are by no means the only valuation we see as relevant for characterizing human linguistic performance, and using a seven point scale $s = \{0, \ldots, 6\}$ is clearly arbitrary. Be it as it may, similar scales are standardly used in the measurement and modeling of all sorts of psychological attitudes since Osgood, Suci, and Tannenbaum, 1957, and there is an immense wealth of experimental data linking linguistic expressions to valuations. Perhaps the simplest of these would be the GOOD/BAD scale we discussed in 5.2. For improved modeling accuracy, we may want to consider this a three-point scale *good, neutral, bad*, since most things, in and of themselves, are neither particularly good nor particularly bad.

Another valuation of great practical interest would be TRUST. For this we can assume a set of fixed (or slowly changing) sources like people, newspapers, etc., and a set of nonce propositions coming from these. Sometimes the source of a proposition is unclear, but quite often we have information on which proposition comes from which source. By a trusted source we mean one where we positively upgrade our prior on the trustworthiness of the propositions coming from them, and by a distrusted one we mean one that triggers a downgrade (negative upgrade) in the trustworthiness of the proposition. As (dis)confirmation about particular propositions comes in, we can gradually improve our model of sources in the obvious manner, by backpropagating the confirmation values to them. This can be formulated in a continuous model using probabilities, but the essence of the analysis can be captured in terms of discrete likeliness just as well.

Of particular technical interest is the ACTION POTENTIAL valuation taking values in $A = \{-1, 0, 1, 2\}$, where -1 means 'blocked' or 'refractory', 0 means 'inactive', 1 means 'active', and 2 means 'spreading'. These can be used to keep track of the currently active part of the graph and implement what we take to be the core cognitive process, spreading activation (Quillian, 1969; Nemeskey et al., 2013). Here we will not pursue this development (see 7.4 for further details), but note that we don't see this valuation as formally different from e.g. the probability valuation, except that innateness is plausible for the former but not the latter. To paraphrase Dedekind's famous quip, spreading activation was created by God, the other valuations are culturally learned.

Unlike the lexicon itself, valuations are not permanent. The inputs to a valuation are typically nonce hypernodes 'death at Snowbird' and the linguistic subspace L only serves as a basis for computing the mapping from the hypernodes in question to the scale s. We assume that the activation mechanism is unlearned (innate), but this still leaves open the question of how we know that forces of nature are a likely cause of death in Reykjavík but not in Istanbul? Surely this knowledge is not innate, and most of us have not studied mortality tables and statistics at this level of specificity, yet the broad conclusion, that death by natural forces is more likely in Reykjavík than in Istanbul, is present in rational thinking at the very least in a defeasible form (we will revise our naive notions if confronted with strong statistical evidence to the contrary).

Part of the answer was already provided in 5.2, where we described the mechanism to compute these values. Aside from very special cases, we assume that such valuations are always computed afresh, rather than stored. What is stored are simpler building blocks, such as 'volcano near Reykjavík', 'volcano isa danger' from which we can easily obtain 'danger near Reykjavík'. A great deal of background information, such that *danger* is connected to *death*, must be pulled in to compute the kind of valuations we described in Table 5.1, but this does not alter the main point we are making here, that inner models are small information objects (the entire mental lexicon is estimated to be about 1.5MB, see Mollica and Piantadosi, 2019).

From the foregoing the reader may have gathered the impression that learning of nodes is relatively easy, learning of edges is harder, and learning of valuations is the hardest, something doable only after the nodes and edges are already in place. The actual situation is a bit more complex: any survey of the lexicon will unearth nodes that are learnable only with the aid of valuations. The strict behaviorist position that learning is simply a matter of stimulus-response conditioning has been largely abandoned since Chomsky, 1959. Whether the alternative spelled out by Chomsky, an innate Universal Grammar (UG) makes more sense is a debate we need not enter here beyond noting the obvious, that lexical entries are predominantly language-particular. This is no doubt the main reason why Chomsky places the lexicon in the "marked periphery", outside "core grammar".

Children acquiring a language acquire its lexicon, and there is no reason to believe that this process relies on innate knowledge of concepts (nodes) for the most part. In keeping with our approach to consider the entire lexicon, we begin with a brief survey of the semantic fields used by Buck, 1949:

1. Physical World
2. Mankind
3. Animals
4. Body Parts and Functions
5. Food and Drink
6. Clothing and Adornment
7. Dwellings and Furniture
8. Agriculture and Vegetation
9. Physical Acts and Materials
10. Motion and Transportation
11. Possession and Trade
12. Spatial Relations
13. Quantity and Number
14. Time
15. Sense Perception
16. Emotion
17. Mind and Thought
18. Language and Music
19. Social Relations
20. Warfare and Hunting
21. Law and Judgment
22. Religion and Beliefs

The list offers whole semantic fields like 4, 12, 14, and 15, where we have argued (see 3.1) that the best way to make sense of the data is by reference to embodied cognition, a theory that comes very close to UG in its insistence of there being an obviously genetically determined component of the explanation. The same approach can be extended to several other semantic fields: we discuss this on 6: Clothing, Personal Adornment, and Care.

We start from an embodied portion, 4, and proceed by defining *shoe* as 'clothing, worn on foot'; *leggings* as 'clothing, worn on legs'; *shirt* as 'clothing, worn on trunk'; etc. We begin by noting that *clothing* 'the things that people wear to cover their body or keep warm' is already available in 4lang as `cloth, on body, human has body, cause_ body[warm]`. Using this, a good number of Buck's keywords fit this scheme: 6.11 *clothe, dress*; 6.12 *clothing, clothes*; 6.21 *cloth*; 6.41 *cloak*; 6.412 *overcoat*; 6.42 *woman's dress*; 6.43 *coat*; 6.44 *shirt*; 6.45 *collar*; 6.46 *skirt*; 6.47 *apron*; 6.48 *trousers*; 6.49 *stocking, sock*; 6.51 *shoe*; 6.52 *boot*; 6.53 *slipper*; 6.55 *hat, cap*; 6.58 *glove*; and 6.59 *veil*. For some of these case our definition of clothing would need a bugfix to include the 'modesty' aspect (which is actually culture-specific) by merging in our definition of *cover* `=agt on =pat, protect, cause_[lack{gen` cover `see =pat}]` to yield an additional clause e.g. for *veil* `cause_ [lack{gen see face}]`.

This analysis illustrates the point about highly abstract units we made in 1.2: obviously *boot* means different things for different cultures, and the Roman legionnaire would not necessarily recognize the caligae in the skinhead's DMs. But the conceptual relatedness is clearly there, and as we discussed above, the word can be learned as a node in a network composed of abstract units such as `cover` and *foot* `organ, leg has, at ground` which we need anyway.

foot

This is not to say that the 54 main headings covered in Chapter 6 of Buck, 1949 are all automatically covered in 4lang, especially as many are listed in this chapter only because the lexicographer had to put them somewhere, and this seemed the best place. In addition to the core entries discussed so far, we have a wide variety of clothing materials: 6.22 *wool*; 6.23 *linen, flax*; 6.24 *cotton*; 6.25 *silk*; 6.26 *lace*; 6.27 *felt*;

`wool`   6.28 *fur*; 6.29 *leather*. For the most part we treat these as genus `material`, e.g. *wool* `material, soft, sheep has,` but sometimes we place them under other genera.

`fur`   e.g. *fur* `hair/3359, cover skin, mammal has.`

Buck also lists here some professions (6.13 *tailor*; 6.54 *shoemaker, cobbler*); activities characteristic of cloth- and clothing-making (6.31 *spin*; 6.33 *weave*; 6.35 *sew*; 6.39 *dye (vb.)*); professional tools (6.34 *loom*; 6.32 *spindle*; 6.36 *needle*; 6.37 *awl*; 6.38 *thread*). We have discussed professions like *cook* in 2.2, and for a typical tool we of-

`needle`   fer *needle* `artifact, long, thin/2598, steel, pierce, has hole, <sew ins_>.`

More challenging are the 'accessories' or 'adornments' which are not, strictly speaking, items of clothing in and of themselves (6.57 *belt, girdle*; 6.61 *pocket*; 6.62 *button*; 6.63 *pin*; 6.71 *adornment (personal)*; 6.72 *jewel*; 6.73 *ring (for finger)*; 6.74 *bracelet*; 6.75 *necklace*; 6.81 *handkerchief*) as well as culture-specific items that are associated to clothing and adornment only vaguely (6.82 *towel*; 6.83 *napkin*; 6.91 *comb*; 6.92 *brush*; 6.93 *razor*; 6.94 *ointment*; 6.95 *soap*; 6.96 *mirror*). First, we need to consider what is an *accessory* 'something such as a bag, belt, or jewellery that you wear or carry because it is attractive'. This is easily formulated in `4lang` as `person wear, attract`. Similarly with *adornment* 'make something look more attractive by putting something pretty

`attract`   on it'. The key idea is to define *attract* as `=agt cause_ {=pat want {=pat near =agt}}`. Once this is done we are free to leave it to non-linguistic (culturally or genetically defined) mechanisms to guarantee that nice-smelling ointments and pretty jewelry will be attractive. The example highlights the need for a realistic theory of acquiring highly abstract concepts. In S19:2 we wrote:

> the pattern matching skill deployed during the acquisition of those words denoting natural kinds cannot account for the entirety of concept formation. People know exactly what it means to betray someone or something, yet it is unlikely in the extreme that parents tell their children "here is an excellent case of betrayal, here is another one". Studies of children's acquisition of lexical entries such as McKeown and Curtis (1987) have made it clear that natural kinds, however generously defined so as to include cultural kinds and artifacts, make up only a small fraction of the vocabulary learned, even at an early age, and that children's acquisition of abstract items "but not concrete word learning, appears to occur in parallel with the major advances in social cognition" (Bergelson and Swingley, 2013).

While our remarks on the subject must remain somewhat speculative, it seems clear that *attract* is learned together with *attraction, attracting, attractive* i.e. without special reference to the fact that the root happens to be verbal. In fact, there is every reason to suppose that abstract terms are root-like, and it is only the syntax that imposes lexical category

`responsible`   on them. Consider *responsible* `has control, has authority, has blame.` The Hungarian version proceeds from a verb *felel* 'respond' through an adjective *felelős* 'is responsible' to a noun *felelősség* 'responsibility'. In Chinese, we begin with a noun *ze2ren4* 责任, form a verb *fu ze2ren4* 负责任, and proceed to the adjectival *fu4 ze2ren4 de* 负责任的.

We also have roots that are neutral between verbal and adjectival forms, for example *open, cool, warm*. Common to them is the ability to treat the adjective as the result state of the verb, for example *open* `move[can/1246]`, `move through, lack` `shut/2668` and `after(=pat open/1814)`; *cool* `temperature, normal` `er_, er_ cold` and `after(cold)`; *warm* `temperature(er_ gen)` and `after(warm/1655)`. The nominal forms *warm/warmth, join/joint, cool/cold, heat/hot,* ... are remarkably close, but perhaps not close enough without making recourse to the kind of stratal morphology that makes a strong distinction between stem-level and word-level morphology (Kiparsky, 2016).

```
open/1814
open/1815
cool/1103
cool/1101
warm/1655
warm/878
```

The use of `=agt` and `=pat` in the definition of *attract* makes clear that it is essential to have two items in attraction, i.e. the relation is binary. Our theory of learning must start with an elementary act of recognizing attraction, just as we recognize nearness, one thing being on top of another, and a host of other relations. Clearly, there is huge evolutionary advantage to recognizing nearness to us, as this will be a primary signal of whether something can attack us and/or whether we can manipulate it to our advantage. We consider proximity marking (`near`, see 3.1) to be a reasonable candidate for universality.

We also consider the other two components of our definition, a naive theory of needs and wants (6.2), and a naive theory of causation (2.4) as evolutionarily highly motivated: clearly, being able to model what other actors in the environment are likely to do, based on their needs/wants, will hugely improve our own chances for surviving and thriving, and a theory, even a naive theory, of causation has a similar salutary effect. The challenge here is to put together three highly abstract theories to produce a fourth one for *attraction*.

While our remarks must remain somewhat speculative, what we believe is that the putting together is driven by valuations. We recognize attraction by first seeing increasing nearness, `after(=agt nearer =pat)`, next by attributing this change to the desire of the patient, `=pat want {=pat near =agt}`, and finally the desire itself as being aroused by the property of attractiveness lodged in the agent. At some point, the `after` clause is converted to the `cause_` by means of the naive analysis of causation.

The micro-analysis of how these steps are built on one another during language acquisition requires further study, and will clearly involve polarity: nearness is GOOD for good things, but BAD for bad things. Since we expect all beings, not just `self`, to want good things, we must assume that attracting something is itself a good thing (as it is, unless `=pat` is bad). In certain cases, we can expect these valuations to play out on primary linguistic and sensory data: to the extent something is pleasant to the touch a baby may want to touch it and thereby bringing it near, or even put it in their mouth, to bring it even nearer. Signaling of attractiveness is already evident in flowers, and signaling of badness, aposematism, is common in the animal kingdom. Where earlier generations of researchers may have seen the theory of causation as something only the properly trained mind is capable of (Kant even assumed some innate human capacity), we see a perfect Darwinian continuity connecting humans to far simpler life forms.

6

Modality

Contents

The notion of *modality* is almost inextricably intertwined with metaphysics, some kind of theory of what is real, what exists, and why (a theory of 'first causes'). At the center of the commonsensical theory is the real world, but the idea is that there exist, or at least there can exist, other worlds. This idea is most clearly supported by the common-sense notion that the world existed yesterday and will exist tomorrow, even if it will be slightly different from what it is like today. In 3.2 we already discussed that the current world V_n requires two modal versions: V_b for the past and V_a for the future, and in 6.1 we will considerably refine this idea by describing a a nonstandard theory of the standard temporal modalities.

A central question of metaphysics is the existence of the things that are co-present in all worlds, the things that do not change. Are these collected in another world? Are there other worlds to begin with, and if there are, in what sense are they real? In 6.2 we use the same technique to introduce an ideal world V_d where rules are kept, and investigate the real world in relation to this. In 6.3 we use an even simpler technique to bring epistemic modality in scope, and in 6.4 we reanalyze defaults.

6.1 Tense and aspect

In 3.2 we introduced the naive theory of time, and described how it requires at least one, and possibly two somewhat imperfect copies of V_n to explicate word meaning. When we say that statements about these subworlds, and especially statements that involve more than one of these, have modal import, we rely on the broad family of theories known collectively as modal logic. (For a terse discussion, see S19:3.7, for a book-length one see Blackburn, Rijke, and Venema, 2001.)

© The Author(s) 2023
A. Kornai, *Vector Semantics*, Cognitive Technologies, https://doi.org/10.1007/978-981-19-5607-2_6

Fig 6.1 depicts part of a two-way infinite graph, the nodes of which correspond to (full or partially specified) *possible worlds* and the directed edges depict the *accessibility relation* that obtains among these. The left-to-right direction corresponds to the flow of time, with nodes to the left (right) of the vertical line in the past (resp. future), and the vertical line itself marking *now*. The bulleted node is the real world, as of now, those above and below are alternative worlds at the same time. There is only limited movement across the horizontal timelines: not every past world is compatible with the present world, and not every future world is accessible from it.

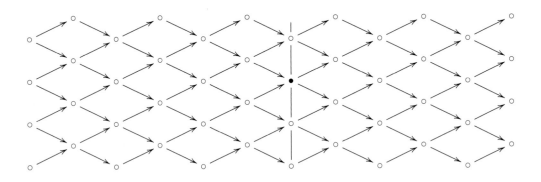

Fig. 6.1: Modal accessibility

The sophisticated reader can find all kinds of faults with Fig. 6.1. How can we have a single (absolute) time that flows synchronously in all worlds? How do we know there are alternative timelines, rather than a single, fully deterministic one? What makes us think we could move across timelines? In response, let us reiterate the disclaimer we already made at the beginning of Chapter 3: linguistics and cognitive science, our primary tools here, are highly unlikely to contribute to contemporary cosmology, precisely because in physics we already assume things such as a continuous timeline (possibly with a singularity at the beginning) which go beyond the scope of the naive theory. No matter how much the study of the naive system contributes to a better understanding of early natural philosophy, especially the Presocratics, Zeno in particular, there is no reason to suppose that it can contribute to contemporary natural philosophy (physics, chemistry, biology, and so on) since the naive system is already insufficient to sustain integer arithmetic, let alone real numbers, functions, and higher mathematical constructs essential to the practice of science.

In fact there is no reason to believe that the commonsense theory is in any way trivial. A fuller mathematical reconstruction of the temporal aspects requires more sophisticated tools from mathematical logic and analysis than are generally employed in mathematical physics. (Readers who wish to refresh the standard modal concepts may find S19 Ex. 3.17-19 helpful.) Time instances are best seen as infinitesimal neighborhoods (called monads in nonstandard analysis, and *timelets* in smooth analysis by Bell, 1988). Such a formulation goes a long way towards capturing some of the seemingly paradoxical properties of time instances: that they are discrete (monads corresponding to different time instances have no overlap), that they have no tangible beginning or end, that they have nonzero length yet ordinary time is suspended with a single timelet. For reasons that we will discuss shortly, we consider Lawvere-style smooth analysis, as opposed to Robinson-style nonstandard analysis, to offer a better formulation for the dynamic aspects of commonsense temporality.

Since normally it is verbs that carry tense marking, the expectation is that it will chiefly be verbs that require explicit temporal clauses. Perhaps surprisingly, temporal modality is already required to treat suffixes like `-th` (*half, third,* and *quarter* are irregular and not discussed here) which we must define by some unit that is getting divided (or broken up, as in Skt. *bhinna*): `-th -ad -ias -ta part, in whole,`   `-th` `before(divide)`. To conceptualize a fraction we must assume a preexisting unit. Again, this does not imply, or even suggest, that quarks, with fractional charges, must have originated by breaking electrons up – what the evidence shows is that the *words* for fractions post-date the words for integers, not that fermions post-date bosons.

A more typical entry, with extension in all three temporal slices, is *pause*, defined as `lack action, before(action), after(action)`. Similar to frac-   `pause` tions, which are tied to their units, pauses are inescapably tied to some action that is being paused. There are many relational nouns that only make sense only if some other entity is invoked, but these are typically in the same V_n temporal slice, whereas the act of pausing requires the presence of activity both before and after. A particularly interesting case is provided by *through*, which we analyze as an adposition of fictive motion: `before(=agt on side), in =pat, =pat has side,`   `through` `after(=agt on other(side)), =pat has side[other]`. To understand this word, we need to invoke a full story about its object, a `body` (in the sense of 3.1) with two sides, and virtual movement that starts and ends outside this body, but is inside for a period.

To build a more refined formal language describing such cases, we introduce further notation, with π_i defined as the projection of the polytope in V_i, for $i = b, n, a$. Since the default is `now` (event time), this is left unmarked in the lexicon, where only `before` and `after` are overtly marked. Typically these three projections are identical, especially for nouns, where we expect the realizations π_i to be isomorphic in the spaces V_i. Only 15.5% of the defining vocabulary contains overt reference to `before/after/cause_`.

It is worth emphasizing, particularly to the reader familiar with the standard theory discussed in S19:3.7, that projecting from the joint polytope (or system of polytopes) in $V_b \times V_n \times V_a$ to one of the components $V_b, V_n,$ or V_a is *not* the same as the extension of a concept in the past, present, or future. Take a stable noun such as *food* : `material,`   `food` `gen eat`. It is quite possible (and historically common) for various things that were not considered food to become generally accepted as food, and conversely, for materials that were earlier considered suitable for eating to get off the menu. But such changes are slow, adiabatic, and do not affect V_b or V_a which are best thought of as *sharing the timelet* with V_n.

Within the adiabatic approximation food is an eternal noun, a proper member of the deontic world V_d that we will discuss in 6.2. This is true even though it is a common-sensical law of nature that food is perishable, i.e. no *instance* of food lasts long (though special efforts to salt/smoke or otherwise preserve it may extend its usable lifetime). This is one point where the modern theory, capable of distinguishing between *instances* and generics, is arguably superior to the naive theory. That said, we can still add a rule *food is perishable* by nonstandard methods without triggering a contradiction.

Since *perishable* means 'likely to decay quickly' (LDOCE) and decay is given as change[slow], after(lack health), we need to consider the timescales in more detail. When we say the magnetic field of the Earth decays, this is on a hundred thousand year or longer timescale, and can only be measured with sophisticated instruments. But even for ordinary decay processes from decaying buildings to decaying teeth a multi-year timescale is implied. When the LDOCE posits quick decay the implication is that for food, the process is quick relative to the multiyear timescale inherent in decay, just as the *enormous flea* discussed in 4.3 that is enormous only on a flea scale. As a practical matter, the timescale is weeks, possibly days or even a few hours, but certainly not minutes or seconds as we have with ordinary action verbs.

On the near-instantaneous timescale embodied in before/after, food does not change, or changes only with imperceptible speed. This is something well captured by smooth analysis, which reconstructs derivatives with the Kock-Lawvere Axiom:

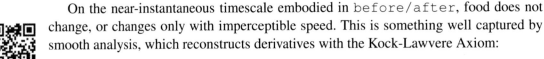

If D is the timelet around 0, and f is any function from D to $\mathbb{R}$, there is a unique real number a such that for all $d \in D$ we have $f(d) = f(0) + d \cdot a$

The axiom guarantees that within a single timelet *all* functions are linear 'linelets' with a unique tangent a. This lets us define derivatives at every instant t as the unique real number $f'(t)$ that satisfies $f(t + d) = f(t) + d \cdot f'(t)$ for any d in the same timelet as t. Notice that there are no higher derivatives that could be obtained by closer inspection of an infinitesimal neighborhood, in fact it is a characteristic property of smooth analysis, as opposed to nonstandard analysis, that infinitesimals d (called *minims* in this theory) satisfy $d^2 = 0$. What we have in smooth analysis is a theory restricted to continuous functions, embodying the famous Leibnizian principle of *natura non saltum facit* 'there is no jump in nature'. To handle the case of food perishing, all we need to add is that the process is akin to the movement of the hour hand on a clock, possessing a derivative, but one that is too small to be perceptible. It is not that the derivative itself is a minim, it is measurable by instruments of sufficient precision to be an ordinary non-infinitesimal quantity, it's just that our senses do not offer this level of accuracy.

Where does this leave us in regards to pauses? The commonsensical, if ahistorical, answer is that we equate movement with nonzero derivative and conversely, we equate zero derivative with lack of motion. This is perhaps best illustrated by a ball bouncing against a hard surface: the idealization that distance from this surface is measured by $|t|$ is simply not available, since at $t = 0$ this function only possesses a left derivative -1 and a right derivative $+1$. Rather, we have to assume that either the ball or the surface is not entirely rigid, that for the time of the impact the center of the ball actually approaches, and departs from, the surface in a smooth fashion. This function will have zero derivative at $t = 0$ from either side. Therefore, the ball is `pausing` at the surface.

The geometrical picture associated with smooth analysis includes not just a clear picture of derivatives as the tangent function of the linelet angle, but also the idea that the entire curve is built from such linelets, just as a circle can be conceptualized as an infinite-degree polygon. The reader interested in how the entire apparatus of multivariate calculus can be built on smooth analysis is referred to (Bell, 2008). Here our goals are more modest: we concentrate on linking the naive theory embodied in language to early theories of natural philosophy, and consider the task of reconstructing modern physics entirely out of scope. A key element of the medieval theory of dynamics is the notion of impetus, what today we would define as speed times mass. It is impetus that endows objects with an intention to keep moving in the direction that they are already moving in. The use of impetus resolves the Aristotelian quandary of why a rock, once thrown, does not fall to the ground as soon as the hand is no longer supporting it, but rather follows a parabolic trajectory.

This much, while clearly insufficient for planetary motion (which will have to wait until Newton) is quite sufficient for Oresme, Buridan, and the great scholastic thinkers whose line actually goes back to Aristotle (via very significant Arab contributions that go well beyond mere transmission and commentary). To the extent that the picture provided by smooth analysis is highly intuitive, we begin to see the intellectual leap that separates second derivatives from linelets. Our intuition, grounded both in everyday experience and in linguistic cues, may readily supply the idea of local linearity (sometimes called *micro-straightness*), but this does not extend to second derivatives. Indeed, if two points of a curve are in the same timelet, they are also on the same linelet, so to build second derivatives we need to reify the derivative as a function on its own. (The very idea of time-distance diagrams goes only back to Oresme, Aristotle didn't have these at his disposal.)

The first person to wrestle with the issue was Zeno, and his paradoxes demonstrate quite clearly that certain commonsensical assumptions about time and space, if held jointly, will result in contradictions. The linguistic conception of time is discrete, but this immediately leads to the paradox of Dichotomy: a discrete system of instances cannot be dense, with a halfway point between any two instances. The standard solution is that time instances are indeed point-like, but we have infinitely many – nonstandard analysis endows these points with monads surrounding them while being disjoint from one another. In Robinson's version, monads have rich internal structure, in Lawvere's version, they are just tiny lines that can be characterized by their centerpoint (position) and direction (impetus). That we need some kind of continuous time for conceptual semantics is evident: in S19:3.3 we wrote

The key temporal notion in the prolepsis is not so much the idea of *time* itself as the idea of a *process*. It seems that humans (and in all likelihood, all mammals) are endowed with a perceptual mechanism that inevitably makes them perceive certain sensory inputs as processes. Try as we might, we cannot perceive the flight of the arrow as a series of states: what we see is a continuous process. The compulsion to do so is so strong that even truly discrete sequences of inputs, such as frames of a movie, will be perceived as continuous, as long as the frame rate is reasonably high, say 20/sec.

More debatable is the concept of space, whether we see it as composed of small discrete voxels, or as a continuum. The egocentric coordinate system we discussed in 3.1 is actually neutral on the issue, presenting space as being composed from a few discrete regions like *inside* and *outside* but without any implication that movement within a single region is imperceptible. But Zeno's argument against the reification of space is still worthy of consideration: if everything has a place, what is the place of place? If there are voxels, can they be occupied by (parts of) physical bodies? If we suppose they can, where are the voxels themselves going to be? Again, it is the background assumptions behind the paradox that matter for us: clearly, Zeno is making the commonsensical assumptions that (i) *two things cannot occupy the same space at the same time* and (ii) *a thing cannot be at two places at the same time*. We will discuss each in turn.

`place` We begin our analysis by defining *place* as `point, gen at`. This is an atomic, point-like entity 'where things can be' as opposed to the conceptual `{place}` schema we defined in 3.1, but it makes little difference whether we consider just the point or the voxel, the 3-dimensional monad surrounding it: the import of (i) is still that no two objects can occupy it. This is considerably stronger than what our primitive negation element, `lack`, is capable of expressing, since `lack` is most natural in situations where some default expectation is not met (see 4.2), whereas in (i) we wish to express an absolute negative. How is this negative enforced? Everyday experience shows that this is done by one object either forcing the other one out, or by not letting the other one in.

Since an exception is provided by 'shapeless' objects such as liquids and gases which, according to everyday experience, can in fact mix in the same place, instead of objects we will insist on *solids*, both in the `4lang` sense of `firm/2215` and in the original LDOCE sense of 'having no holes or spaces inside'. The requisite sense of solidity is easily defined by the uroboros vocabulary as `lack empty(place/2326) in`.

(This will also help with firm objects like Matryoshka dolls which can in fact occupy the same place as far as their center of gravity is concerned.) This notion of solidity, actually quite close to the idealization of the *convex rigid body* that we rely on in classical mechanics, is sufficient for restating principle (i) as

$$\{\texttt{solid at place, other(solid) at place}\} \texttt{ cause\_ move} \quad (6.1)$$

Eq. 6.1 is a far cry from modern dynamics, where impulse is conserved in each dimension, so the resulting movement can be computed with great accuracy: here what moves, the object, the other object, or perhaps both, is left underspecified. The reference to `move` in Eq. 6.1 makes clear that it is a constraint over different time instances that is responsible for (i), making what appears to be a static principle into a principle of (proto)dynamics. Principle (ii), to which we turn now, will be different: no movement is implied.

Since everyday (solid) objects extend over a volume, they actually *can* be at two places at once: the bridge is at the left side of the river, and also at the right side. To go further, we therefore need to restrict principle (ii) to point-like objects and to point-like places. This departs from the general sense of *place* which permits overlapping or even containment: the tree is in the forest, the forest is in the country, so a single point, say the point of an arrowhead, can be both in the tree and in the country. Since 4lang defines *place* as `point, gen at` and *point* as `place, lack part_of` we get the restriction both to point-like objects and point-like places for free. The knowledge engineer would probably state principle (ii) as

place
point

$$\text{point}_1 \text{ at point}_2, \text{ point}_1 \text{ at point}_3 \Rightarrow \text{point}_2 = \text{point}_3 \qquad (6.2)$$

While very familiar to contemporary thinking, Eq. 6.2 is much farther than we are prepared to go. Several notational conventions used in Eq. 6.2, though typical of the uniqueness statements used elsewhere, are beyond our formal language. First, the main connective $\Rightarrow$. We never offered a non-causal theory of implication, and in fact deduction in a system that carries smooth analysis is of necessity a weaker, intuitionistic type (Moerdijk and Reyes, 1991). Second, and perhaps more important, we don't have a theory of indexing, as will be clear for readers of 3.3. Finally, the equality symbol '=', while close to negating our primitive `other`, is subtly different (more powerful). 4lang makes a distinction between `equal/191 azonos idem identyczny` and `equal/565 egyenlo3 aequalis ro1wny` and being `other` really means 'non idem' rather than 'non aequus'. To appreciate the distinction, consider Pappus' proof of Euclid I Prop 5 known as *pons asinorum*: given a triangle ABC with equal sides AB and AC, the angles at B and C must also be equal. Pappus simply considers the triangles ABC and ACB: in didactical terms, he 'lifts up' ABC, 'flips it over' to make ACB, and lays it down on the original. This kind of subtle leveraging the notions of identity and equality bothers many people precisely because of (ii): how can the same triangle be in two places at once?

Staying within the limitations of 4lang , we need to say `gen lack {thing at place(two)}`. The number *two* is defined as `number, one in, other in, follow one`. Here we are free to ignore the ordinal aspect, which is irrelevant for our reconstruction of (ii):

two

$$\text{gen lack \{point at point, point at other point\}} \qquad (6.3)$$

The syntactic mechanism will automatically identify the first three occurrences of `point` in Eq. 6.3, but not the third and the fourth since this is precisely what `other` means (see 1.6). One way to think about this is to consider principle (ii) part of the meaning of being *at* some location.

It is worth emphasizing that a trivial extension of (ii), "n things cannot be at $n + 1$ places at the same time", is impossible to formulate within the limitations of 4lang. It is of course quite possible to build better resourced systems, but sooner or later these will also run out of steam. An interesting case is provided by the obverse of (ii), the Pigeonhole Principle, which asserts that $n + 1$ objects cannot be placed in n boxes without putting at least two in the same box. Proving this using only a polynomial size proof (the total number or symbols of the formulae in the sequence) and keeping each formula at constant depth (unlimited fan-in) is simply not possible (Ajtai, 1994).

other

Here the question is not just why we cannot have n objects in $n + 1$ places at the same time, but rather why this general truth is hard to demonstrate already for $n = 1$? Eq. 6.3 highlights the difficulty, not being at place(other). Since *other* is defined as different, which in turn was defined in 4.2 with the aid of Leibniz' Principle of Indiscernibles, it is true for any X that it cannot unify with other X. This is guaranteed not so much by the semantics of unification (where overriding certain values, especially defaults, is possible) as by the semantics of *other* – this is what being other means. The other schema depicted in Fig. 1.3 demands the coercion (as discussed in 3.3) of X and other X to one and other respectively. To see this

passenger

word in action, consider *passenger* person, person[travel], person in vehicle, other(person) drive vehicle.

There is, perhaps, a more general lesson here: it is the lexical meaning of core elements, such as other, that drives the large-scale behavior of complex operations, such as unification, and not the other way around. Given that specific lexical behavior cannot be deduced from general principles, we may as well adopt a 'generators and defining relations' style of description.

6.2 The deontic world

Almost all (perhaps all) languages and cultures have some notion of another world, generally populated by powerful anthropomorphic beings, ranging from gods and angels to evil spirits and devils. The major exception is the

> kind of disembodied use of higher forces that is taken for granted in Chinese metaphysics: "Thus to say that 'High heaven shook with anger' by no means implies that there is a man up above who shakes with anger; it is simply that the principle (li) is like this [that is, that crime deserves anger]." (Graham, 1958), p24, cited in S19:3:10.

It is this kind of Heaven, populated by principles (eternal laws) rather than by ghosts and spirits, that is closest to the deist metaphysics common in occidental philosophy starting with Edward Herbert's De Veritate. Broadly speaking, we use the same methodology relying on common notions *(notitiae communes)* (innate, universal semantics) as all rationalists from Herbert onwards (at least according to Chomsky's 1966 presentation of rationalist thought), but we don't undertake to faithfully reconstruct any of these systems here. In S19:3 we offered an automata-theoretic reconstruction of the kind of patterns (rules, regularities) that we may consider central both for elementary guidance of behavior 'don't immerse your hand in boiling water' and for stating laws of nature 'unsupported objects will fall'. Here our main interest is explicating such statements in vector semantics, and we start with Yi-ch'uan's example

$$\textit{crime deserves anger.} \tag{6.4}$$

For this purpose, it will be quite sufficient to define *crime* as `action, illegal` and trace `illegal` through `bad for_ law` to obtain `action, bad for_ law`. In fact, by tracing further *bad* as `cause_ hurt` we end up with an even more compact definition of crime: `action, hurt law` – this has the advantage that we don't have to get sidetracked with the issues of experiencer subjects (see 2.4) that the use of `for_` would bring in tow.

The modal element is from *deserve*, defined as `before(=agt DO <good>), should[after(=agt GET/1223 =pat)]` where *should* means not just an option, but the `right/1191` option, an idea that is central to the deontic modality. We could also analyze *anger* further as `feeling, bad, strong, aggressive,` but this would not take us further toward our goal, which is to express the pattern at hand as a normative rule.

In the world of norms, it's not just that the right thing *should* happen, but that it *does* happen, `after(gen angry)`. It's not that there is a man above who shakes with anger, but rather that everybody, the generic subject, will. Putting all this together we obtain the `4lang` translation

$$\texttt{before(action hurt law), after(gen angry)} \tag{6.5}$$

Let us pause for a moment and observe that we have done a great deal more than translating Eq. 6.4 to Eq. 6.5 in that the target semantics is quite universal: for example Hungarian *törvénysértés* 'crime' is a compound that would be literally translated as *law-hurting*, and since *hurt* is defined as `cause_ {=pat has pain}, offend` we also obtain a semantic reading for English *offence*, namely that it is the law that is getting offended.

Given the pre- and post-conditions in Eq. 6.5, we can recognize this as an instance of the `cause_` primitive discussed in 2.4, so we obtain

$$\texttt{(action hurt law) cause_ (gen angry)} \tag{6.6}$$

This lends itself to further generalization on both sides. First, it is not just actions, but any form of hurt that causes anger, and second, `gen angry` is just a restatement of `anger`. Therefore, we obtain

$$\texttt{hurt cause_ anger} \tag{6.7}$$

as a truly general principle that has Eq. 6.4 as a consequence. By the definition of `cause_` this amounts to the co-presence of `hurt` in V_b and `anger` in V_a or, in the two-state temporal model we have `hurt` in V_n and `anger` in V_o. (Observe that in such a model not only does hurt cause anger, but anger also causes hurt, reminiscent of the unbreakable cycles of violence we see all too often in tribal societies, where new generations simply take up the grievances of their forbears and keep on fighting like the Hatfields and McCoys.)

It is precisely the lack of a long-term sustainable commonsense temporal model that leads to the metaphysical search for first causes. Commonsensical cause_ is simply the existence of a cause in one temporal model (V_b or V_n as the case might be) and an effect in a subsequent model. Once this is coupled with some Parmenidean principle of *out of nothing nothing comes* we must continue the backward search for a sufficient reason: what caused the cause? What caused the cause of the cause? If the underlying temporal model is cyclic, as with Ecclesiastes, we end up with simple, irreducible cycles of eternal change: day causes night, night causes day, there is no beginning and no end.

The more general three-state model of past, present, and future depicted in Fig. 6.1 does not cycle back from future to past. Rather, the underlying commonsensical temporal model includes a notion of a (discrete) timeline with successive states (time instances), with *present* always being mapped on a single state, *past* as a half-line extending with the previous state, and *future* another half-line gradually consumed by the passing of time, much as a wheel rolling over a surface only touches a single point of the ground, leaving a half-line trace.

There is nothing in language that is probative about the precise nature of this timeline, and commonsensical reasoning must be replaced by some form of more structured philosophy to be able to reason about infinities. If the timeline is finite, at the first step we must suppose some self-necessitated being, or unmoved mover, giving us the Kalam cosmological argument, whose attraction lies precisely in the fact that it hews so closely to commonsensical notions.

god

This is not to say that any commonsensical system will *prove* the existence of god, but it is quite obvious that the existence of *god* or gods being, religion has, has power, er_ nature is compatible with common sense. As we have argued elsewhere (Kornai, 2010a), "once the names of major religious figures and the titles of sacred texts are treated as pointers to the encyclopedia, there remains nothing in the whole semantic field of Religion that is not definable in terms of non-religious primitives". In other words, religion is *possible*, but by no means *necessary* for a commonsensical worldview.

With this, we have arrived at a central problem of modal logic: what is possible, and what is necessary. Let us begin with a standard example of a non-existent being, the *unicorn* 'an imaginary animal like a white horse with a long straight horn growing on its head'. We know of no law of biology that would make unicorns impossible – after all, both white horses and animals with long straight horns exist. We may search for unicorns in the past (perhaps remains can be found) or in the future (perhaps genetic engineering will produce some), though not every world is accessible to such a search. Even under the most cautious definition of 'possible', we find things such as dinosaurs which we know to be possible, and at the same time know to be nonexistent in the real world. Mathematics furnishes many examples that we know to be absolutely impossible: no search in past worlds or alternative timelines will ever produce an algorithm for squaring the circle by ruler and compass.

Altogether, non-existence has an absolute form, impossibility, and existence has a weaker form, possibility. It is therefore reasonable to look for an absolute form of existence, which we can call 'necessity'. Things that enjoy only a weaker form of existence, or what is the same, a weaker form of non-existence, are called *contingent*. Altogether, we can build a three-point scale of existence, with necessity at the top, impossibility at the bottom, and the contingents in the middle. As long as we assume the accessibility relation to be reflexive, everything that is real (here and now) is either possible or necessary, and everything that is not real (here and now) is either possible or impossible. Looking through the lexicon we see that having a name like *unicorn* is no guarantee of possibility, unless imaginary worlds are also accessible, in which case of course it is (see S19:5.6).

Having a proper name is no more a firm guarantee of possibility than being named by a common noun. Here we take the position that the entire linguistic subspace L is composed of possible things, be they part of the basis (defining words) or be they expressed by more complex expressions. Where does that put us in relation to linguistic expressions that are known to designate impossible things, such as squaring the circle? The answer is that such expressions are also rigid designators (they mean the same thing in every possible world) it's just that they cannot be realized. The fact that something can be expressed in language is no guarantee that it's true or real.

In the naive theory, the type distinction between *true* and *real* is absent (cf. *a true friend/a real friend*). To bring this in line with contemporary logic where only well-formed formulas can be true/false and only objects can be real (there is never any doubt about the reality of a formula) requires a special predicate EXISTS whose type-theoretic signature is from matters[1] to truth values. In 4lang we define *real* and exist as synonyms: we have `real igazi verus prawdziwy 1126 A exist` and also `exist van exsto byc1 2587 V real`. Similarly, we define *true* as `igaz verus prawdziwy 1125 A fact` and *fact* as `telny factum fakt 2323 N has proof[exist]`, something for which proof exists.

real
exist
true
fact

What constitutes proof? Again we eschew the modern proof-theoretic statement in favor of the naive theory, where *proof* is given by `prove`, and is a conjunction `after(other(people) know =pat[true]), real ins_`. Whatever a proof is, it is something that convinces others that its object is true and, moreover, it is an instrument (most effective means) to truth/reality. The most immediate proof of existence is supplied by our senses, compared to which scientific proof is effective only to the extent we consider scientific instruments more reliable than our senses. While our contemporary worldview considers scientific evidence to be superior to our senses, it is worth noting that in the age of science, many people insist that the Earth is flat, preferring the direct sensory evidence to more sophisticated theories.

proof
prove

---

[1] Recall that we use 'matter' as a cover term for objects and events/relations

Bodily feeling, proprioception, is the ultimate reality. This extends not just to strong signals of joy and pain, but to much weaker afferent signals pertaining e.g. to the status of our extremities: if my leg fell asleep it means it went numb. The physiological mechanism (restricted circulation, lack of oxygen, blocking of nerve path, etc.) is relevant only to the extent that a reasonable person *must* conclude that this is what is happening, *no other conclusion is available*. It is precisely this lack of choice in terms of explanatory mechanism that makes torture such an effective weapon: pain is real, as is joy. If 4lang we have *pain* as `bad, sensation, injury cause_` and *injury* as `damage, body has`, so pain/injury is tightly coupled to bodily sensation.

pain

All of this is not to deny the existence of higher sentiments of joy and pain such as moral triumph and outrage. The claim here is that such higher sentiments are directly modeled on the direct, bodily sentiments, even if they are stronger, as they can be, when people withstand torture in the service of higher moral ideals. Again we follow the commonsensical notions that this is a matter of willpower, and `will/132` is simply a synonym of *want* which is defined simply as `=agt[feel[=agt need]]`. We have already seen that *feel* is bodily feeling, proprioception, and 4lang skirts the issue of differentiating between needs and wants. (There is a culturally widespread theory that the two are subtly different, what you want is not necessarily what you need, and that you can't always get the former.)

want

The Swadesh list is a rich source of concepts with high universal presence: it is hard to imagine a language that doesn't have a word for *river* or *mother*, and perhaps even harder to imagine a world without rivers or mothers. There are no doubt such worlds exist, e.g. Mars is one, but the commonsensical theory that we are trying to model here loses its grip over these. There are two views of the deontic world, populated by eternal entities and governed by eternal laws, what we will call the *large* and the *small* view. Under the large view, every world of Fig. 6.1 is part of the deontic world, it is merely the case that some of them are inaccessible from the real world. The central deontic world V_D is one where rules are kept, children have parents, sugar is white, caramel is brown, and so on. We will compare this to the *default world* in 6.4.

Under the small view, the only accessible ones are where the concepts are instantiated, making them truly rigid designators in a way proper nouns are not (see 8.1), in spite of the original Kripkean intent. Consider the louse. We can well imagine a world where lice are absent, in fact most of us desire to live in such a world. There is nothing in the definition 'a small insect that lives on the hair or skin of people or animals' that suggests that a world without lice is beyond our ken. A world without insects is a bit harder to imagine, certainly this would result in an unimaginable ecological catastrophe with the very survival of humanity in doubt. Compare this to *mother* 'parent, female' and notice that the absence of parents implies the absence of children (indeed, 4lang defines *parent* by `make child`) and a whole different idea of human life pattern than what we have now, with concomitant loss of sense-making concerning a very broad swath of human languages (as spoken until this science-fictional future becomes reality) and concerning human cultures.

parent

The closer we get to the core (defining) vocabulary the more we see the necessity of the concepts they name for meaningful discussion of any sort. In the vectorial perspective we can express this observation by saying that the special predicate EXISTS must be true for the basis vectors, but not necessarily for the entire space they span. The small view assumes that only things expressible in terms of this basis can exist. Most of them, like unicorns and lice, exist only contingently, but some of them, like *mother*, are necessary, if not for maintaining human life, at least for making sense of human discourse. The question must be asked: how do concepts like *river,* a Swadesh near-universal,   river but a defined word in `4lang`, behave? We have `river folyol fluvius rzeka 848 N stream, has water, in valley` but the defining terms don't all appear in the core vocabulary: *stream* is `<water> flow` and *valley* is `land, low,`   valley `between <mountain>, between <hill>`. Altogether, this gives 'water flows between hills or mountains' where *hill* can be eliminated in favor of `on land,`   hill `high, mountain er_` to yield 'water flows on land between more high land'. Concepts of this kind are possible, but not necessary. Their necessity, such as it is, comes from the fact that their shared knowledge is a precondition of communication.

The large view also permits existents that are not definable in terms of the basis. We are very aware of such existents, e.g. Fourier series, but we have trouble conceptualizing them as being present in the same world where we find mothers. The problem is not that Fourier series are contingent: to the contrary, mathematicians are as convinced of their existence as they can be. The problem is that they don't obey ordinary laws of nature, e.g. they have no weight, color, energy, shape, position, etc. We may say, together with all mathematicians who find a realist ontology convenient, that such object exist in a Platonic world, one that is even accessible to the human mind, but the commonsensical theory, which is our object of inquiry in this volume, has no grip over such worlds.

Even so, the commonsensical view remains useful in understanding the commonsensical concepts of probability discussed in Chapter 5 and, remarkably, the concept of instrumentality. We begin with *prior probability* of a matter X, defined simply as the proportion of worlds within our experience (think of the 'reverse light cone' terminating in the bulleted world of Fig. 6.1) where X obtained. To some extent, this is shrouded in lack of information, but if $l(X) = 0$ this means that EXIST(X) is false in all prior worlds, and $l(X) = 6$ means it is true in all of them. *Current probability* is a more mysterious notion precisely because we are attempting to estimate the proportion of existence across the vertical line of Fig. 6.1, those worlds that share the same time but, at the same time, worlds we don't really have access to. Much better to deal with *future probability* or *rational expectation*, which measures the proportion of existent Xs among the accessible future worlds.

As we shall see in 6.4, if no effort is made to change the outcome, matters continue on their default path: if $l(X) = 5$, X will continue to obtain, and if $l(X) = 2$, X will continue to not obtain. Further, no amount of effort can change $l(X) = 0$ or $l(X) = 6$ outcomes. This latter case, referred to as an act of God in the Anglo-American legal tradition, is delimited precisely because no human act could prevent it (when $l = 6$)

or bring it about (when $l = 0$). The real scope of human intentions is in the middle, particularly at $l = 3$, the broad domain where things are neither particularly likely nor particularly unlikely.

While the real world may be deterministic, the commonsensical world is certainly not: chance plays a big role, things can go many ways, and a key part of the human condition is that we don't know, some would say we cannot know, key events from the future. Accurate prediction of rare events is the hallmark of science, and Thales' predicting the eclipse of 585 BCE (this is now disputed, but see Couprie, 2004) is often described as the beginnings of science, as opposed to common sense. Be it as it may, *John hoped to win the competition* is different from *John won the competition*, and clearly the difference is lodged in *hope*, which is defined as `desire, want, =agt think =pat[possible]`. Next, *desire* is defined as `feeling, want`, and *want* as `=agt[feel[=agt need]]`. After all these substitutions, *hope* is still a feeling (by the agent) of needing something, obviously still a modal, and by looking up *need* we are not getting any closer, since this is defined as `=agt want`. However, the condition on hope `=agt think =pat[possible]` is helpful, as *possible* is given by `gen allow, can/1246`. We defer the analysis of *can* to 6.4, because it is simply given as `<do>`, and it is the optionality of doing, marked by the `< >`, that does the real work there. However, we can continue with *allow*, which is defined as `=agt[lack[=agt stop =pat]]`. Here the agent is `gen`, so *possible* means nothing (can) stop `=pat`, the object of the matrix agent's hope.

hope
desire
want
need

possible
can

allow

It is at this point of the analysis that the nondeterministic world-view comes into play: the mere fact that something *can* happen (there is no general force stopping it) doesn't in any way imply that it *will* happen. Fortunately, we are not restricted to hoping: we can improve the odds. Compare *John hopes to win* and *After a year of relentless training, John hopes to win*, or compare *John hopes to keep the wolves at bay* to *With his rifle, John hopes to keep the wolves at bay*. Altogether, *instruments* are tools that improve the chances of the desired outcome. Hope is good, but being prepared is better. Using our naive probability model, we see the immediate successor worlds of the current real world (see the arrows starting from the bullet in Fig. 6.1) as containing a distinguished default state, but for $l(X) \neq 0, 6$ also many other states where the desired outcome X obtains, and many where it does not. What we want to say is that the likeliness of the desired outcome is increased by instruments. Needless to say, it has to be an instrument fit for the purpose: *With his hacksaw, John hopes to keep the wolves at bay* is as dubious as *With his rifle, John hopes to trim the beam*.

instrument   We have *instrument* as `object, work ins_, gen use, has purpose, at hand`, and note that in the `ins_` relation the instrument need not be a physical object, cf. *John won by cheating*. The prototypical instrument is a hand-held device with a specific purpose and this, seemingly accidental, aspect will be relevant for 6.3: what is near us is epistemologically certain. Our sense of touch is considered more reliable than our vision, which in turn is more reliable than our hearing. Our most reliable sense is is our proprioception: if we *feel* something, this overrides our auditory and visual

perceptions. By definition, *feel* is `=pat in mind, =pat at body, =agt has` `feel`
`body, =agt has mind,` something that brings body and mind together in an act of
perception. To feel something means that it is something right here, within the body
schema, that is being brought to mind. Unlike things we hear, and even things we see,
what we feel is something that cannot be denied.

To summarize, instruments are simply goal-oriented likeliness-increasing devices.
This again illustrates a point we already made at the end of 6.1: it is the lexical semantics
of the elements such as the instrumental case marker `ins_`, defined as `=pat make` `ins_`
`=agt[easy]` that drives the way instruments are referred to in language, not some
top-down theory (such as hierarchical ordering of thematic roles). This is not to say
that conceptual definitions such as Fillmore's "The case of the inanimate force or object
causally involved in the action or state identified by the verb" or Pāṇini's "most effective
means" are useless. To the contrary, these are both powerful paraphrases for trying to
get to the meaning of the instrumental marker, and for the analytically minded, they
provide excellent guidance in trying to sort out what (if anything) can be considered an
instrument in a given situation. Our own definitional attempt differs from these chiefly
in being provided in a fully formalized language, in keeping with the overall plan of the
work.

6.3 Knowledge, belief, emotions

We now try to articulate some fundamental assumptions about knowledge and belief.
First, these are things in the head. Gordon and Hobbs, 2017 trace back the standard Theory of Mind (ToM) to Heider and Simmel, 1944, and here we follow in their footsteps to
the extent feasible, but concentrate on how ToM is reflected in `4lang`. In this particular
case, the definition of as *thought* as `idea, in mind`, relies on two notions we will `thought`
analyze further, *idea* and *mind*, but readers of 3.1 will know that the spatial `in` relation is used in earnest: the mind is a `{place}`, and thoughts are in it. This gets further
specified by the longest definition in the entire core vocabulary:

```
mind tudat conscientia umysll 2457 N
     human has, in brain, human has brain, think ins_,
     perceive ins_, emotion ins_, will ins_,
     memory ins_, imagination ins_
```

We will not do justice to the complex discussion that followed Premack and Woodruff,
1978 whether the the proper definition should include `animal has` rather than `human`
`has`, but note that the tendency to typecast animals, machinery, and even simple household objects as 'having their own mind' is strong not just in children but adults as well.

We obtain our starting point, that thoughts are in the head, by transitivity of *in*: if
thoughts are in the mind, the mind is in the brain, and the brain is in the head, thoughts
are in the head. We use `idea` as a near-synonym of *thought*, defining it as `in mind,` `idea`
`think make.` More interesting is the relation of the nominal *thought* and the transitive
verb *think*, defined as

```
think gondol cogito mysllecl 907 U
     =pat in mind, =agt has mind
```

There is a subtle intransitive/transitive alternation often seen in psych verbs: if *John thinks* it is not necessarily the case that he is thinking of something – anybody who has ever struggled with putting thoughts into words will recognize the cases when the object cannot be formed easily, or at all. In English, the object of thought is typically expressed in a prepositional phrase, the agent thinks *of* something, or *about* something. This can easily be encoded by `"of" mark_ =pat` or `"about" mark_ =pat`, but the cross-linguistic variability is such that we refrained from doing so.

Second, the thoughts in the head are ontologically just as well established as the objects/events/qualities in the real world. We follow Meinong (see Parsons, 1974 for a clear modern exposition) rather than Frege, who places thoughts in a 'second realm', the internal world of consciousness. We would like to strongly discourage the reader from thinking about this in New Age terms, how consciousness creates reality, etc. Rather, this is a straightforward explanation of the human capability to *model* all kinds of things, from alternative outcomes of actions (as required for weighing the fitness of instruments for this or that purpose) to predicting the behavior of other agents. Further, the evolutionary advantage conferred by modeling ability is overwhelming: in any competition for resources if A can model B but not conversely, A is far more likely to obtain the resource.

Third, the assumption of thoughts in the head being real inevitably leads to the conclusion that other things in the head, such as feelings, emotions, desires, … must also be treated as real. This, of course, is everyday human experience, and the commonsensical theory of emotions views them as humors flowing through the body. To better articulate the commonsense theory we have already gone one step further, endowing feelings with direct, non-negotiable reality in proportion to the reliability of the sense that conveys them. At the top of this hierarchy stands proprioception, followed by touch, vision, smell, and hearing in this order. Thinking is generally considered less reliable than our senses, and this includes discounting our own thoughts in relation to the words of the sages. Whether we like it or not, this is precisely the advantage that traditionalism and revealed teachings have over rationalism.

At this point, the reader may wish to revisit the discussion of grammatical moods and logical modalities in S19:7.3, but for greater convenience we summarize the 4L logic approach used there, which relied on the introduction of two more truth values in addition to the standard T (true, ⊤) and F (false, ⊥), called U (unknown) and D (unDecided). Negation, as standard, makes F out of T and T out of F. In 4L the negation of U is U, cf. Codd's 'missing data'. The modal operator K will mean *known*, or rather *learned*, and will be given by `after(T or F)`.

The other nonstandard truth value, D, maps out the scope of agentive decisions and free will in terms of `before` and `after`. At any given time, the truth of a statement may depend on our own decision. Tomorrow morning I may drink tea, or I may not; this matter X is unsettled in all theories of free will (except in the denialist version, which takes all such matters to be deterministically set in advance). In 4L the negation of D is D: if I am undecided about something I must perforce be also undecided about its negation. D means a nondeterministic transition `after`, to T or F, but not to both, and in this regard it is not at all like the 'both' value of Belnap (1977). The operations ¬, ∧, ∨ are defined by the truth tables given in Table 6.1 below.

```
              ∧ | T U D F          ∨ | T U D F
              --+--------          --+--------
              T | T U D F          T | T T T T
  | T U D F   U | U U D F          U | T U D U
¬ | F U D T   D | D D D F          D | T D D D
              F | F F F F          F | T U D F
```

Table 6.1: Boolean operations in 4L

In 6.4 we will refine this simple theory of decision-making with a key observation: matters cannot stay undecided forever, not making a decision generally amounts to making a definite choice of letting the default operate. Certainly, if I defer the decision whether to drink tea until noon, this is for any observer quite indistinguishable from having made the positive determination not to drink any in the morning. We will use the modal operator S to describe the process of *settling* on a decision, meaning `after(T or F)`, where *or* carries the full force of the logical primitive `"_ or _" mark_ choose.` or

How is, then, the modal operator K, the act of learning, different from the modal operator S, the act of decisionmaking, especially as both satisfy `after(T or F)`? The most salient difference is in the frequency of the outcomes: if no learning takes place, we generally assume positive statements to be false, whereas if no decision is made, we generally assume that the default will carry the day (be true). Since the everyday experience that we are surrounded by an ocean of falsity, with truth being a rare find, seems to extend even to scientific studies (Ioannidis, 2005), we see no need to argue the point about K in detail here, S will be discussed in 6.4.

Clearly, emotions are as real as other things in the head, and in fact electrocardiogram-based emotion recognition systems can reach remarkable accuracy (Hasnul et al., 2021). Subjectively there doesn't seem to be a significant difference between bodily sensations like *feeling hot* and emotions like *feeling angry*, and most of the 4lang definitions for emotions eventually go back to *feel* `=pat in mind, =pat at body, =agt` feel `has body, =agt has mind.` This is true both for basic emotions listed in 4lang such as *anger* `feeling, bad, strong, aggressive` or *desire* `feeling,` anger `want,` and for abstract categories such as *feeling* `mental, other cause_,` joy desire `is_a, sorrow is_a, fear is_a, anger is_a` and *emotion* feeling `state/77, in mind, feeling.` emotion

Furthermore, the same holds for the entire emotion vocabulary, very much including words not explicitly listed in 4lang such as *grief* 'extreme sadness, especially because someone you love has died' (LDOCE); 'very great sadness, especially at the death of someone' (Cambridge). To reduce this to the core system, we first note that *-ness*, a clear deverbal and deadjectival noun-forming suffix (ignoring lexicalized cases like *business*) is not essential. We can use *sad*, already defined as `emotion, bad` in the definition of *grief* as `sad, <{=agt love person die} cause>`. For the 'extreme, very great' part `4lang` actually offers *sorrow,* `emotion, ER sad`, suggesting a better definition such as `sorrow, <{=agt love person die} cause>`. The naive theory of emotions embedded in `4lang` is not very sophisticated, but the links between sorrow 'dolor' and badness `cause_ hurt` are laid bare.

(margin: sad)

(margin: sorrow)

As in other semantic fields (Buck, 1949 devotes an entire chapter to emotions), we resist the temptation to offer a full taxonomy. Many words that Buck considered key are removed during the uroboros search, for example *pity* has the following definition: `sorrow, {other(person) suffer} cause_`, but we see no reason to trace these exhaustively, let alone to trace all emotionally loaded words one may wish to consider. Broadly speaking, the naive theory treats feelings along the Hippocrates/Galenus lines as vapors or liquids (humors) flowing through the body, and we see traces of this in the free use of several motion verbs with emotions as subjects *joy flooded him,* or *his blood boiled* etc. We offer a mechanism for uncovering such taxonomies by tracing the definitions to the core, but we do not offer a policy.

(margin: pity)

6.4 Defaults

Perhaps the cleanest statement of defaults comes from programming languages such as C++, where function arguments can be equipped with default values. Other familiar examples include standard unix/linux utilities like `ls`, which will, when invoked with a directory path argument, list the contents of the directory in question, but will list the contents of the current working directory if invoked with no argument.

Natural language offers many similar examples, where a default object is assumed if no overt object is specified. Often the default object is highly unspecified as in *eat <food>*, other times it is highly specific, as an *expect <give birth>*. On the unspecified end, we often find cognate objects as in *sing a song, think a thought, ...* and very weak subcategorization as in *prove <something>*. Neither of the extremes poses a great challenge to a modal treatment of defaults invoking the large deontic world V_D or the small deontic world V_d respectively. As a practical matter, over 6.3% of LDOCE definitions contain defaults encoded by the keyword *especially*, as in *admit* 'to say that you have done something wrong, especially something criminal' or *rat-a-tat* 'the sound of knocking, especially on a door'. `4lang` relies on defaults even more heavily, with 13.8% of the core definitions containing clauses demarcated by $\langle\rangle$ (Rule 6 in 1.6).

We begin with a simple case, where the ambiguity is caused by predicating the default. Consider *-ist* as in *pianist, receptionist, scientist, tourist, violinist*. There is something of a slippery slope between characterizing a person for whom the notion expressed by the stem is important, as in *Calvinist, Marxist, Unionist, abolitionist, activist, ...,* and naming a profession, as in *archivist, anaesthetist, artist, ...* `4lang` provides

(margin: -ist)

```
person<profession>, think stem_[important], "_-ist" mark_
stem_.
```

In many cases, we are not sure whether the person is professional: *arsonist, bal-loonist, philanthropist, ...* That the default is *profession* `job, before(educate` `profession` `for_)` is clear from the fact that in these cases we tend to treat them as such, e.g. we assume that the arsonist is a career criminal, the balloonist has undergone rigorous train-ing and flies balloons for a living, etc. We even have a word, *amateur* whose main use is to defease this implication. In the cases where it is hard to distinguish professionals from amateurs, the default `profession` takes precedence over the more general `person`.

Profession descriptors are a subset of person descriptors (as long as we don't in-sist on strict InstanceOf typing, see 4.5), so the lexical rule for *-ist*-suffixation oper-ates the same way as the more static entries we quoted from LDOCE above. More challenging are those cases where there are two, seemingly disjunct defaults, as in *bake <cake, bread>; drink <water, alcohol>;* or *can/1427* `cylinder, metal,` `can/1427` `contain [<food>,<drink>]`. Such entries resist the kind of analysis based on `is_a`, since neither *cake/bread* nor *food/drink* has a superordinate member that the sub-ordinate (more specific) member could override.

To compare this to the Pāṇinian idea of "habitual, professional, or skilled" actors noted in 2.2, we need to analyze what the three-way disjunction between *habitual, pro-fessional,* and *skilled* amounts to. For *habitual,* LDOCE offers 'usual or typical' in one sense, and 'as a habit that you cannot stop' in another. The distinction is carried back to *habit* 'something that you do regularly or usually, often without thinking about it because you have done it so many times before' versus 'a strong physical need to keep taking a drug regularly'. It appears we can do away with the compulsive sense, especially as the formations where it is most prevalent *(chain-smoker, pill-popper)* are synthetic to begin with. This leaves something like 'usual, typical, regular, done without thinking, done may times before' for *habitual.* For *professional* LDOCE offers 'doing a job for money rather than just for fun' and 'a job that needs special education and training, such as a doctor, lawyer, or architect'. Finally, for *skilled* it provides 'has the training and experience'.

It seems quite hard to disentangle the senses of *professional* and *skilled* as both require the `before(educate for_)` aspect that we used in the definition of *pro-fession.* In modern society, 'need special education and training' really means educa-tion/training that provides a license: practicing law, medicine, or architecture without a license is criminalized, no matter how skilled the practitioner. This means we can col-lapse the second and third terms of the Pāṇinian disjunction (no doubt distinguishable back in his day) to just *habitual* or *professional,* perhaps adding to the latter an optional default clause `<licensed>` which must be omitted for *employer, farmer, manager, ruler, waiter,* etc. at least until regulations are further tightened.

It is clear that the range of the remaining two options overlaps greatly, but perhaps differently for nominals obtained by deverbal zero-suffixation (a device we have no need for, given the type-free nature of `4lang`); by deverbal *-er* suffixation; and by denominal -*ist* suffixation. More important, the identification of these sub-meanings is *post hoc,* relying on the subdirection (see 2.2) rather than on the parts themselves. A *habitual offender* is simply a person who has offended many times before, there is no implication that they get paid for it, or that offending required any education or training, let alone licensing. If *for our next outing, Jim will be the cook,* this does not make him a professional cook, or even a skilled one, just one who assumes the role, quite possibly without the benefit of special education or training. It is precisely because of the post hoc nature of the choice between the habitual and the professional reading that the rule lacks productivity: we don't have a notion of the *??habitual sleeper* not because nobody is trained in sleeping (some people with disorders actually are, but we don't consider them professionals for that) but because everybody is on the habitual branch of the definition of *sleeper, eater, breather,...* to begin with.

Returning to dual defaults, it is intuitively quite clear that we would want to follow Pāṇini and permit disjunction e.g. in *bake* `cook/825, =pat[<bread>, <cake>], =agt cause_ =pat[hard]` whose default object is either bread or cake, but not both. One way to resolve the issue would be the introduction of some abstract supercategory such as 'dough-based baked food' or 'victuals'. We call this the KR-style solution, as it is seen quite often in systems of Knowledge Representation. This is unattractive for most languages (cf. 5.3 for 'doors and windows' in Hungarian), especially as the first paraphrase sneaks in *bake* on the right-hand side of the definition, and the latter (together with its less current synonym 'aliments') defaults to food, whereas 'refreshments', at least in current usage, defaults to drinks. The KR-style solution also goes against the lexicographic principle of reductivity (see 1.2) that the definiens should be simpler than the definiendum.

> bake

The key to the treatment of defaults is to see them as triggers for spreading activation. We will discuss the activation process in greater detail in 7.4, but the general picture should already be clear: if a default is present in a lexical entry, it is active unless it gets defeased. At the discourse level, such activity is easily tested by the immediate, felicitous availability of definite descriptions. Compare *I went to a wedding. The minister spoke harshly* (Kálmán, 1990) to *I went to a restaurant. #The minister spoke harshly.* The wedding script comes fully equipped with a slot for *minister,* but the restaurant script does not. In fact, one need not resort to the full conceptual apparatus of Schankian scripts or Fillmorean frames to see this, the lexical entry for *wedding* 'a marriage ceremony, especially one with a religious service' (LDOCE) already carries the religious service and its officator by default, whereas the entry for *restaurant* does not.

Under the view presented here, a *restaurant* is not fully defined by 'a place where you can buy and eat a meal' (LDOCE) because the same test *I went to a restaurant. The waiter spoke harshly* shows *waiter* to be available by definite description. The existence of a specific negative, *self-service restaurant* also points at the conclusion that waiters are present in restaurants by default, as are chefs, maître d's, busboys, tables, etc. The Oxford definition, 'a place where people pay to sit and eat meals that are cooked and served on the premises' shows the slots for cooks/chefs and servers/waiters, and *sit and eat* does seem to imply the presence of a chair and a table. Whether the maître d' hôtel is a default feature of a restaurant seems very much culture-dependent, but a *real restaurant*, as we shall see in 7.2, can hardly do without.

Let us return to conjoined defaults. Consider *ash* `powder[<grey>, <white>,` `ash` `<black>], {<wood> burn} make`. What is the default color of ash? The word *ashen* suggests 'pale gray' but an *ashen face* 'looking very pale because you are ill, shocked, or frightened' (LDOCE) is actually not grey, just pale. The larger encyclopedia, https://en.wikipedia.org/wiki/Shades_of_gray is already overwhelming, and a broader search leads to sites such as https://simplicable.com/new/ash-color which call into question even the Knowledge Representation-style solution relying on some technical term (in this case, *grayscale*).

There are cases like *broadcast* `signal, <radio,television> receive` `broadcast` where a KR-style solution is easy. Unlike *aliments* discussed above, where the defining word is lexicographically unreasonable, here we could use *antenna*, not just as something common to TVs and radios, but also as the instrument of both broadcasting and reception. But there remain cases like *opponent* `person, oppose, <compete>,` `opponent` `<in battle>` where the defaults are rather contradictory between friendly competition and adversarial battle. In the spreading activation model we don't have to make early choices between polysemous senses or pretend that these involve a single abstract sense. Rather, the system can resolve later on which of the adjacent polytopes is meant.

We began S19:1.1 with two Fregean principles, the better known Compositionality, and that of *Contextuality*:

> Never ask for the meaning of a word in isolation, but only in the context of a sentence.

In computational linguistics, relating word senses to contexts is known as the problem of Word-Sense Disambiguation, see Agirre and Edmonds (2007) for the state of the art before 2010. Perhaps the greatest step forward in solving the WSD problem was the introduction of *dynamic* embeddings that produce a word vector based on context. Unfortunately, this is a black box solution, and part of our goal here is to understand the mechanism of disambiguation. Defaults, contradictory defaults in particular, offer an important insight into the structure of lexical entries: while the basic structure is conjunctive, their joint activation, by spreading, is disjunctive. The broad agentive `-er` `-er` `stem_-er is_a =agt, "_ -er" mark_ stem_` is simply 'one who stem-s' (cf. *buyer, sleeper, . . .* rather than 'one who habitually stem-s', so there is no disjunction to consider. The more narrow agentive *-er*, and *-ist* are, perhaps just like in Sanskrit, ambiguous between the habitual and the professional readings (cf. *smoker, exhibitionist* for the former and *plumber, pianist* for the latter) but we see no supercategory that connects these two: rather, we see these as disjunctive by virtue of being defaults. The work is done by the `person[<profession>]` clause which defaults to `profession`. We have to do extra work to escape this conclusion in order to fall back on the default `person`, and this extra work is unrelated to any notion of habituality, since the pros obviously 'do stem' habitually. In synthetic compounds *teetotaler, navelgazer,* ... we assume the work is done during the formation of the compound, in other cases we may have to bring in the compulsive aspect we chose to ignore above.

The entire network of lexical entries is remarkably tight. We have seen that from the uroboros core every word can be reached in three steps via the LDV and LDOCE. Three is the maximum: those familiar with the use of `4lang` can often write a one-step definition that relies only on the uroboros core. (By now, most readers will have seen enough examples and will understand the principles well enough to try themselves.) Since the average number of clauses within the V2 uroboros set is 2.66, if we let spreading proceed through any undirected 'associative' path, we may activate the entire vocabulary in 5-6 steps starting from the words of any sentence. Consider *colorless green ideas sleep furiously. Color* immediately activates `sensation, light, red, green, blue`; *-less* activates `lack`; *green* activates `has, plant`, and the already active `color`; *idea* activates `in, mind, think, make`; *-s* activates `more`; *sleep* activates `rest, conscious`, and the already active `lack`; *furious* activates `angry, er_, gen`. Only *-ly*, a pure category-changing affix, does not activate any element, as it is semantically empty. This is not to say that it entirely lacks a categorial signature: for English, *-ly* is clearly [AN]\D, but in `4lang` we wish to avoid the claim that operators turning adjectives or nouns into adverbs are universal.

In one step we have already activated 20 elements (2.86 per morpheme), and only four of these, `er_, gen`, and `lack` are primitives that resist further spreading, while `in` will invoke the entire `place` conceptual scheme (3.1). In fact, the morpheme count is somewhat arbitrary, as we should clearly add a nominative and an accusative marker, 3rd person singular, present, and perhaps other unmarked operators such as *I declare to you.* To limit combinatorial explosion we need to constrain spreading activation in various ways. First, it is clear that permitting activation in the other direction would be unwise, since one in seven words involve spatial `in`, almost one in three involve possessive `has`, some 40 involve comparative `er_`, some 60 involve negative `lack`, and the same number involve generic `gen`. Second, we need to enforce some condition of locality, in that it is cognitively implausible that the negative element explicit in *colorless* could

sleep   reinforce the negative element implicit in *sleep* `rest, lack conscious`. We will return to spreading activation in 7.4, where we discuss how to implement locality by island parsing, but we note in advance that the key building block will be the *construction* in the sense of Berkeley Construction Grammar.

For the synthetic compounds in *-er* this superficially takes the form $(N\ V\ -er)_N$, e.g. in *navel.gaze.er.* Remarkably, the spreading analysis often leads through the unattested intermediary that we use use for agentive *-er*, $(V\ -er)_N$. Once this pattern is activated, the very frequent noun-noun compounding pattern $(N\ N)_N$ can be spread to. In S19:6.4 we wrote

> The algebraic approach (...) largely leaves open the actual contents of the lexicon. Consider the semantics of noun–noun compounds. As Kiparsky (1982) notes, *ropeladder* is 'ladder *made of* rope', *manslaughter* is 'slaughter *under-*

gone by man', and *testtube* is 'tube *used for* test', so the overall semantics can only specify that $N_1 N_2$ is 'N_2 that is V-ed by N_1', i.e. the decomposition is subdirect (yields a superset of the target) rather than direct, as it would be in a fully compositional generative system.

This applies to entries like *teetotaler* which we analyze with an unattested agent noun *totaler* who totals (does always) the V-ing of *tee* (tea). Unsurprisingly (though not exactly predictably) the verb in question is *drink*, so we obtain 'one who always drinks tea'. While still a bit off the actual target 'one who abstains from drinking alcohol', this is close enough for memorization, and offers considerable economy relative to memorizing the entire definition.

7

Adjectives, gradience, implicature

Contents

Adjectives are present in most, though not necessarily all, natural languages. In 7.1 we begin by discussing the major properties of adjectival roots and the vector semantics associated to the base, comparative, and superlative forms. We discuss the logic associated to these, and extend the analysis to intensifiers.

Starting with Bloomfield (1926), semantics relies not just on the idea of identity (equality) of meanings, but also their similarity. Logical semantics offers *implicational equivalence* for defining identity of meaning, and vectorial semantics offers cosine similarity for defining meaning similarity. But there is a third notion, *strength*, which remains somewhat elusive in both frameworks. In logical semantics, we speak of *implicational strength* e.g. that 'run fast' implies 'run', but not the other way around. In 7.2 we discuss how this kind of scalar intensification can be implemented both for adjectival and for verbal predication using voronoids.

In 7.3 we turn to implicature in a broader sense. From the vantage point of algebraic semantics, implicature is also a kind of intensification process, except what gets intensified is not some direct aspect of the meaning but rather the degree of indirection (number of substitutions) we need to carry out. We will present implicatives as satisfying simple inequalities imposed on their agents and patients, and show how the force dynamics-like analysis offered by Karttunen (2014) can be taken on board in vector semantics.

Finally in 7.4 we summarize how substitution or, what is the same, spreading activation, is operating during parsing and generation. This puts in a new light the perplexing lack of transitivity in implicature, that we may conclude B from A, and C from B, yet we are often reluctant to get from A to C. When someone tells us *It can hardly be disputed that X*, what this means that it is hard to dispute X, which only happens if X is obviously true. From this it follows that X is true, self-evidently so. Yet once our sus-

A. Kornai, *Vector Semantics*, Cognitive Technologies, https://doi.org/10.1007/978-981-19-5607-2_7

picions are aroused, we are reluctant to draw this conclusion. What needs to be made explicit in this regard is that sentences whose meaning cannot be tied to truth conditions (other than truth conditions pertaining to the mind-state of the speaker and the hearer) actually demonstrate that truth-conditional semantics is a blunt instrument, incapable of assigning meaning to sentences.

7.1 Adjectives

The traditional understanding of adjectives is that they are content words that freely attach to nouns as modifiers. Categorial grammar expresses this by assigning the signature N/N (something that takes N input and returns N) to adjectives, and dependency grammar sees them as dependents of nouns (nmod). Many adjectives are morphologically distinguished from their nominal counterpart: *angle/angular, desire/desirable, habit/habitual* ... and the phenomenon is not at all restricted to the Latinate segment of the English vocabulary: *anger/angry, fool/foolish, help/helpful,* .... Time and time again it is the adjectival form that is seen as basic, and the nominal/verbal as derived: *hard/hardness/harden, obese/obesity,* ....

This basic picture is complicated by two facts: first, that many adjectives have a nominal reading e.g. *purple* 'the color purple', *heavy* 'a criminal', *safe* 'reinforced metal box', etc. Second, in certain languages (Mandarin Chinese, Acehnese, Puget Salish) the distinction between stative verbs and adjectives is very hard, perhaps impossible, to make. From an Indoeuropean perspective it is hard to imagine that some language would say *Mother kinds* instead of *Mother is kind* to express the same idea, yet in many languages we find affixes that turn adjectives into verbs without any change in meaning, and we also find light verbs constructions as *Mother acts kind* that achieve the same effect.

The primary domain of adjectives seems to be qualia, which in the ontology of (Jackendoff, 1983) correspond to Properties or Amounts (see 2.1). In our more sparse ontology, qualia are just *matters*, as all unary functions. What seems to distinguish them from other matters, Things and Events in particular, is the systematic ability to invoke them in comparisons, such as *The weather is colder today*, and the equally systematic ability to seek extrema, as in *This was the coldest day of the year*.

That adjectives tend to be amenable to comparative and superlative degree modification receives a natural explanation in vector semantics, where qualia correspond to the simplest polytopes, (affine) half-spaces. Moreover, if nouns are more complex polytopes, while adjectives are just half-spaces, the semantics we associate to adjectival modification will simply be conjunction, i.e. intersecting the polytope with the half-space as assumed in 1.3. Altogether, this makes typical Things (we defer Events to 7.3) just bundles of typical Properties, but it would be rush to say that there is a one-to-one mapping between syntactic and geometrical types in any single language, let alone a universal correspondence across all languages. Rather, qualia/adjectives *typically* represent half-spaces, and it *typically* takes bundles of qualia to represent things. But there can be sporadic mismatches, words that are treated as adjectives even though they do not naturally correspond to a half-space; and words that are syntactically nouns or verbs, even though semantically they are elementary half-spaces, rather than the intersection of such.

Recall that a half-space H is defined as a side of a single hyperplane, which in turn is defined by a normal vector $\mathbf{n}$ so that $\mathbf{x} \in H \Leftrightarrow \langle \mathbf{x}|P|\mathbf{n}\rangle > 0$ i.e. iff the scalar product of $\mathbf{x}$ and $\mathbf{n}$ in the prevailing metric P is positive. (More precisely, this is an *open* half-space – we obtain a *closed* half-space by letting the scalar product to be ≥ 0.) An *affine* half-space is obtained by shifting an ordinary half-space by some fixed vector $\mathbf{s}$, but clearly only the component of this vector falling on $\mathbf{n}$ matters, which will be $\|\mathbf{n}\|$ times some scalar c, so in the affine case we have $\mathbf{x} \in H + c\mathbf{n} \Leftrightarrow \langle \mathbf{x} - c\mathbf{n}|P|\mathbf{n}\rangle > 0$ or, what is the same,

$$\mathbf{x} - c\mathbf{n} \in H \Leftrightarrow \langle \mathbf{x}|P|\mathbf{n}\rangle > c\langle \mathbf{n}|P|\mathbf{n}\rangle \tag{7.1}$$

where $c\langle \mathbf{n}|P|\mathbf{n}\rangle$ is a *bias* that depends on P. Increasing this term (either by changing c or by changing P) will shift the boundary hyperplane in the positive direction, where things display the qualia more strongly. For any vector $\mathbf{x}$, the length of its component in the $\mathbf{n}$ direction serves as a natural measure of H-ishness: the larger this number the more $\mathbf{x}$ displays the qualia/enjoys the property of being in H. In other words, affine half-planes come naturally equipped with a numerical scale that makes comparison and seeking extrema easy.

When testing this on numerals, a special subclass of adjectives that are supposed to make reference to a scale by their very nature, the results are confounded by a variety of phenomena that philosophers of language, starting with Grice (1975), consider 'pragmatic'. For example, with *I have three children* the expectation is that I don't have four, and the logical implication that I have two must be carried. This is in contrast to *I am tall*, where it is quite possible that I am in fact very tall, and nobody considers me a liar for omitting *very*, and the implication that I am only a tiny bit tall, say 180 cm as opposed to the adult average of 177.6, does not carry.

This is due to the fact that numerals, while syntactically very much nominal modifiers, are hard to represent as half-spaces. Arguably, *two* is the intersection of the ≥ 2 and the ≤ 2 half-spaces, but among the integers there is no scale of 2-ishness: either something is two or it isn't. This makes it possible to define '2' in an extensional manner as the class of all sets that have exactly two members, but offers no possibility of a scale, so no *two-er* or *two-est*. It is precisely because we lack clear guidance which half-space to take as basic that the upward and downward entailment issues we discussed briefly in 4.5 are so prominent among numerals and numeral-like elements like *many* or *few*. Under the 4lang analysis *many* is quantity, er_ gen and *few* is amount (gen er_).

Since *amount* is simply defined as quantity, at first blush *many* and *few* are perfectly symmetrical. But the implicational depth is different, since *quantity* is gen count, gen measure, <much> is defaulted to *much*, but *amount* does not carry

many
few
amount
quantity

`much` this implication. The use of *much*, which is simply defined as `many`, does not change this asymmetry.

7.2 Gradience

For the implicational account, negation causes difficulties, since in negative contexts the direction of implication reverses. Whereas *John runs* follows from *John runs fast, John doesn't run* doesn't follow from *John doesn't run fast*. This is something of an idealized example, in that the primary reading for *John runs fast* is habitual 'John is a fast runner' whereas the primary reading for *John runs* is episodic 'John is running now', so on the most natural readings the implication doesn't even hold!

If we just look at strength as this term is ordinarily understood, at least in subject position the effect is rather clear: from *A red car is overtaking us* it follows that *A car is overtaking us* but not conversely. This much, however, is easily obtained from the vectorial account as well: since the polytope corresponding to *red car* is the intersection of the *red* and *car* polytopes, it is contained in the latter. The discussion can be extended to negative polarity contexts along the standard lines (Giannakidou, 1997), but we call attention to another phenomenon whose explanation has hitherto been lacking: it is precisely in the case of non-intersective adjectives that the implication fails *A former president will give the commencement talk* ⇏ *A president will give the commencement talk*.

`-er` To see how all this plays out for comparatives, we need to define the comparative morpheme *-er*, for which `4lang` provides `er_, =agt has quality, "_-er"` `mark_ stem_[quality], "than _" mark_ =pat, =pat has quality`. Most of this definition just serves to pin down A, B, and C in the *A is B-er than C* construction: A is the agent, B is the quality marked by the stem, and C is the patient. The only critical element is the relational `er_`, which we take to mean ordinary numerical comparison '>' between the stem-ishness of A and C. `er_` is a primitive only under the algebraic view: in the geometric view we can replace it by

$$\langle \mathbf{A}|P|\mathbf{B}\rangle > \langle \mathbf{C}|P|\mathbf{B}\rangle. \tag{7.2}$$

As we have already discussed in Chapter 6, non-intersective adjectives like *former* actually shift P (the projection that falls on V_n is replaced by the projection to V_b), but the inequality 7.2 will otherwise remain homogeneous in the basis of comparison: *cold-er* is obtained from *cold* and *-er* the same way *blue-er* is obtained from *blue* and *-er*. In other words, the semantics of *-er* is perfectly compositional, and stays the same for intersective and non-intersective adjectives alike.

`-est/3625` Turning to superlatives, we can define *-est* as `er_ all`. Since *all* is defined as
`all` `gen, whole`, we obtain

$$\langle \mathbf{A}|P|\mathbf{B}\rangle > \langle \mathbf{C}|P|1/n, \ldots, 1/n \cap \text{whole}\rangle \tag{7.3}$$

where we have used the fact that `gen` is a fixed vector with $1/n$ on all coordinates, and that the semantics of conjunction is intersective. This can be further improved by substituting the definition of *whole* which is `all member`, and also *member* `group has, in group`. This brings to sharp relief the essence of *-est*, that we have some implicit comparison group, and *A* is *-est* means that for every other group member Eq. 7.2 holds. Again the analysis is entirely compositional, and leaves implicit exactly what needs to be left implicit, the comparison group. Note that this group is not entirely supplied by the noun that the superlative attaches to: *the tallest boy* is not the boy who is tallest among all boys, just the one who is tallest among all relevant boys (Moltmann, 1995).

`whole member`

Eq. 7.3 can be faulted for using $>$ instead of $\geqslant$. This can be easily fixed by replacing `all` by `other` in the definition of *-est*, but this of course implies a unique maximum. We have arrived at a situation that is fairly common in formulaic semantics, where the correctness of an analysis must be evaluated based on the felicity of readings in somewhat contrived situations. Suppose there are two twins of the exact same height in a class, Bill and Dave. Can Bill be called the tallest, why or why not? If we think *-est* does not imply a unique maximum, the definition of `-est/3625` is along the right lines (we'd still have to make provisions for the fact that no thing is strictly larger than itself). Since superlative plurals like *the strongest boys, the most beautiful paintings* are common, this lends strength to the proposal, as long as we assume that the strongest boys are equally strong.

However, if we are committed to the idea that extrema are unique, we can use a different definition of *-est*, `er_ other`. This can again be further analyzed by substituting the definition of *other*, which is simply `different`. Recall that *other* is a procedural keyword that prevents unification, and when we define it as `different`, we rely on Leibniz' Principle of Indiscernibles, i.e. we bring in a property that distinguishes the two: *different* means `=pat has quality, =agt lack quality, "from _" mark_ =pat`. As unification operates silently, the simplest assumption is that this property must be the one marked by the stem. Other properties could also be invoked, as exemplified by Kornai, 2012 as follows.

`-est/1513 other`

`different`

> In the case of () *She promised immunity for a confession* (), we assume the promissor p is in a position to cause some suspect s to have immunity against prosecution q for some misdeed d, and that it is s who needs to confess to d. Yet the sentence is perfectly compatible with a more loose assignment of roles, namely that the actual misdeed was committed by some kingpin k, and s is merely a witness to this, his greatest supposed crime being the withholding of evidence. This d', being an accessory after the fact, is of course also a misdeed, but the only full-force implication from the lexical content of *immunity* is that there is some misdeed m that could trigger prosecution against which s needs immunity, not that $m = d$ or $m = d'$. The hypothesis $m = d$ is merely the most economical one on the part of the hearer (requiring a minimum amount of matters to keep track of) but one that can be defeased as soon as new evidence comes to light.

That said, the most natural (default) assumption is that the distinguishing property is indeed the one supplied by the adjectival stem, which implies that the noun modified by *stem-est* is indeed the one that has the property signified by *stem* in the greatest measure among all candidates.

`best`   `4lang` has only two definitions, *best* 'optimus' `good, -est`; and *main* 'primus'
`main`   `er_ other, rank, lead/2617` that rely on the superlative morpheme. In the former case, we left the choice of superlative between `3625` and `1513` unresolved, since *best* inherits the ambiguity, but in the latter case we resolve it in favor of `1513`, as it is commonly assumed that there can only be one main city, main thoroughfare, etc.

In a weaker form, gradience phenomena are observable not just on adjectives but also on nouns. Many languages have diminutives like English *-ette* (cigar/cigarette, kitchen/kitchenette, pipe/pipette, . . .) and augmentatives like Italian *-one* (minestra/minestrone, provola/provolone, spilla/spillone . . .) but these are rarely productive, whereas comparatives and superlatives are so productive that the existence of such forms is often taken as diagnostic for the adjectival status of the stem.

Another set of examples comes from syntax: English and many other languages have a fully productive construction with *true/real* in the noun modifier slot. A *true Scotsman* is one that enjoys all properties Scotsmen are supposed to have, a *real Colt* is a revolver actually made by Colt's Manufacturing Company, and so on. Since this 'prototypicality' reading of the construction is non-compositional (all `4lang` definitions of *true, real, fact* revolve around existence and proof) we must supply the semantics based on the word *prototypical*, which means 'very typical' (LDOCE). *Typical* means 'having the usual features or qualities of a particular group or thing', so a *true/real X* must be something that has the usual features/qualities of X in large measure. This can be implemented using the same idea. If we model a word by a polytope that is the intersection of some half-spaces H_i, we can form the intersection of H_i' defined by the same normal vectors $\mathbf{n}_i$ but higher biases $b_i' > b_i$.

Cross-linguistically, intensifier morphemes can apply to all categories, as in Russian *pre-* (*predobryj* 'very kind' from adjectival *dobryj* 'kind'; *premnogo* 'very much' from adverbial *mnogo* 'much'; *preizbytok* 'large abundance' from nominal *izbytok* 'abundance'; *preuspet'* 'succeed in' from *uspet'* 'manage') though not with equal productivity (Endresen, 2013). Importantly, quantifiers are no exception, they can be intensified *anything at all* and serve as intensifiers *so I don't work or anything* (Labov, 1984).

The ease of creating homogeneous semantics for intensification processes, coupled with the very visible heterogeneity of their productivity across lexical categories, provides yet another argument in favor of Bloomfield's rejection of 'class meaning' we cited in 2.1. It is not that we need to reject the very notion of lexical categories, in fact Lévai and Kornai (2019) demonstrates that word vectors in different syntactic categories show different behavior, a matter we shall return to in 8.2. Rather, we have to draw the line between inflectional and derivational morphology following the dictum of Anderson (1982): *inflection is what is relevant for syntax.* Syntactic constructions and inflectional regularities can be most economically stated over lexical categories, and when we see limited productivity, as we do in most derivational processes, no explanation in terms of the semantics is likely to make sense.

Finally, we note that negative intensification, `lack` in particular, involves a shift away from the broadest sense of a predicate i.e. the *largest* half-space we countenance in terms of containment, which is defined by the *smallest* positive bias. Clearly *blinder than the bat* means having even *less* sight than a bat, to which we grant some vision (echolocation).

7.3 Implicature

As in 4.3, we assume a scale, and here we will first discretize it the way we did with probabilities in Chapter 5. Let us begin with simple positive/negative pair such as adjectival *good/bad* or verbal *approve/disapprove*. H_g for *good* is a half-space, H_b is another halfspace for *bad*, and we need not assume that the two hyperplanes that limit them are parallel. There may even be regions of the space where they intersect, and there is definitely a large neutral region between the two.

Working with *very/somewhat/slightly* as a 3-point augmentative/neutral/diminutive scale we obtain two shifted versions of H_g, H_g^+ *very good* and H_g^- *slightly good*. Similarly, we have H_b^+ *very bad* and H_b^- *slightly bad*. When a verbal base is used, there are some syntactic complications: in English we must say *I approve very much* or *very strongly* instead of **I very approve* though expressions like *?I slightly approve* are attested. On the negative side, we don't see the same asymmetry: both *I slightly disapprove* and *I disapprove very much* are common. There is a great deal of intra-speaker variability in the interpretation of *I approve somewhat* – does this imply I also disapprove somewhat or not?

We take it for granted that adverbs operate on verbs essentially the same way as adjectives operate on nouns, and that intensifiers and moderators operate on these, generally with equal ease on adverbs and adjectives (see S19 Ex 8.8). Now, if verbs were geometrically just like nouns, polytopes defined by the intersection of half-spaces, we could readily deploy the same intersective mechanism for adverbial modification that we already have in place for adjectival modification. But verbs, transitive verbs in particular, appear more complex than nouns, in that they often make essential reference to participants and event structure. Also, using the same semantics for nouns and verbs is cross-linguistically very suspicious, as it is much easier to find cases where adjectives and verbs, or nouns and verbs get conflated, than cases where nouns and verbs are conflated. In fact, the latter does not seem to exist, as the best known putative example, Eskimo (Thalbitzer, 1911) has been shown to have a sharp category distinction between nouns and verbs (Sadock, 1999).

To see what makes verbs verbs, let us consider a few prototypical examples: *eat*   `eat` `=agt cause_ {=pat in mouth}, swallow, <=pat[food]>, <chew>, <bite/1001>, =agt has mouth`. According to this definition, biting and chewing are optional, but swallowing is an obligatory feature of eating. It is perhaps debatable that it is the agent that causes the food to get into their mouth (maybe someone is feeding them), but that food somehow gets located in the agents mouth and swallowed by them is a feature under any definition. We highlight three aspects of the definition: that it makes reference to =agt and =pat; that it involves a coercive aspect: whatever is the patient, it is *by definition* food; and that it involves temporal marking.

move   Recall from 3.2 that our primary tool for handling temporal constraints are `before` and `after`: for example *move* is given as the conjunction `before(=agt at place)`, `after(=agt at other(place))`. Here the motion element is the movement of food into the mouth, and more important, the *swallowing* which moves it from mouth

swallow through throat to stomach: `=agt cause_ {=pat[move]}, after(=pat in stomach), =pat in mouth, =pat in throat, =agt has stomach, =agt has mouth, =agt has throat.`

kill   Next, let us consider *kill* `=agt cause_ =pat[die]`, which displays two of
die these three verbal features, but not coercion. It is true that *die* means `after(=agt`
dead `[dead])` and *dead* means `still, lack live, before(live)`, so kill *implies* that the object of killing was live before, but it is not evident that such a chain of implications is truly coercive, e.g. that from *the lawyers killed the proposal* or from *John killed the time chainsmoking* we actually conclude that the proposal or the time were alive before the action took place. This is in contrast to the implicatives like *dare* that we will turn to shortly: whatever is the object of *dare*, it is by definition dangerous.

see   Finally, let us consider *see* `perceive, ins_ eye`. Here there is no temporal marking, no coercion, and the linkers are brought in only by a deductive chain, via *per-*

perceive *ceive* `know, ins_ sense, hear is_a, smell is_a, see is_a, ...`,
know which brings in *know*, which in turn is defined by making explicit reference to =agt and =pat as `=agt has information, information connect =pat`. None of the three verbal characteristics we started out with are directly manifest, making this a lexical entry that is neutral between *see, seeing,* and *sight*. The `has` appearing at the end of the deductive chain is worth special attention, since it is one of the handful irreducible binaries (see Rule 15 in 1.6), which must, on any theory of semantics, be treated as relational. There is no act of possession that does not involve both a possessor and an object possessed, and this clearly goes back to *perceive*, for which again both agent and patient are obligatory.

On the one hand, *see* is obviously a verb of perception, and one cannot perceive without perceiving something, since the act relates some qualia, the object, to some mental state of the subject. On the other, *see* has many intransitive uses, ranging from a patient recovering their eyesight after surgery *I can see again!* to simple assent *I see* which has at best a dummy zero object. Whether the object truly percolates up from the possessed `information` through `know` and `perceive` is unclear. In the case of assent, we have good reason to suppose that it was the information conveyed by the speech of the first speaker that the second speaker now acknowledges to have, but it the case of eyesight it is not at all evident what information is relevant.

After these preparations let's turn to some typical implicative verbs. The lexicon defines *dare* as 'to be brave enough to do something difficult or dangerous' (Cambridge Dictionary of English); 'to be brave enough to do something that is risky or that you are afraid to do' (Longman); 'to have enough courage or confidence to do something, to not be too afraid to do something' (Merriam-Webster). This is a special case of the general analysis that Karttunen (2014) offers for the whole class of verbs: 'overcoming an obstacle' with the obstacle being *fear* for *dare*, *indifference* for *bother*, *empathy* for Finnish *hennoa,* which he illustrates with *Hennoitko tappaa kissan?* 'Did you overcome your pity to kill the cat?', akin to 'overcome your fear' for *dare*.

John dares VP is taken to mean *John does VP* in conjunction with *VP is risky*, and for the moment we leave open the issue whether it's risky for John, or really risky for everybody. What we did here was to incorporate a hidden element, 'object being risky' for *dare*, 'object being boring' or 'subject being indifferent' for *bother*, and 'subject being humane' for *hennoa*. It is helpful here that vector semantics is detached from part of speech, and we make no distinction between the adjective *risky*, the V' *is risky*, and the noun *risk*, but this lack of morphosyntactic typing will not be relied on heavily in what follows, except for making it easier to draw the graphs and talk about their nodes.

Let us consider how an ordinary sentence, such as *John dared to criticize the mayor* will be analyzed. The matrix verb is *dare* and it is John who does the daring, so we have John $\overset{1}{\leftarrow}$ dare, and it is also John who does the criticizing, so we have John $\overset{1}{\leftarrow}$ criticize. How the subject equi is effected during the parsing process is something we leave to the phenogrammar, the point here is that few grammarians (including LFGers who would use an XCOMP here) would seriously doubt that the subject of both verbs is the same John. The object of the criticism is no doubt the mayor, and the object of daring is the entire *criticizing the mayor*.

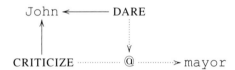

Figure 1. *John dared to criticize the mayor*

As we said, dare is_a overcome, and whoever *dares* actually has power, or at least he thinks he does, to overcome risk. In fact, we can rely on a dictionary definition of *bravery, courage* as 'power to overcome risk' or 'power to overcome fright caused by risk', or even 'power to overcome one's own fright caused by risk'. Neither the final cause nor the precise application site of the power ends up being very relevant for the task at hand, which is to explain certain implications, whose non-fulfillment makes sentences using *dare* infelicitous, but not outright false. With *criticizing the mayor* it is rather clear that mayors are powerful people, and criticizing the powerful is dangerous. But when we say

#John dared to chew gum   (7.4)

we need to abductively infer some theory that makes chewing gum dangerous. Perhaps John had throat surgery, and the wounds haven't quite healed. Perhaps he is in the presence of some superior who considers this disrespectful. Perhaps he was told the gum could be laced with poison. There are many theories that would make the use of *dare* felicitous, and we need not choose among them. But we do need to draw the implication from `dare` to `risk` or `danger`. As with `obstacle` versus `difficulty` above, we need not be very precise which of these terms is operative here. What matters is the se-

`risk`   mantic concept, defined both for `risk` and `danger` as `can/1246(harm)`, not the
`danger` (English) printname we assign to the concept.

A significant advantage of relying on such hidden conditions on `=agt` and `=pat` is that they all fit in the category of `difficulty`. As a result, they will participate in a larger frame involving `virtue overcome difficulty`. The virtue, be it *bravery* as in the Cambridge and Longman dictionaries, or *courage/confidence*, as in Merriam-Webster, comes for free here, in that bravery, diligence, or decency are obviously virtues, This may even be made part of their lexical definition, but we will not pursue the matter here for the following reason. It is plausible that it is a piece of lexical knowledge that fear is a difficulty (an obstacle, as Karttunen (2014) has it), but this does not work for the other cases: it is rather unlikely that `indifference` or `empathy` are lexically specified for genus `difficulty/obstacle`. Such a conclusion, if not available lexically, must somehow be derived by a process of typecasting. The essence of the force-dynamic analysis is that *dare* and *hennoa* both `is_a overcome`. There is every reason to suppose that *overcome* subcategorizes for a *power* subject and an *obstacle* object. To make some action an instance of *overcome*, its subject must be typecast to *power* and its object to *obstacle*. That *virtue* `is_a` *power* is hard to deny, and by transitivity of `is_a` we can treat all implicative subjects under this heading. With the implicative objects, this is less trivial: *empathy* gets to be an obstacle only because *hennoa* `is_a` *overcome*, and outside this frame we cannot draw the usual conclusions, e.g. that obstacles are bad things and therefore being humane is a bad thing.

`dare`   For *dare* we can use a lexical entry `do, =agt[brave], =pat[danger]` which includes the selectional restriction on the object that it is dangerous. Similarly, the object of *deign* is low status (or perhaps the subject is high status), that of *remember* is hard to memorize, and so on. *Manage* has an object that is simply difficult, manner unspecified. Since it is the relationship of the subject to the object that is getting characterized by the implicative verb, we often have a choice between alternative framings: e.g. with *deign* we may be describing (i) the subject as high status; (ii) the object as low status; or (iii) the subject as higher status than the object. These alternatives are logically equivalent, since by default things are neither high nor low status. Yet it is quite conceivable that different speakers have different lexical entries for *deign*, and different lexicalization options may play differently with negation. Translational near-equivalents in different languages may also differ only in the choice between (i)-(iii).

In the algebraic theory, what implicative verbs bring to the table (working memory) are little lexically prespecified hypergraphs that demand abductive inferencing (type-casting) over and above the normal inferencing process, We assume, as is standard, that verbs can subcategorize for their arguments. To reuse an example from S19 4.2, "*elapse* demands a time interval subject. If we read that *A sekki elapsed* we know that this must refer to some time period even if we don't know the details of the sekki system." This is part and parcel of knowing what *elapse* means: people who can't make this inference are not in full possession of the lexical entry. In the representation of *elapse* there is thus a direct prespecification =agt[period], which contrasts with the prespecification inherited from a nominative that subjects are agentive (active, causing, volitional).

There is little reason to suppose that time intervals are inherently active, that they causally contribute to their own elapsing. To the contrary, time periods are abstract objects and will inherit several features of these, in particular lack of volition and lack of physical capabilities, which are hard to square with agentivity. But once *sekki* appears in the subject slot of *elapse*, it is_a period and the inheritance of non-volitionality from any supercategory of abstract objects is blocked, since the specific, lexically prespecified case will block the general inheritance mechanism as a matter of course. Similarly, in the representation of *dare*, the object of daring is_a risk, and this will block the general assessment that e.g. chewing gum is not normally considered a risk.

Another aspect of the analysis that is relatively easy in the framework presented here is adding *overcoming* to the small set of preexisting force dynamic primitives *letting, hindering,* and *helping.* We can dispense with the force-dynamic diagrams entirely in favor of analytic statements such as *overcome* 'the agent, initially weaker than the object, is subsequently stronger'. In 4lang we can say before(force(=pat er_ =agt)), after(force(=agt er_ =pat)). As usual, it matters but little whether we call the basis of the comparison *force, power, might, heft, momentum,* or something else, there is a single concept here, and we chose to call it force mainly to make clear our indebtedness to Talmy (1988) and Jackendoff (1990). As a matter of fact, the 4lang of *force* power, and that of *power* as cause_ change works reasonably well for the naive physics we assume throughout, very much including the 'metaphorical' cases like *The rain forced them to seek shelter* or *She changed his mind by force of thought.*

force
power

Putting this all together, in 7.4 we start from highly skeletal lexical entries such as for *dare,* which stipulates only that this is an act do and some selectional restrictions =agt[brave], =pat[danger] and obtain, by simple algebraic means, implications such as *John chewed gum* and *chewing gum was risky (for John).* This inferencing is accomplished by the same process, substitution *salva veritate,* our rule of expansion (Rule 1 in 1.6). In the subject position we obtain =agt has brave. Now what about *brave?* Substituting the definition will er_ fear yields 'subject has greater willpower than fear' and it is by yet another substitution, that *fear* is not just any old sensation, but sensation, danger cause_, <anxiety> that we finally de-

brave
fear

rive the conclusion that the object of *dare* is indeed dangerous, with the danger being the same instance that appears on the lexical prespecification of the object.

As in many parts of the lexicon, there can be serious disagreement over how much of this is precomputed and already stored in the lexicon, see e.g. Pinker and Prince (1988), and defending one analysis over the other would take as far afield from the central claim, that this weak (proto)logic calculus is sufficient for deriving the meaning of the whole 7.4 from the meanings of its parts, and further, that the semantics directly accounts not just for the strict meaning that John chewed gum, but also for the 'pragmatic' portion that this was dangerous/took courage.

7.4 Spreading activation

Implicatives offer a rich area for comparing various approaches to semantics. In the formulaic approach, providing for implication is easy, but creating the right meaning postulates is relatively hard, especially if the intention is to go beyond a single fragment, and to create representations that are (re)usable everywhere. In a standard system of logic such as first order predicate calculus (FOPC) we would need at least two variables x and y, a one-place predicate *force* and seven two-place predicates *SubjectOf, ObjectOf, Before, After, IsA, Has,* and $>$, to express the meaning of force-dynamic *overcome* in a conjunctive formula:

(2) *Before((x IsA force & SubjectOf(x,overcome) & y IsA force & ObjectOf(y,overcome) & y > x),overcome) & After((x IsA force & SubjectOf(x,overcome) & y IsA force & ObjectOf(y,overcome) & x > y),overcome)*

even with the dubious expedient of reusing x and y in the *Before* and *After* subformulas. To simplify matters, notice that the variable x is just a paraphrase 'the force of the subject of overcome' and similarly y is 'the force of the object of overcome', so in order to put the formulaic system together what we need to handle first are binary relations that have a first and a second argument.

In 4lang there are only a few intrinsically binary relations (16 out of 775, or about 2%) and for the technical reasons discussed in 2.3, we need to model these as matrices. This is one place where 4lang perceptibly differs from more standard conceptions of vector semantics, which offer perfectly reasonable algorithms to assign just a vector to every word, including to intrinsic binaries like has. But the totally homogeneous treatment of all words as belonging in the same semantic type 'vector' is actually highly problematic, as it makes such models (static and dynamic embeddings) very hard to interpret, and renders the search for declarative knowledge very hard.

has   Consider the possessive relation, given simply as *has* =agt control =pat, =agt has =pat. As the circularity of this definition makes clear, we have no alternative but to treat has as a primitive. In the 4lang lexicon, almost a third of the entries refers to this element, in clauses like state has government, sheep has wool, etc., and in LDOCE over 10% of the senses contain *has, have* or *had*. In the approach taken here, if M_{has} is the matrix in question, we have several hundred equations of the form $\mathbf{y} = M\mathbf{x}$ and M_{has} is one of the solutions (it is underdetermined which one).

Many (indeed, most) of our binaries express spatiotemporal relations, and again all indicia point to the conclusion that these are primitives, definable only in terms of a language-independent body schema (3.1). We start with a primitive relation `under` that obtains, by definition, between anything below the `ground` and the figure (ego) of Fig. 3.1. Everything else gets to be *under* something else by being coerced into the basic schema that has the prototypical instance. This means that we need to do a great deal more than simply finding *x under y* trigrams (or broader contexts containing these) to fully interpret this term: we need to parse the text, obtain abstract semantic representations (typically vectors) and use these to solve for the matrix corresponding to *under*.

It should be clear from 7.3 that parsing in algebraic semantics is relatively easy: after some morphological analysis, we just look up the words in the lexicon and apply spreading activation and unification algorithms in the same manner as we analyzed *colorless green ideas sleep furiously* in 6.4. This is the approach taken e.g. in Unification Grammar (Shieber, 1986); HPSG (Pollard and Sag, 1987); and more recently in Extended Dependency Unification Grammar (EDUG) (Hellwig, 1993); and Abstract Meaning Representation (AMR) (Banarescu et al., 2013). Broadly speaking, all 'algebraic' approaches (among which we count not just the classic AI models originating with Quillian (1967), but also Pāṇini) and Generative Semantics proceed from meaning representation to surface form directly, without any reliance on Logical Form, and all view interpretation as the inverse task, analysis by synthesis.

The natural domain of such parsing and generation is (hyper)graph manipulation. The linearization of meanings as formulas is a relatively new development: Frege (1879) actually used 2D notation, as does Generative Semantics, which used tree structures on the semantic side, and almost all theories listed above. The exception is Pāṇini, whose notational conventions were geared toward the pronuncability of the sūtras: linear, but with special indicatory letters (it, anubandha) interspersed.

Starting perhaps with Yngve (1961), linguists have long wrestled with assessing the impact on sentence processing of the limitations of short-term or working memory (Miller, 1956). The bulk of this work concerns syntax and takes it for granted that the central issue is dealing with the linear succession of words. Island parsing techniques, based on the idea that a full parse may be built from well-understood subgrammars, came two decades later (Carroll, 1983), and it was only under the impact of Ken Church's famous declaration, *parsers don't work,* that interest in partial parses, such as offered by light parsing (Abney, 1991), was beginning to be seen as legitimate.

If our focus is with semantics, the defining data structure of sequential processing, the tapes common to FSA, FSTs, and Turing machines, appear neither relevant nor particularly useful. Clearly, humans have huge long-term memory, but there is no reason whatsoever to suppose that this memory is sequentially organized outside of procedural/episodic memory. In particular, the bulk of linguistic information is stored in the lexicon, a device that is best thought of as random access. The classic model of graph transduction best suited to random access of this sort is the Kolmogorov Б-complex (КБС), originating with Kolmogorov (1953) – for a more accessible English-language introduction see Ch. 1 of Uspensky and Semenov (1993). There are more modern concepts, in particular the Storage Modification Machine of Schönhage (1980), the Pointer Machine of Shvachko (1991), and the Random Access Computer of Angluin and Valiant (1979) – for a good discussion, see Gurevich (1988). We are in no position to make a compelling case for one over the other, but the key issue, as emphasized by Gurevich, is that all these models are "more appropriate for lower time complexities like real time or linear time" than the standard Turing Machine. We are, in the relevant sense, at the (sub)regular portion of the Chomsky hierarchy, since it is evident that humans perform both parsing and generation near-real time.

Theories of 'algebraic conceptual representation' Kornai and Kracht (2015) share much of the tools, concepts, and formal underpinnings of 4lang because they take all lexical entries, and the knowledge representations, to be (hyper)graphs, just as the КБС family of models. But vector semantics demands a new set of parsing and generation techniques. These can be *direct*, assigning vector output to each sentence input, or *indirect*, proceeding first to create a hypergraph, and computing the vector based on this. A simple, and broadly used, direct method is simply ignoring all sentence structure and adding the word vectors together. Here we will discuss the indirect method, not so much as a suggestion about the architecture of the grammar but rather as a means of filling the vacuum left by the deprecation of the Eilenberg machines used in S19. We emphasize at the outset that there is no theoretical claim of 'psychological reality' attached to the hypergraphs we use, nor is there a practical claim that indirect methods will turn out to be the best possible way to organize computation. Hypergraphs are not seen as Logical Form any more than binary strings are 'logical form' for numbers. They are merely a perspicuous shorthand, and receive their only justification in the economy of the rules that can be stated with their use.

In earlier chapters we have already covered the main ingredients of the functorial parallelism between hypergraph structure and polytope structure. The typical atoms are word- or morpheme meanings, which correspond to polytopes given by the intersection of a few half-spaces (in a sparse overcomplete basis). Occasionally the intersections involve other, temporally shifted versions of the basis, so their projection on the current instant *now* may appear non-intersective, but these are also intersective, one just to have to view them from the broader basis to see this.

A 0-link (is_a) between A and B corresponds to containment of polytope structures, see Eq. 1.1. A 1-link (subject), denoted B(A) or A[B] does the same, forcing containment of the subject in the predicate, see Eq. 2.8. A 2-link (object) merely enlarges the active structure, forming the union (disjunction) of the two components with the head (typically a verb or a preposition) staying in head position, so that *eat fish* is_a *eat* and *on the hill* is_a *on*, see Eq. 2.9. Coercion is creating not just subsumption, but equality between two terms A and B, see 3.3.

The key, most compelling notion these theories have since Quillian is *spreading activation*. This is island parsing writ large, beginning with nouns, named entities, NPs and PPs, detection of case marking, assembly of clausal structure, and verbal slot filling. At every stage, morphemes, words, or larger lexical entries are active, and by spreading activation so are their links. A structure is detected whenever two such spreading waves of activation meet. Pragmatics, in the sense relevant to our understanding of *dare* and other implicatives, is simply an effort to find paths where none initially exist. There is clearly no link, at least initially, from *chewing gum* to `danger`. But the post-verbal position really compels a reading whereby *chewing gum* is the object of *dare*, so we make the link, and now *chewing gum* is risky. Speakers are of course very aware that such sense-making activity is under way. They use it to eliminate redundancy, and abuse it to set up semantic traps like *when did you stop beating your wife*.

Geometric approaches require extra work to get to spreading activation, even though the very notion was inspired by neural nets in the first place. As we emphasized from the beginning, Rule 1 (1.6) is *not* a rule of the grammar. Rather, it is our way of implementing spreading activation both in algebraic and in geometric semantics. How far the activation is spread (how many substitutions are made) is obviously related to short-term memory limitations: we, as humans, can only keep so many balls in the air simultaneously. In fact, the motivating experiments of the spreading activation model established the fact that humans respond faster to specific questions like *Is a robin a bird?* than to more general ones *Is a robin an animal?* (Collins and Quillian, 1969). In `4lang` terms, there are 0-links from `robin` to `bird`, and from `bird` to `animal`, and it simply takes more time to traverse two links than one.

At the lexical level we do consider the entries to have psychological reality, but to establish the micro-mechanism of spreading and adjustment of thought vectors will require a great deal more psycholinguistic experimentation before we could claim reality for these. What seems clear is that our linkers =agt and =pat are equalizers, with coercive effect. When we say *John sleeps* we are committing not just to the idea that the small polytope corresponding to *John* (see 8.1) is located inside the half-space corresponding to *sleep*, but (by implication) also to the idea that there is nobody else participating in this particular sleeping event. This can be relevant in establishing the proper implications, particularly in cases like Hungarian, where focus positively carries the meaning `lack other` (Szabolcsi (1981) and Onea (2009) are good entry points into the voluminous literature on the subject).

In the theory developed here, the appropriate data structure is not just a thought vector $\Psi(t)$, but also a dynamically updated transition matrix $P(t)$ that combines background knowledge and current context (see 2.4 for notation). With a theory of lifelong learning one could assume that P is built from immediate contexts incrementally, taking everything into account that went on before. Our view is that the lifelong mechanism must be far more complex, because we are perfectly capable of sentence understanding within fiction, where much of the culturally shared background knowledge must be disabled for the story to make sense. Rather, we follow the same adiabatic approximation as in earlier chapters, and have little to say about learning beyond learning the lexical entries themselves (see 5.3).

To put the adiabatic hypothesis in vector terms, we assume the basis, and thus the entire linguistic subspace L, to be fixed for the duration of parsing. The transition matrix P is altered every time we encounter a new predication or even sub-predication such as a relative clause or a selectional restriction. The key driver of changes in P is the coercion process, which in turn is governed by =agt and =pat, very much including prepositional subjects and objects.

Let us see how our starting example, *It can hardly be disputed that X*, works out under this analysis. According to LDOCE, *hardly* is 'used to mean 'not', when you are suggesting that the person you are speaking to will agree with you'. This gives us two propositions: *(speaker says) X cannot be disputed* and *speaker expects hearer to agree*. Who is the subject that 'cannot be disputed' declares to be incapable of dispute? The speaker expects hearer to fill the slot, and under the default unification mechanism we discussed for the 'kingpin' example in 7.2 this would follow. But again a different interpretation, whereby it is the speaker who is incapable of disputing X is also possible. Doubt as to the plain interpretation, triggered by this and similar rhetorical phrases like *we all know that* or *as we all know*, etc. leaves ample room for this second interpretation, and to choose between the speaker-intended and the skeptical reading we would have to inspect the mind-state of the speaker.

To see how island parsing can provide locality constraints on spreading, let us return to our earlier example, *colorless green ideas sleep furiously* (see 6.4). By definition, 'tactics' regulate the linear succession of elements. To do syntax (or morphotactics, which will be handled by the same tools) we therefore need some notion of a *pattern* that imposes linearity constraints (immediate precedence). A well-known example would be the standard English SVO pattern, which we give as =agt V =pat and the 'nmod' pattern A N. Both assume linearity, i.e. that in SVO the subject will precede the verb, and the verb will precede the object, and that in English noun modification the adjective precedes the noun.

We make no apologies for using lexical categories in the specification of patterns, since these (together with inflectional morphology) are essential for the economical statement of syntactic regularities, as we argued in 7.2 above. But we have to apologize for not stating the patterns fully, e.g. that in the English SVO (and intransitive SV) patterns the subject agrees with the predicate in number and person, and that other critical phrase-level features such as bar level (Harris, 1951; Jackendoff, 1977) will also be ignored. To include these would cause no technical difficulty, but would require more complex graphs (feature structures in the GPSG/HPSG tradition) and would just complicate the presentation.

In the 'radical lexicalist' paradigm (Karttunen, 1989), tactic patters are also lexical entries, though considerably less contentful than the lexemes we investigated so far. The first islands are built on the morphologically preferred *color+less* and *furious+ly*. We can try to match the nmod pattern A N to *green ideas*, to *ideas sleep*, and to *sleep furiously*. The standard approach to narrow this down is to invoke the lexical category of the elements. Though `4lang`, being a universal theory, lacks the resources for this (see 2.1), the English binding will contain a better developed system of English lexical categories, very much including the fact that *-ly* is a deadjectival adverb-forming suffix, and *-less* is a denominal adjective-forming suffix. Since *colorless* is an adjective, it can match the first member of the A N pattern, while *furiously*, being D, cannot match the second member.

At this point, we have three active 'nmod' candidates: *colorless green*, *green ideas*, and *ideas sleep*, each carrying the category N or N̄. The first two could in fact be brought together by repeat application of the nmod pattern, giving us *colorless green ideas* as N or N̄. We could also coerce *ideas sleep* to this pattern the way we deal with *beauty sleep*, but the presence of the adverbial *furiously* calls for a sentence (or verb) modification pattern S D or V D. Spreading activates both, so we must search for something that would fit these. Certainly, *sleep* is trivially matched on V, so we can join *sleep furiously* together in a V or V̄ island. At this point we have N V (we ignore the proper assignment of bars) which actually corresponds to a major pattern of English syntax, the intransitive 'SV' pattern we represent as =agt V. This provides a strong locality constraint, the subject must precede the verb, and absorbs both islands we now have. Subject-predicate agreement (which we haven't formalized here though it is obviously part of the SV pattern) also works, since *ideas* inherits the plural from *-s*, and this is compatible with the zero person marking on *sleep*.

What about extramorphological formatives acting as separate words or clitics? A good example is the English subordinating particle *that*, which can act as a device for activating either the subject of the subordinate clause *the flood that engulfed me* or its object *the vote that I cast*. The `4lang` semantics of *that* thing, is rather thin, in fact `that` it is little more than a shorthand for N or N̄ or even NP. The pattern (construction) that contains *that* is reasonably simple: N that =agt V =pat. In the first example, the N to the left of `that` is trivially identified as *the flood*, the V as *engulfed* and =pat as *me*, leaving the =agt to unify with the N. In the second example, *the vote* is trivially identified as N, *I* as =agt, and *cast* as V, leaving =pat unify with N.

To summarize, we need only one operation, spreading activation, to handle all forms of sensemaking as long as we have a low-level unification primitive that enforces well-formedness at all times, somewhat analogous to autosegmental spreading, resyllabification, and similar processes maintaining phonological well-formedness. Under this view syntax, much like morphology, is about matching patterns specified in lexical entries (constructions). The conceptual similarity of this view to classical (Lambek-style) Categorial Grammar and modern Combinatory Categorial Grammar is evident, as is the relation to the more lexically inspired Berkeley Construction Grammar.

Where does this mechanism leave us in regards to the autonomy of syntax issue (2.5)? Since we rely on lexical categories, and these are not universal, the strongest hypothesis we could make is that the syntax of individual languages is autonomous. Since this is not at all supported by recent fMRI studies such as Fedorenko et al., 2020, we are left with formal universals pertaining to the makeup of the patterns (linearized (hyper)graphs) and their substantive elements (nodes and edges). On this level `4lang` is fully universal, with huge scope for cross-linguistic variation in the form of language-specific formatives, categories, and constructions.

8

Trainability and real-world knowledge

Contents

Until this point, we concentrated on the lexicon, conceived of as the repository of shared linguistic information. In 8.1 we take on the problem of integrating real-world knowledge, nowadays typically stored in knowledge graphs as billions of RDF triples, and linguistic knowledge, stored in a much smaller dictionary, typically compressible to a few megabytes. We present proper names as point vectors (rather than the polytopes we use for common nouns and most other lexical entries), and introduce the notion of *content continuations*, algorithms that extend the lexical entries to more detailed hypergraphs that can refer to technical nodes, such as Date, FloatingPointNumber, or Obligation (see 9.1) that are missing from the core lexicon.

In classical Information Extraction, our goal is to abstract the triples from running text, and a great deal of effort is directed toward *database population*, finding new edges in a knowledge graph. After a brief discussion of this task, in 8.2 we deal with the inverse problem: given that we already have real-world knowledge, in fact orders of magnitude more than we have lexical knowledge, how can we bring it to bear on the acquisition problem? As we shell see, there are some striking successes involving not just one-shot but also zero-shot learning that rely on dynamic embeddings instrumentally.

In 8.3 we turn to dynamic embeddings. We briefly outline the four major ideas that, taken together, define the modern, dynamic embeddings: the use of *vectors*, the use of *subword units*, the use of *neural networks*, and the use of *attention*, linking the latter to the idea of the representation space we introduced in 2.3. We propose a semi-dynamic embedding, DilBERT, which occupies a middle ground between fully static and fully dynamic embeddings, and enables careful study of the representations learned while sacrificing very little of the proven usefulness of dynamic embeddings.

8.1 Proper names

Psychology traditionally divides declarative memory into *episodic* and *semantic* components (Tulving, 1972). Since episodic memory stores individual events and impressions that are autobiographical, episodes are not expected to be part of the shared cultural heritage. This is not to say that they are not expressible in language, very often they are, but they are rich in non-linguistic qualia and it is only the projection to the linguistic subspace L of the episodic thought vectors that are captured in vector semantics.

More pertinent to our concerns is *semantic memory*, the contents of which we called *rules* in S19:3.2-5. These are regularities ranging from general rules like *when it is cold, water freezes* or 6.7 `hurt cause_ anger` to highly specific (singular) rules such as *Shakespeare married Anne Hathaway in 1582*. There is no clear-cut boundary between the general and the specific, but there are some broad characteristics signaling high specificity such as the use of proper names and explicit temporal marking.

The situation is further complicated by what is known as *stage-level* v. *individual-level* predication since (Carlson, 1977). In one case, e.g. *operators are standing by* we are reporting something that is true of some group of operators at this specific time, while in the other, e.g. *operators are underpaid* we are reporting on a general property associated with being an operator. We can find the same distinction when the subject is a proper name: *John is at home* refers to a particular temporal stage of John, while *John is a coward* refers to John the individual. The standard solution (Kratzer, 1995) to the problem of making the stage/individual level distinction is to assume that it is lodged in the predicate by means of it having (resp. lacking) a temporal argument slot.

Temporal modality is often relevant for restricting the generality of some statement, but individuals are divisible not just into temporal stages but also as being composed of behaviors: *John is a coward when it comes to shorting stocks*. Other modalities can play the same role: *operators are underpaid as long as they don't unionize* is about the default state (see 6.4). In both of these examples, the syntax indicates the modal restriction by means of conventionalized pseudotemporal clauses *when it comes to, as long as*, but this is not necessary: John is a coward *in regards to* shorting or *about* shorting would work just as well, operators are underpaid *because* they don't unionize, operators must unionize lest they remain underpaid, and so on.

For this reason, we will avoid speaking of stages, and speak of *modal qualifications*. Everything is general, unless it gets specified further. Even the example we started out with about Shakespeare's marriage is quite a bit less specific than *Shakespeare married Anne Hathaway on November 27, 1582*. Obviously the wedding didn't take a full year, so the original sentence displays an epistemic limitation as to the precise date. Kratzer (1995) already calls attention to the effect that background knowledge has on the stage/individual distinction:

> If I dyed my hair every other day, my property of having brown hair would be stage-level. Usually we think of having brown hair as an individual-level property, though, since we don't think of persons dying their hair capriciously.

Here we argue that in natural language it is next to impossible to make specific statements, further qualifications can always be added. This is true even of natural language paraphrases of scientific knowledge. The square of the hypotenuse of a right triangle will equal the sum of the squares of the other sides *in Euclidean geometry*. Water boils at hundred degrees centigrade *at standard atmospheric pressure*. Most people are aware of the italicized qualifiers after high school, but few will know that the boiling point of water, with pressure kept constant, will decrease after magnetization (Wang, Wei, and Li, 2018).

Natural language offers a rich set of devices to denote temporal effects, but no similar mechanisms exist for signaling pressure- or magnetism-driven effects. In analyzing generality versus specificity or, what is the same, invariance under changes in conditions versus dependence on conditions, the stage/individual distinction is special only because language (or, at any rate, standard theories of language following in the footsteps of Davidson, 1967) offers a filler/slot mechanism for handling it. A direct extension of Kratzer's theory to cases like property bundles would require slots for stock trading, union organizing, and whatever activity is involved in the qualification. Only an indirect extension, whereby the qualifiers are coerced into the temporal slot, makes sense, and this much is clearly supported by the superficially temporal nature of expressions like *when it comes to*.

Whether the temporal slot attaches to the subject, turning it into a stage, as Carlson supposes; or to the predicate, making it stage-level, as Kratzer argues, is a question that is, perhaps, easier to settle in vector semantics than in the logic-based approach. Consider John, the cowardly investor. Since the predicate 'being an investor' obviously has no direct slot for bravery/cowardice and, being stative, has no temporal slot either, we must follow Carlson and assume that cowardice is predicated of John the investor, what we can call the investor *facet* of John (to avoid the temporal association that *stage* brings). Facets have clear intersective semantics, so sentences like *As a boss John is very forgiving, but as a father he is very strict* give rise to no interpretative challenges.

Earlier (see 4.5) we found it advantageous not to make a category distinction between classes (typically common nouns, such as *poet*), and their instances, such as *Allen Ginsberg*. Qualified versions inherit this: compare *as a teenager, Ginsberg wrote letters to the New York Times* to *as teenagers, poets often write letters to the press*. Clearly, common nouns can have stages, and in fact we have lexical entries that refer to such, e.g. *juvenilia* are 'compositions produced in the artist's or author's youth' (Merriam-Webster). Since the use of a common noun already implies a certain amount of generality, while the use of a proper name endows the thing named with a great deal of specificity, it is tempting to view these as endpoints of a single scale.

Remarkably, the standard theory (Kripke, 1972) considers proper names to be maximally general, designating elements that are invariant even under choice of possible world, except perhaps temporal (stage level) changes. As we noted in S19:3.8, this view is hard to reconcile with the fact that proper names can be adjectivally modified:

(direct reference predicts) that the word *Shakespeare* refers to the unique histor-ical person William Shakespeare. But if this is so, who is *the Polish Shakespeare* that the *"Looking for the Polish Shakespeare" Contest for Young Playwrights* wants to find? Clearly, not some British subject born in Stratford-upon-Avon but a brilliant playwright who is a Polish national.

Such examples, virtually impossible to handle in mainstream formulaic semantics, are trivial from the geometric standpoint. The geometrical picture of proper name is that of a single vector, a point in the semantic space L. There is a strong intuition, first articu-lated by Russell, 1905, that the same can be said of definite descriptions, noun phrases that uniquely determine their denotation, but we treat the two as different: definite de-scriptions are polytopes, and even if very small, they cannot be treated as a single point without losing an important conceptual distinction. The *Polish Shakespeare* is simply the projection of the *Shakespeare* vector $\mathbf{S}$ on the *Polish* half-space $\mathbf{v}^{\perp}$(Polish). This works, even though the original $\mathbf{S}$ had no Polish component, since projection guarantees the Polishness of the result. In fact, the result is the one that preserves the original $\mathbf{S}$ maxi-mally among the candidates. When the adjectival property is met by the original vector, as would be the case with *the brilliant Shakespeare*, the result of the operation is just the original $\mathbf{S}$.

There appears to be something of a contradiction between the formulaic view, which takes proper names to be highly general (invariant under all context changes except time), and the geometric view, which takes them to be the most specific (least subsump-tive) of regions, a single point. Kripke, 1972 articulated the formulaic view in modal terms: proper names are rigid designators that keep their meaning across possible worlds, whereas definite descriptions like *the current president of the EU* may refer to unique individuals, but not necessarily the same unique individual. Different designations are possible not just across the temporal modality, but also across possible election out-comes.

In contrast to this, the naive theory (see 2.5) considers *naming* an act of free will (every point can be given a proper name), which connects a name not just to something but to the essence of that thing. Under this 'magical' theory, things are composed of essence and appearance, and usage referring to the latter is 'in name only'. Dictionary definitions always strive at capturing the essence, but 4lang undertakes the task only for common nouns, leaving proper names to the encyclopedia (1.2). It is, indeed, quite questionable what constitutes the essence of a proper name. When we hear the word Jena, we know that this is a medium-size city in Germany, that an important battle of the Napoleonic wars took place there, that the optical manufacturer Zeiss is there, and so on. But which of these facts are essential to an understanding of the word *Jena*?

Since our theory of modality is far less sophisticated than that of contemporary model logic (in the large deontic world of 6.2 we may have only 2-3 modal alternatives, depend-ing on our theory of time, see 3.2), our proper names will not be fully rigid in the sense Kripke intended. In S19 Ex. 5.10, repeated below, we left it to the reader to develop their own theory of the matter, here we start with a specific solution closely tied to the geometric picture.

> What are the truth conditions of *the river Garonne*? Does it, or does it not, mean the same as *the Garonne river*? In what sense are *Garonne* and *Garumna* identical, especially if we agree with Heraclitus (DK 41) that we cannot step into the same river twice? Is *the summer Garonne* the same as *the winter Garonne*? Is *the languid Amazon* the same as *the cruel Amazon*? (S19:Ex° 5.10)

Starting at the end, an expression like *the cruel Amazon* asserts that the vector $\mathbf{v}$(Amazon) is in the positive halfspace limited by the affine hyperplane $\mathbf{v}^{\perp}$(cruel). Whether this assertion is direct, as Russell (1905) would have it, or indirect, presuppositional (Frege, 1892), is from the geometrical point of view a distinction without a difference: in both cases we have a point in a halfspace by the same mechanism of Eq. 2.8 that we use for intransitive predication everywhere.

This makes clear that *the summer Garonne* is the same as the *the winter Garonne* in the sense of occupying the same point, but not in the algebraic sense, since it occupies the same point in a different structure. For a simple example, consider the element 2 in Z_7 and in Z_8. It is the "same" element, defined as $1 + 1$ in both rings, but it is a quadratic residue mod 7, and a quadratic non-residue mod 8. This is exactly like the winter Garonne, which we assume to be cold, and on rare occasions (as in 1956 and 1985) actually frozen, as opposed to the summer Garonne, which never freezes.

The first part of the problem is much easier: we treat *Garonne* and *Garumna* identical as per our lexicographic principle of Universality (see 1.2), and of course we treat the expressions *the river Garonne* and *the Garonne river* as meaning the same thing. To fully reconstruct the naive theory that assigns the actual river as the reference to these expressions would take us far afield, as this requires *grounding* points of the vector space in real-world objects (see S19:3.7). For our purposes it is sufficient to trace the meaning of linguistic expressions to vectors and matrices. This way, we can remain silent on the usual bestiary of strange objects that are the center of attention in philosophical logic: those that have no denotation such as *the present king of France, the gold mountain, the method of squaring the circle*, those that are the referentially the same but are spoken of differently such as the *Morning star* and the *Evening star*, and so on.

We assume that the central object of study in semantics are the thoughts in the head, and it is these that we wish to capture in thought vectors (see 2.3). The theory of vector semantics that is the subject of this book also includes matrices and other ancillary constructs from linear algebra, but not the standard model theoretic constructs like ultraproducts (for which see S19:2.3). To formalize what happens beyond these stages, how the vectors or model elements are mapped on real-world objects would require a formal theory of the codomain, something we obviously don't have.

Of particular interest in this regard are direct deictic gestures: clearly, by pointing from some distance at the Garonne the speaker can disambiguate the subject pronoun in *This is the Garonne*, but the exact same pointing gesture can be accompanied by saying *cold* or *steamboat* in which case the (implicit or explicit) pronoun will be resolved differently. There is something of an implicit visual ontology which treats homogeneous patches of an image as objects, but this does not appear to be human-specific. Indeed, computational vision systems such as YOLO9000 (Redmon et al., 2016) have great success in identifying objects in pictures, and even actions such as 'jumping' or 'running' Karpathy and Li (2014) which are not at all object-like.

Instead of developing a fuller (naive) theory of geographical features, we will sketch a less ambitious theory of *content continuations* which connect the lexical entry *Garonne* 'a river in SW France, rising in the central Pyrenees in Spain and flowing northeast then northwest into the Gironde estuary. Length: 580 km (360 miles)' (CED) to different knowledge bases. As this and the earlier examples already show, knowledge about proper names is predominantly relational, with simple predicates connecting proper names to one another *Shakespeare married Hathaway, the Garonne starts at the Pyrenees*, or more rarely, a proper name to a common noun, adjective, or verb.

Since the 1990s, a great deal of computational linguistic effort focused on the detection of proper names (called named entities since Grishman and Sundheim, 1996), in particular names of PERsons, ORGanizations, and LOCations. In theoretical linguistics a distinction is often made between proper names such as *Shakespeare* and proper name phrases such as *William Shakespeare*, but in computational linguistics, the Named Entity Recognition task by definition includes the phrases as well. For locations, the average phrase contains about 1.3 word tokens (counting *Georgetown, Guyana* as three words because the comma token gets included in the phrase). In early work, NER was viewed as being composed of two tasks: *entity segmentation*, finding the boundaries in the text, and *entity classification*, deciding on the type of the segment found (Collins and Singer, 1999). Remarkably, systems that are specifically designed for segmentation (as opposed to classification) have found their way into contemporary neural systems (Xiao et al., 2019) in spite of the overarching end-to-end design philosophy where division into subtasks is generally rejected by a near-religious intensity. Early on, the extraction of *numerical expressions* (NUMEX) such as we discussed in 3.4 was often lumped together with the extraction of named entities (ENAMEX) but, as we shall see, the lexicon can be made to carry named entity information far more easily than it could be made to carry arithmetic.

Given a common noun such as *city*, which is represented by a polytope, actual cities form a point cloud inside this polytope, but without filling it: there is plenty of room for imaginary/nonexistent cities. Dictionaries will list a handful of cities, e.g. (Guralnik, 1958) lists Hyderabad, but not Chennai, while purpose-built gazetteer databases list thousands, or even tens of thousands. Exactly how many is depending on the definition of *city* which normally involves various thresholds for size, population density, available services, etc. The problem of arbitrariness in definitions is not restricted to nouns: in S19 3.7 we wrote "What makes a person *obese*? Insurance companies might agree that payment for medical treatment may be justified if and only if the weight of a person (expressed in pounds) divided by the square of their height (expressed in inches) exceeds

0.04267, but this is hardly a definition of obesity that makes sense outside a very limited healthcare context, and even there its applicability is dubious, as it is easy to imagine some committee of learned doctors and actuaries moving the threshold to 0.0393. We will favor a definitional style where *obese* is defined as 'very fat, overweight' in accordance with the everyday meaning."

The `4lang` definitions of `city` as `town, buy in, society in` and `town` as `artifact, many(people) in, many(house) in` lack this kind of artificially imposed specificity, and very much leave open the possibility that different people will have different opinions on whether a particular place is a city or not. In fact, the discrete scale offered by the dictionary, defining *metropolis* as 'a very large city'; *city* as 'a large town'; *town* as 'smaller than a city and larger than a village'; *village* as 'a very small town'; and *hamlet* as 'a very small village' (all definitions from Procter, 1978) provides us with a five-point scale that is very similar to the generic very large > large > medium > small > very small scale we discussed in Chapter 7. In practice, detailed gazetteers have largely given up on this level of subcategorization, using PopulatedPlace for all, given the cultural relativity of these terms: what is a small town in China may be considered a sizeable city in Rwanda, even though population density is much higher in Rwanda than in China.

Looking at dictionaries it is evident that there are other, similarly rich sources for proper names: rivers, mountains, famous people, trademarks, institutions/organizations, first names, and so on. There are two main drivers of the mechanism whereby `4lang` can add such entries: indicative lists and content continuation. *Indicative lists*, such as found in the definition `color sensation, light/739, red is_a, green is_a, blue is_a`, simply list conjoined is_a clauses. These are merely indicative: there is no claim that green, red, and blue are the *only* colors, or the only colors worth listing in the lexicon, but the explanatory function (for human readability) is clear. The device is used sparingly in `4lang` – less than 2% of the definitions rely on it, with the longest lists devoted to *season* `season/548 period[<four>], part_of year, spring/2318 is_a, summer is_a, fall/1883 is_a, winter is_a, has weather;` *furniture*; and *emotion*. In principle, indicative lists could be eliminated entirely, since the relevant entries for the seasons all begin with `season/548`, the relevant entries for furniture (`bed, chair, cupboard, table`) all begin with `furniture`, and so on.

As is typical of taxonomies, we often find abstract group terms in higher positions of the hierarchy: for example we may define `geographic_feature` as `feature, geographic, city is_a, river is_a, mountain is_a, country is_a, ocean is_a, lake is_a`. Are the names of waterfalls our prominent rock formations part of this list? This depends on the particular gazetteer we wish to link to, and it will be the task of the interpreter, a custom-built piece of software, to fit the precise list to the preexisting structure of the knowledge base in question. Different knowledge bases will require building different interpreters. The key observation here is that none of the domain-specific (generally numerical) knowledge, such as the longi-

city
town

color

season

tude and latitude of a geographic feature, will actually belong in the dictionary: for this we require external pointers (see 1.3) that appear in addition to the definition head that provides the genus.

For other classes of proper names we may be able to encode a great deal more non-numerical information: for example a trademark such as *Fanta* designates a product from a given class 'soft drink' and is owned by an organization, in this case, the Coca Cola Company. These statements are naturally expressed as `Fanta is_a trademark`, `Fanta is_a soft_drink`, `Coca_Cola_Co. has trademark` using the same hypergraph structure (see 1.5) that we use to describe lexical entries. If we are prepared to add numerical types such as Integer, Float, ... and semi-numerical types such as Date as common nouns (this was done routinely and without much reflection in early Knowledge Representation work), we may even capture in hypergraph format further fields in the database record such as FilingDate or RegistrationNumber.

The overlay between the lexical (hypergraph) and the knowledge base (record) structure is sufficiently similar to analytic continuation that we will talk about *content continuation* among domains. This is the second, and in a sense more complex method for integrating the lexicon with an encyclopedic database. Indicative lists are suitable when we wish to incorporate a few dozen records from a database, but when we have thousands of database records, there is no alternative to writing a content continuation interpreter that converts the records to hypergraphs. These are typically formatted as attribute-value lists, often in standardized syntax as offered by the World Wide Web Consortium's Resource Description Framework or Wikipedia's infoboxes.

Just as in the case of numbers and counting we discussed in 3.4, a full account will have to invoke some external theory, analogous to the equation solver used there. For the geographic case, the solver will have to consolidate knowledge about the World Geodetic System and the methods whereby the positions of geographic features are encoded (e.g. as polygons, bounding boxes, or centerpoints). Without such a solver, it will be impossible to disambiguate e.g. between the two cities named Hyderabad in India and Pakistan. However, the use of a strict system of geographic coordinates is not always appropriate. The Getty Thesaurus of Geographic names, assembled for the broader purpose of identifying cultural artifacts, will often list the relevant cultural period, or religion. Perhaps more important from the linguistic standpoint, well-crafted named entity detectors will generally identify fictional heroes as PER, fictional places as LOC, and fictional organizations as ORG, a matter we shall return to in 8.2.

Needless to say, content continuation is a considerably weaker notion than analytic continuation, as there is nothing for dictionaries like the identity theorem that complex functions enjoy. In fact, it often takes a nontrivial amount of work to establish the identity of points across content continuations, especially if the names associated to the object have spelling variants. We still consider the Hyderabad of (Guralnik, 1958), 'a city in South central India, population 1,086,000' to be identical to (a stage of) the Hyderabad of Wikipedia, having "an average altitude of 542 metres" and "a population of 6.9 million residents within the city limits, and a population of 9.7 million residents in the metropolitan region" according to this much more detailed, and more up to date, data source. Further, both of these are identical (as lexical entries, or points in the *city* polytope) to the Hyderabad of (Gumma et al., 2011), so in this case the key Kripkean insight of rigid designation is preserved.

That the Polish Shakespeare is *not* identical to the better known **S** is guaranteed by the converse of Leibniz' Principle of Indiscernibles: since the two are separated by the hyperplane limiting the *Polish* half-space, they cannot be the same. Remarkably, we know this without knowing who actually won the contest, just as we know that none of the candidates for the description Venice of the North can be Venice.

In all continuation work, the central (lexical) domain is kept fixed, and the knowledge base containing the encyclopedic information is regarded the codomain. For each mapping there are three major error sources. When the entry exists in the lexicon but not in the database we speak of 'lack of coverage'. With the English Wikipedia now well over 6m entries this is rarely an issue. The converse, there being a target entry in the knowledge base that has no source in the lexicon is very common, and is likely to remain so, since back-filling such entries would very soon overwhelm the lexicon.

To get a sense of the size of the problem, about 112m of tokens (words and punctuation) of the BNC correspond to about 777k word types, of which 29.8% are proper names. Frequency weighted (not counting punctuation) this is only 1.4% of the entire corpus. Adding these proper names would increase the size of the OED by over a third and that of Webster's 3rd by over half, and a fuller encyclopedia may very well overwhelm the lexicon: the US Board of Geographic Names database alone contains over 7m geographic features and over 12m names for these. Standard lexicographic practice is well summarized on the OED webpage:

> Proper names are not systematically covered by the dictionary, though many are entered because the terms themselves are used in extended or allusive meanings, or because they are in some way culturally significant.

Finally, there is the error of linking the wrong knowledge to a lexical entry, something we could call 'continuation error'. This is significantly different from the error being propagated from the KB itself: the link from the lexical entry may correctly identify Anne Hathaway as Shakespeare's wife, but if the DB has the date of the marriage as October 1st, this does not make the continuation algorithm wrong.

It is not just the sheer bulk of specialized databases that makes it impractical to enlarge the lexicon by back-filling their entries, but also their structure. In the lexicon only hypergraphs are used, no nontrivial theory of implications is required, and only a barely discernible layer of proto-numbers and proto-measures like *a handful* are present. Knowledge bases of various sorts typically go well beyond these limitations, but not in a uniform fashion. What one DB regards a city another one may regard a town. History is full of entire cities being moved from one place to another: some databases will consider these stages of one and the same city, others will assume disjoint individuals. Part of our goal with `4lang` is to delineate the absolute minimum required for imposing a conceptual schema on lexical entries, and a bare-bones logic for inference. It is relatively easy to extend this in many ways, but to extend it uniformly so as to serve the needs of knowledge engineers operating over various domains of knowledge does not seem to be feasible.

Content continuation (interpreting lexical entries in databases) implements what Putnam, 1975 called the 'linguistic division of labor' between ordinary speakers and domain experts:

> Every linguistic community exemplifies the sort of division of linguistic labor just described, that is, possesses at least some terms whose associated "criteria" are known only to a subset of the speakers who acquire the terms, and whose use by the other speakers depends upon a structured cooperation between them and the speakers in the relevant subsets.

The actual difficulties of building such interpreters make clear that the process is far from smooth. Putnam grants that "the "average" speaker's individual psychological state certainly does not fix [the expert meaning], it is only the sociolinguistic state of the collective linguistic body to which the speaker belongs that fixes [it]". Since our interest is with the cognitive state of the individual, the thoughts in their head, it is worth looking in a bit more detail how knowledge flows from the experts to our ordinary speaker, Joe. Let us return to Gauss' Law of Magnetism, $\nabla \cdot \mathbf{B} = 0$. The best we can expect from Joe is to realize that this is a mathematical formula that says something is zero.

Turning to a mathematician will elicit detailed instructions prescribing a semester of vector calculus together with dire warnings that this will only make sense if Joe takes the standard Calculus 1 and Calculus 2 sequence first. The formula cannot be explained to Joe Layman – by the time he understands the expert meaning, he is one of the experts.

This is not to say that such formulas are ineffable. If anything, the opposite is true: they are clearly understandable to any student willing and able to learn. But the conceptual structures involved are only weakly tied to lexical facts, and practically not at all to the basic everyday ontology of Things, Events, Actions, States, Properties, Places, Paths, and Amounts that Jackendoff (1983) articulated (see 2.1 for a brief overview of this system, and (Dahlgren, 1995) for a more computational alternative). Vector calculus has its own ontology, quite distinct from the more cognitively or computationally motivated ontologies, and it is precisely the problem of aligning the different ontologies (Fossati et al., 2006) that makes the problem hard even when the knowledge contained in the DB is not in doubt. In this regard, Gauss' Law presents a simple problem, as it belongs to the undisputed part of classical physics, and has no alternative. Other specialist fields like law or economics have many rival theories, each with its own ontology. As there simply isn't a universally, or even largely consensually, accepted body of knowledge, set of inferential rules, or ontology, linguistic division of labor presents a far more thorny issue than Putnam's 'let's just ask the experts' approach would suggest.

8.2 Trainability

Clearly, much of our knowledge about the world is lodged in proper names. In the computational linguistics and information retrieval literature we find many systems directed at acquiring this knowledge from running text. Typical examples include (Ayadi et al., 2019) in the biomolecular and (Desprès et al., 2020) in the biomedical domain, but similar highly specialized systems exist for all kinds of tasks, e.g. for the extraction of biograpic information for famous painters. In fact, each domain ontology and each target language require a new system. We have already discussed the major steps, segmentation and classification, that are generally used in NER systems, but said very little about the computational methods in use. In fact all kind of methods, from the traditional list-based and rule-based to the more modern conditional random field (CRF) and long short-term memory (LSTM) techniques, which are more easily combined with word vectors, are still in use both as standalone systems and as part of larger systems like BERT (Devlin et al., 2019).

Perhaps more important than the methods themselves is the fact that computational linguistics has standardized on key figures of merit. Here we will use *precision*, the probability that an entry returned by the system is actually an instance of the category sought; *recall*, the proportion of instances actually found; and F-measure, the harmonic mean of precision and recall (see Jurafsky and Martin (2022) 4.7). Harmonic, rather than arithmetic mean is used because actual system strength shows hyperbolic, rather than linear, tradeoff between precision and recall.

In addition, for most problems there are *shared tasks* such as SemEval which offer standardized, human-verified 'gold' data to measure system performance. Initially, while systems don't perform too well, the value of these datasets is considerable, but as time goes on and the systems improve, the value of the original datasets is decreased by too much targeted optimization even if the authors working on these systems never cheat (use the test data for training). We will discuss one example from our own practice shortly, for a more systematic study see Manning, 2011.

All methods combine an encyclopedic listing (in our case, a gazetteer) and a pattern matching component, the latter close in style to the rudimentary syntax mechanism we discussed in 7.4. Just as Berko, 1958 could demonstrate the need for rules that are sensitive to the phonological class of the stem-final consonant by means of using nonsense words, we could demonstrate the power of context by considering sentences such as *A severe battle erupted between militants and Pakistani troops in Wana and Shakai that continued for several hours*. Few readers will have heard of *Wana* or *Shakai* before having seen this sentence, but most would assume that these are LOCs, likely in Pakistan or a neighboring country. Once they find that these places are in South Waziristan, they will infer, chiefly on the strength of the suffix *-istan*, rather than based on actual lexical knowledge, that *South Waziristan* is likely a province of Pakistan.

In earlier work (Kornai, 2006) we have constructed a small "Tier 1" gazetteer (2,171 entries) that contains only large cities, states, provinces, seas, oceans, and other entries found in any school atlas and expected to be known to any college-educated person. This can be taken as a generous upper bound on the lexical, and perhaps even the encyclo-pedic, geographic knowledge of the average speaker of English. A superset "Tier 2" list from a professionally edited gazetteer (66k entries) will cover many names that will be known only to inhabitants of the area and experts in geography. At that time we also constructed a large "Tier 3" gazetteer (5.1m entries) based on data from the USGS and the Board of Geographic Names. Today, a well curated public domain gazetteer is avail-able at https://www.geonames.org (11m entries) and for replicating some of the work presented here we suggest using this in place of the Tier 3 list. As our example above shows, we are reasonably certain of the geographic status of *Wana, Shakai*, and *South Waziristan*, though neither of these appear on Tier 1. Domain experts will know Wana and South Waziristan (Tier 2), but not necessarily Shakai, of which the Tier 3-level geon-ames server offers five different resolutions (four populated places and a named grave).

In fact, our primary interest is with the ability of the system to capture those proper names that are not on the list yet. These are called *out of vocabulary* (OOV) entries, and dealing with these is known as the OOV problem. As Chen and Lee, 2004 put it:

> Among all OOV words, named entities are one of the most important sorts. It is impossible to list them exhaustively in a lexicon. They are the most productive type of words. Nearly no simple or unified generation rules for them exist. Be-sides, they are usually keywords in documents. Named entity recognition thus becomes a major task to many natural language applications, such as natural language understanding, question answering, and information retrieval.

So far we illustrated the main points using LOC rather than PER or ORG data, but we believe our conclusions extend to these as well. Judging from the size of bibliography databases such as Marquis' Who is Who, there appear to be several million notable persons. This is far more than necessary for characterizing the lexical knowledge of speakers of English or any other language – the considerable body of empirical work surrounding Dunbar's number suggests that people don't maintain models for more than 500 'Tier 1' people in their lives, with Dunbar himself putting this number at 150. The actual number of people tracked in databases is much larger: many organizations will track their customers, and some will simply track everybody they can.

With the increased availability of open data, the primary target nowadays is the cus-tomer list or, in a more sinister fashion, everybody everywhere. We encourage the reader to try this at home, e.g. with Stanford's DeepDive. It should be emphasized that such an undertaking is by no means restricted to nation-level organizations and megacorpo-rations. At this point one can buy an 8TB disk in any computer store for less than $200, and this is sufficient for storing a megabyte of information for each and every living person on Earth, far more than the average Who's Who article (the compressed source of this book is about a quarter megabyte). In Kornai and Halácsy, 2008 we presented an algorithm, runnable on an ordinary laptop of desktop PC, that will fetch a third of a ter-abyte from the web per day. With the availability of Common Crawl, the task is further reduced to building an infrastructure such as described in Chapter 4 of Nemeskey, 2020 for filtering, deduplication, etc. Again we advise the reader to try this at home, if only to get a better understanding of the impact publicly available web pages have on personal privacy. Organizations are an even richer source of proper names: the widely used Dun and Bradstreet database contains over 225m businesses, and there will be many non-profits and non-governmental organizations not listed there. With hundreds of millions of organizations collecting data on billions of people, the number of datapoints, even the number of publicly available datapoints, is massive. Can this world knowledge help us understand language better?

Following the influential work of Brill, 1994, pattern lists with entries such as `stopped at X` or `traveled to X` are often used for adding entries to the gazetteer, and similar methods leveraging not just the context but also the internal syntax of named entities (e.g. `Mrs. X, The Right Honourable X, ...` for persons, `X Co.` or `Institute of X` for organizations) are well known and widely used. Word vectors offer a novel method for adding new entries, because named entities will concentrate in a small cone, within small cosine distance from the average named entity vector. The brown line on Figure 8.1 shows the expected proportion of vectors within a certain cosine distance from the center of gravity of the group for randomly chosen vectors: this is 70% for a half-space (a cone of rotational angle $90°$) because the word vectors themselves are non-randomly distributed (more coordinates have positive value than would be the case for truly random vectors). If we tighten the cone to a smaller rotational angle, $77°$, practically no random vector is that similar to the average, i.e. random word vectors are nearly orthogonal.

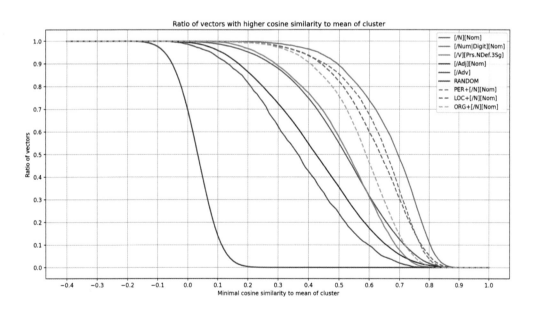

Fig. 8.1: Proportion of in-cone vectors with given cosine similarity to average

Major lexical categories already show more coherence than would follow from the non-random distribution of the vectors: among adjectives, 70% falls within a much tighter cone of $71°$, among nouns and verbs even tighter cones of $61°$ and $60°$ are sufficient. The most coherent lexical category is provided by numerals, requiring only a $46°$ cone to fit 70% of the vectors (Lévai and Kornai, 2019). As can be seen from the dashed curves in Figure 8.1, named entity vectors are almost as coherent as numerals with $58°, 56°$, and $54°$ for ORG, LOC, and PER respectively.

Perhaps unsurprisingly, static word vectors already provide strong hints as to the lexical category and subcategory of words. More striking are the results of Tshitoyan et al., 2019, who used word vectors to discover knowledge that is latent in a corpus (zero-shot learning). By inspecting the angle of word vectors that represent the chemical names of materials such as Bi_2Te_3 to the vector corresponding to the word *thermoelectric*, they found several novel (hitherto unknown to materials science) thermoelectric conductors.

Integrating dynamic embeddings with NER systems is not a solved problem (Bommasani, Davis, and Cardie, 2020). Using averaged dynamic embeddings, the currently best results on the standard CoNLL-2003 named entity shared task (Tjong Kim Sang and De Meulder, 2003) were obtained by Akbik, Blythe, and Vollgraf, 2018, whose system performs at $F = 0.93$. But the remarkable recall and precision of these systems is not probative when it comes to the OOV problem, since the huge corpora used for training BERT (Devlin et al., 2019), GPT-2 (Radford et al., 2019), RoBERTa (Liu et al., 2019), and other major systems are not publicly available and therefore cannot be checked for test-on-train effects. We therefore compare systems (only on the LOC domain) where the size and the quality of the gazetteer can be explicitly controlled.

Using the Tier 1-3 gazetteers discussed above we compared four systems: a rule-based (Brill-style) tagger (Rauch, Bukatin, and Baker, 2003) set for high recall, BR; the same system set for high precision, BP; a multichannel Hidden Markov Model, HM, used at TERN 2004 (Kornai and Stone, 2004); and a Maximum Entropy tagger, ME, based on Apache OpenNLP. While the performance of the more rule-based (and therefore more explainable, see 9.4) systems improved with the size of the gazetteer, the statistical HM and ME systems do not show the same effect. In fact, the true error pattern can be masked by trivial data normalization effects: a tiny change in the normalization of some frequent item can amount to a great deal more, than systematic changes affecting many low frequency items. To cite a specific example, a slight modification in the tokenization of items like U.S. vs US may account, in and of itself, for a full half percent change in F-measure, more than the total improvement of moving from the Tier 1 entity list (2,171 elements) to Tier 3 (5,108,239 elements).

To correct for this problem, we recomputed the results with the Tier 1 placenames excluded. At the high precision end, accepting all and only those entries that appear on the list will have very high precision. Still not 100%, since usage that leverages the salient associations of a proper name as in *Baby Einstein Toys* or *New York steak* does not actually refer to the person or place in question. The list-based approach always suffers from limited recall, since many named entities appear very infrequently. This cannot be easily fixed by larger lists, not just because of the power law distribution of entity frequencies guarantees diminished returns, but also because as the list grows in size, the number of false positives is increasing with it. There is a village called Energy in Illinois, yet few occurrences of the string "Energy" in random text will refer to it. In the example we cited above, a statistical system supported by the small Tier 1 list finds *Wana* and *Shakai* in spite of the fact that neither appears on this list, while the same system operating with the much larger Tier 3 list that actually includes *Wana* (but not *Shakai*) misses both. Manual inspection of false positives such as *In his Friday sermon, Shaykh Khamis Abidah emphasized the importance of national unity* reveals that it is the overlap of the large Tier 3 gazetteer both with person names (*Shaykh* is a Tier 3 city) and with ordinary dictionary words such as *energy*, that causes the discrepancy.

We started by saying that much of our knowledge about the world is lodged in proper names, but we urge the reader not to lose sight of the difference between knowledge and understanding. There is a great deal of attraction to the the idea that knowing the *who, what, when* is tantamount to understanding what is going in. We think this is patently false, real understanding actually comes from understanding the essence, just as the naive theory would have it.

> La historia era increíble, en efecto, pero se impuso a todos, porque sustancial-mente era cierta. Verdadero era el tono de Emma Zunz, verdadero el pudor, ver-dadero el odio. Verdadero también era el ultraje que había padecido; sólo eran falsas las circunstancias, la hora y uno o dos nombres propios.
>
> The story was unbelievable, yes - and yet it convinced everyone, because in substance it was true. Emma Zunz's tone of voice was real, her shame was real, her hatred was real. The outrage that had been done to her was real, as well; all that was false were the circumstances, the time, and one or two proper names.

8.3 Dynamic embeddings

We begin by a brief historical overview of the four main ideas we see as defining the the modern, dynamic embeddings.

Vectors Both Harris, 1954 and Firth, 1957 are frequently cited as early precursors of vector semantics, but contemporary readers will be greatly disappointed if they read these works with the goal of understanding these key elements of modern systems. Harris' work is clear, well reasoned, and forms the basis for much theoretical and computational linguistic work in the subsequent decades, but neither vectors nor cooccurrence statistics play a major role in it. Firth, besides providing the slogan "You shall know a word by the company it keeps" has had practically no impact, and much of what he wrote is quite opaque today.

In truth, none of the key ideas can be fully traced to the early precursors.[1] Rather, the use of vectors appeared first via multivariate statistics, importing standard methods such as Principal Component Analysis and the strongly related Singular Value Decomposition which go back to the late 19th and early 20th century. In S19:2.7 we discussed how Osgood, May, and Miron (1975) and Deerwester, Dumais, and Harshman (1990) used these methods to study human conceptual structure and improve information retrieval. The key conceptual step, parting with the venerable Prague School tradition of discrete features and relying entirely on embedding the discrete elements in a continuous vector space, was taken by Schütze, 1993.

Subword units That OOV expressions are critical to NLP applications has long been known (cf. our quote from Chen and Lee (2004) above). Modern systems such as Fast-Text (Bojanowski et al., 2017) mitigate the problem by employing representations such as character n-grams (see Jurafsky and Martin, 2022 Ch. 3). In 2.2 we discussed how *femin* is just a 'call to associative memory, something that will be matched by *feminine, femininity, feminist, feminism* and perhaps even *effeminate*'. As a character 5-gram, *f.e.m.i.n* matches all these, and also *feminize* 'to change something so that it includes women, is suitable for women, or is considered typical of women'. But for the most part, contemporary systems use a different set of subword units obtained by data compression, either by byte pair encoding or by the slightly more complex but faster WordPiece algorithm (Song et al., 2021).

While the linguistic theory of subword units, morphemes, is well established, and their cognitive status is hardly in doubt (Newman, 1968), ongoing efforts to find the morphemes automatically such as Creutz and Lagus, 2007 are still not quite successful. In S19:4.5 we used the example of *tendovaginitis*, a word whose meaning 'inflammation *(itis)* of the sheath *(vagina)* of the tendon' is obtained quite effortlessly from the meaning of the component morphemes. Given the eminent usefulness of the morpheme-level segments in all kinds of practical tasks, it took significant intellectual daring to give up on this and to decompose the word as ten.do.va.gi.nit.is, where there seems no hope of obtaining the meaning of the whole from the meaning of the parts. Initial support for more 'syllable-like' units came from speech recognition, where subword units were shown to significantly mitigate the OOV problem (Bazzi, 2002). Remarkably, Harris used the same kind of argument in favor of morphemes:

> For example, before the word *analyticity* came to be used (in modern logic) our data on English may have contained *analytic, synthetic, periodic, periodicity, simplicity*, etc. On this basis we would have made some statement about the distributional relation of *-ic* to *-ity*

---

[1] A remarkable exception, Kiss, 1973, was called to my attention by Viktor Tron Ethereum SWARM (pc), and eventually located by Mark Steedman Edinburgh (pc). This work introduces a neurally inspired vector model that limits context to one word following the target, a restriction no doubt necessitated by the limitations of the computers available at the time.

Given the almost complete lack of segmentation in all spoken and signed languages, and given the practical usability of scripto continua writing systems both historically and in contemporary Chinese, Lao, Thai, etc. scripts, it is not particularly surprising that segmentation into words is hard, especially for languages like Chinese or Vietnamese where many words are short, only one or two syllables. The best performing system (Shao, Hardmeier, and Nivre, 2018) achieves only 91.28% on Chinese, 87.95% on Vietnamese, but over 99.9% for many other languages from Ancient Greek to Urdu, suggesting that word boundaries in these languages are highly predictable (contain very little information in themselves).

Segmenting the words further, into meaningful atomic units, *morphological analysis*, is a considerably harder problem. The best unsupervised method, Morfessor (Smit et al., 2014) reaches only about 60%, and the best supervised methods are at 73% (Ács and Velkey, 2017). As our example shows, most of the parts *ten, do, gin, it, is* are perfectly legitimate morphemes. It is quite often the case that the actual morphemes (here *tendon* and *vagina*) get truncated in the morphological composition process, and overanalysis (ten.don, va.gin) is hard to avoid.

In 2.4 we alluded to the possibility of tricking the standard training algorithms such as Gensim (Řehůřek and Sojka, 2010) into producing vectors for =agt and =pat as well. This requires preprocessing the training corpus by running it through a morphological analyzer and produce a 'deglutenized' or 'gluten free' (GLF) corpus where stems are separated from the affixes by whitespaces (Nemeskey, 2017). What we obtain this way are vectors for every stem and affix morpheme, including, at least for Polish, Latin, and Hungarian, for the nominative and accusative case endings, which are reasonable proxies for =agt and =pat respectively. For English the preprocessing has to be more complex, but subjects and objects can be easily identified based on standard context-free parse trees or dependency graphs. Essentially the same deglutenization method is applied for Kinyarwanda in (Nzeyimana and Rubungo, 2022).

The problem is well recognized: for an early summary see Lazaridou et al., 2013, and for recent study comparing alternative solutions see Mager et al., 2022. But for now, this is one place where theoretical and computational linguistics part ways. For the theoretician, the psycholinguistic evidence, starting with Berko, 1958 that children are aware of the complex phonological changes that take place at morpheme boundaries is impossible to ignore. For the computational person, the difficulties of getting the morphemes right are overwhelming and a workaround, any workaround, such as n-grams, byte-pair encoding, or WordPiece are preferred. This situation is unlikely to change until neural net architectures become more capable of global optimization by dynamic programming (see Schwartz, Thomson, and Smith, 2018 and Ács and Kornai, 2020 for some tentative steps in this direction), since building morphological analyzers for the deglutenization remains a complex task requiring a great deal of manual labor.

Neural nets With the Perceptron (Rosenblatt, 1957), the use of neural networks also has a long and venerable tradition, and natural language applications such as McClelland and Elman, 1986 were part of the connectionist revival of the 1980s. Linguists, however, remained keen on discrete units in spite of the promise of such systems, see e.g. Pinker and Prince, 1988, and it was only in the 2020s that end-to-end speech recognizers became competitive with the more established Hidden Markov Models. Neural nets remain central to vector semantics, and this is in no small part due to the fact that on modern GPU and TPU architectures they offer an extremely efficient way for utilizing multiple processors. Today, we have good methods for computing static embeddings by more direct linear algebraic computations, but the breakthrough work that established the multitask usability of embeddings (Collobert and Weston, 2008; Collobert et al., 2011) was neurally inspired.

Perhaps the clearest case where neural models were not just playing catch-up to the dominant statistical approach but actually improved performance is provided by the *Long short-term memory* (LSTM) circuits invented by (Hochreiter and Schmidhuber, 1997). Early successes include very notable advances in the recognition of handwriting (Graves and Schmidhuber, 2009; Graves, 2012) and speech (Graves, Mohamed, and Hinton, 2013). Linguists may consider these tasks *etic* rather than emic, but as we shall see shortly, LSTMs have proven extremely useful for both.

A well-known problem with classical neural nets is that the dimension of the input vectors is fixed once and for all. This forced early connectionist research such as McClelland and Elman, 1986 into using fixed templates. But there are many cases, where input size is arbitrary. These are better described as *sequence labeling*, such as part of speech tagging which asks, given some text, for the lexical category label of each word. Other important cases, such as shallow parsing can be easily treated as sequence labeling (also known as *seq2seq* transduction): all we need to add is open and closing bracket tags at the beginning and end of major (phrase-level) constituents. Even tasks that seem to strongly rely on parse trees or other graph-like representations, such as semantic role labeling, can be oftentimes recast as sequence labeling: for each role (linker) we need to label the beginning and end of the arrow in question. Irrespective of the specifics of the system of lexical categories or linkers adopted, these are clearly discrete, emic tasks.

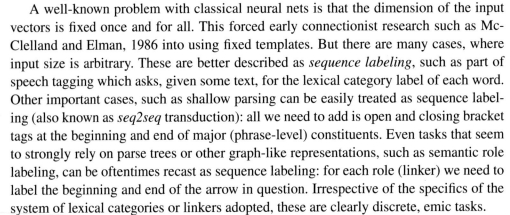

Once we feed a sequence of inputs into a neural net, it is natural to permit some or all of its output to also feed back. Early work in this direction (Jordan, 1986; Elman, 1990) solved the key technical problem of how to train by gradient descent without underflow in the computation, known as the vanishing gradient problem. (There is also overflow, known sometimes as the exploding gradient problem, but the two are not symmetrical: vanishing gradients are a real impediment to learning, whereas exploding gradients are relatively easily circumvented by clipping.) However, the training of Elman and Jordan nets was always relative to immediate context, leaving long-distance dependencies out of scope. LSTMs control both the material that is fed back by an *output gate* and whether it is worth remembering by a *forget gate* – the part that actually serves as memory is known as the *input gate*. When the dependency to be learned is long distance, the LSTM can be trained to ignore the intermediate material, as is necessary for handling classic examples like *The people/person who called and wanted to rent your house when you go away next year are/is from California* (Miller and Chomsky, 1963). As Greff et al.,

2015 demonstrate, "the forget gate and the output activation function [are the LSTMs] most critical components".

A central element of training is the encoder-decoder or autoencoder paradigm which we illustrate on a well-known problem, the compression of English text, for which standardized test sets exist, such as `enwiki8` and `enwiki9` for the Hutter prize. The Hutter competition assumes a simple Minimum Description Length (MDL) two-part scheme, where the first part is the compressed file, and the second the compressor algorithm itself. One of the best algorithms, nncp (Bellard, 2019) uses the text to train the LSTM-based model, but escapes the problem of transmitting the weights by a decoder that works symmetrically. By training incrementally, the encoder and the decoder are always in the same state.

Another good example is machine translation, when the task is viewed as encoding a source-language string by the system, and decoding it in the target language, essentially the same idea as in statistical machine translation (Brown et al., 1993). Dynamic embeddings were first introduced for this task (McCann et al., 2017), and the resulting CoVe system proved superior both on MT and classification tasks. By contemporary standards, CoVe was data-limited, using only 7m aligned English-French and English-German sentence pairs. The breakthrough papers on ELMO (Peters et al., 2018) and BERT (Devlin et al., 2019) have done away with this limitation, as they are taught on far larger (muti-gigaword) monolingual corpora using a Language Modeling objective.

Since the Transformer architecture is by now extremely widely used (at the time of this writing, there are over 48,000 citations to ELMO/BERT), it is important to keep in mind that the leading systems are trained on multi-gigaword (in the case of GTP-3, half a tera-word) corpora. This is simply not feasible for parallel text for lack of data, and even collecting monolingual text in gigaword quantities is a challenge for all but a handful of languages (see Nemeskey, 2020 for a text collection pipeline based on CommonCrawl). On the other hand, generating billions of cloze tests *with known solutions* is trivial. From the statistical viewpoint, cloze tests are just an instance of the *masked* language modeling task where we mask out one word (the cloze target) and provide sufficient two-sided context (see Bengio, 2008; Jozefowicz et al., 2016 for replacing the standard (n-gram) techniques by neural nets).

Multilingual BERT and RoBERTa also make the basic WordPiece vocabulary shared across over a hundred languages (something that would make no sense if morphemes were used as subword units) and injects position markers (relative to word start) in each subword unit, so that the vectors corresponding to *pa* in *pa.ta* and *ta.pa* are not identical. Phonology will sometimes rely on similar positioning of syllables within metrical feet, and even feet within cola (Hammond, 1995), but the computational systems, being entirely orthographic, make no effort to discover or mark metrical structure.

Attention Perhaps the single most important feature of dynamic embeddings is attention (Bahdanau, Cho, and Bengio, 2015; Luong, Pham, and Manning, 2015; Vaswani et al., 2017), whereby correlations are learned between elements that may be separated by some distance from each other. Attention 'heads' are matrices trained to store connections between earlier and later parts of the sequence (see Jurafsky and Martin, 2022 10.4 or Ketan Doshi's blog post for a more leisurely explanation), and these, we claim here, are conceptually related to the dynamically updated transition matrix $P(t)$ we proposed in 7.4.

Attention heads, much like cross-linguistic subword units and artificial prosodic marking, may drive to despair the linguist, even one sympathetic to the general idea of using vectors and neural nets. Yet the idea of *some* dynamically evolving memory of what went on earlier in the sentence and across sentences is commonly accepted in Discourse Representation Theory (DRT) (Kamp, 1981; Heim, 1982), and has proven its worth in handling anaphora, presuppositions, and discourse relations in general. In this book we argue that 'current knowledge state' is obtained from dynamic updating of a basic knowledge state, which we identify with lexical knowledge. The update adds a (softmax) matrix in the manner of Equation 2.6.

Attention, to be sure, is not your father's DRT – there remains a great deal of work in bringing the two in closer alignment, just as there remains much work to be done in replacing WordPieces with morphemes. The dynamic nature of current embeddings is actually an obstacle: rather than there being a single, dynamic vector for *bank*, we follow lexicographic tradition and assume two distinct senses `bank/227` and `bank/1945` (see 5.3). In ordinary static embeddings, a single vector is obtained which is just the log frequency weighted sum of the two sense vectors. But there is no unique way to recover the two sense vectors from their weighted sum, even if we know the weights.

At the considerable risk of making the embedding very data-limited, we can begin with SemCor or some similar sense-tagged corpus (Mihalcea, 2002) and obtain static 'DilBERT' embeddings that contain different vectors for different senses. Here we still have the problem of which sense to employ during the processing of ordinary text that lacks sense-tagging – in effect, the same word sense disambiguation problem that we discussed in 6.4. But the solution is much easier than in the original case, since in most cases all we need to do is to inspect the leading term of the definition, `institution` for `bank/227` and `land` for `bank/1945`, and compute the angle with these: for the right choice we obtain high cosine similarity, and for the wrong choice, low.

The risk of being data-limited is not to be taken lightly. It is well known in computational linguistics that an algorithm which gets more training will outperform much better algorithms for which there is less data, *there is no data like mo' data*. Sense-tagged corpora, besides being orders of magnitude smaller than the gigaword size that is standard today, are generally 'silver' quality: machine tagged, as opposed to human 'gold' data. The idea of training on these is not very attractive, since the model will inherit the errors of the automatic tagger. A better way for obtaining static embeddings is to start with the dynamic ones which already underwent very large-scale training (Akbik, Bergmann, and Vollgraf, 2019; Bommasani, Davis, and Cardie, 2020).

Since our goal is to study linguistic regularities, we expect to rely on standard linear algebraic methods such as decomposition into orthogonal subspaces (Rothe, Ebert, and Schütze, 2016), which lose their grip over the dynamic embeddings. Either we develop special methods for 'untangling' the dynamic embedding (Moradshahi et al., 2020) or we preprocess the material to obtain static embeddings. Since the dynamic embeddings lack cognitive realism (it is not that we have a single *bank* entry in our mental representation that now means one thing now another) we plan on following the second route, a decision also supported by direct measurements such as Dufter, Kassner, and Schütze (2021), who found that on relational triples "static embeddings perform 1.6% points better than BERT while just using 0.3% of energy for training".

While dynamic update of the thought matrix (or whatever data structure we use) is typical for intra- and inter-sentential parsing, the mechanism we employ is already at play in the lexical domain. In 1.6 we already considered *attract* , defined as `=agt attract cause_ {=pat want {=pat near =agt}}`. For the logical semanticist comfortable with the use of VBTOs this is very clear: we define *attract* by reference to its agent and patient arguments. If there is a need to quantify over a role, as would be the case in *attractive* 'can attract'; *descriptive* 'can describe'; *effective* 'can effect'; *explosive* 'can explode' etc. we simply fill the patient slot with the all-purpose proquant, `gen` (see 4.3). This way, *attractive* becomes `attract gen`, or by substitution, `=agt cause_ {gen want {gen near =agt}}`. The modal aspect, 'can cause; can describe, can effect; can explode' is analyzed, as before, by optional (default) `<do>`, so we obtain `<attract>; <describe>; <effect>; <explode>` etc.

Let us briefly consider how the uniqueness (what the linguist would call *coindexing* of the elements) is enforced in the formulaic, the algebraic, and the geometric systems. If we have variables and VBTOs their uniqueness comes for free: no logical semanticist would want to translate *X attracts Y* as *Z causes W to want to be near T*. The formulaic theory makes a sortal distinction between VBTOs on the one hand and formulas on the other: VBTOs are *operators* that operate on the formulas, closing off the open variables, and have no status in isolation. (This is the typical setup, but see da Costa (1980) for an alternative.) In the algebraic theory, the same distinction is reflected by VBTOs used as labels on edges that connect two (hyper)nodes, and having no status without some source and target nodes for the edge. A linguist would say that in both views VBTOs are *bound forms* (S19:5.2) while formulas or hypernodes are *free forms*.

In the (hyper)graph view (1.5) we enforce uniqueness by unifying all atoms that have the same name. This guarantees, even though we treat `gen` as an ordinary noun, that the person/thing/matter that is being attracted is the same as the one that desires to be near the source of the attraction. We only have two variable-like entities, `=agt` and `=pat`, so renaming variables is not an issue. The price we pay for this simplicity is that we lose much of the analytic freedom available to linguists who follow Perlmutter (1978) and analyze certain intransitive verbs as unaccusative with a `=pat` subject. `4lang` currently treats all intransitives as having an `=agt`, even if the subject in question lacks agency. This is yet another place where the lack of derivational sophistication in the overall system is keenly felt: something like Pāṇini's uniform treatment of voices across verbal and nominal constructions, however attractive, must remain the unique achievement it is until the cross-linguistic picture is better understood.

Since 4lang enforces uniqueness by fiat, it also requires a special operator, other,
reproduce   to block unification, as in *reproduce* =agt make other[similar]. Since *other* is
other   defined as different, and this in turn is defined as =pat has quality, =agt
different   lack quality, "from _" mark_ =pat, the correctness of the definition chain
critically relies on the identification of the two instances of quality. As the conceptual
structures depicted in Fig 1.3 are really the simplest ones imaginable, we don't see reliance on other as particularly problematic, and made no effort to eliminate it in favor
of some other primitive.

The uniqueness requirement is observable in any definition that contains a rel-
red   ative clause, such as *red* colour, warm, fire has colour, blood has
colour, resemble anger. The red is_a colour clause that we arrive at
after we undo the anuvr̥tti requires very little in way of explanation. blood has
colour is also nearly self-evident in isolation: most physical things have color, blood
substance   is defined as a liquid, liquid is defined as a substance, substance is defined as has
mass, in space/2327, physical, so the notion that blood has color, some
color, is nearly evident. But this is not what the defining clause means: it means that
red is the color that blood has, an effect only achieved by the unification of the separate
instances of colour in the definition.

A fuller graph is computed as follows. Recall from 1.5 that plain arrows run from the
predicate to the subject, dotted arrows to the object, and is_a links are represented by
dashed arrow running from the subcategory to the supercategory. After the elimination
resemble   of the *resemble* clause by substituting its definition =agt has quality, =pat
has quality, and unifying the multiple has and quality nodes, we obtain the
graph depicted in Fig. 8.2:

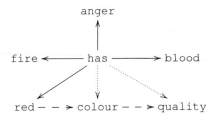

Fig. 8.2: *red*

Turning to the geometric system, relative clauses are a bit more complex, as the notion of
graph node unification has no obvious equivalent in the vector system. We certainly don't
want to claim that the color of fire is the same as the color of blood, let alone the color of
anger. The definition provides these as deictic examples: 'whatever is the range of red,
blood fits in this range', etc. Assuming that the language learner has seen instances of
blood, fire, or even anger, these provide instances (exemplars) of the color in question.
Whether a few instances are sufficient for concept formation is a hard question: on the

one hand, most theories and computational models of learning require many examples, on the other, journal studies in child language acquisition make clear that humans learn many words in one shot.

Relative clauses provide yet another justification for using polytopes rather than point vectors. On the standard view of semantic spaces, red things congregate on one side of the affine half-space demarcated by the red/nonred boundary. Here we say that the color of anything is obtained by projecting on the `=agt has color` polytope, which we obtain by applying the `has` matrix to the `color` vector from the left, i.e. $B_{\text{has}}P_{\text{color}}$. This applies in metaphorical usage *color me skeptical* just as it applies in the core cases: once the imperative is heard, the operation is performed, and the discourse representation includes the clause `<speaker> has color[skeptical]`. What the definition of *red* depicted in Fig. 8.2 does is to enforce several constraints on the `red`, `blood`, `fire`, `anger` polytopes via the B_{has} matrix. This is the same mechanism that we already use to enforce constraints like `blade has edge` or Eq. 6.7. What `blade has edge` means in geometric terms is

$$Y_{\text{blade}} \subset B_{\text{has}}Y_{\text{edge}} \tag{8.1}$$

This means both that blades are inside the set of edge-having things or, if we prefer, that edges are among the things that blades have. Most observations about polarity follow without further stipulation not just for `has` but for all our binaries represented by matrices: if Damascan swords are blades they will perforce have edge, but if Damascan swords never go dull it does not follow that blades in general will never go dull. Longer chains such as `blood has color is_a red` work the same way:

$$Y_{\text{blood}}B_{\text{has}}Y_{\text{color}} \subset Y_{\text{red}} \tag{8.2}$$

and in general, more complex graphs like Fig. 8.2 are translated to conjunctions of such equations.

The case when we have graphs functioning as nodes, as in Fig. 1.4 *Video patrem venire*, works in the same manner. Since `father` is obviously the subject of coming, we need to first express `father ← come` as a geometric constraint, which we do by Eq. 2.6, which has the net result that in the currently prevailing scalar product $P(t)$ we have $Y_{\text{father}} \subset Y_{\text{come}}$. In order to make sure that the entire `father ← come` is the object of `see`, we use Eq. 2.9 which requires adding to the verb `see` the entire object. Since `father` is now a subset of `come`, this requires only the addition of the Y_{come} polytope to the Y_{see} polytope, in effect recovering the 'principle of chain forming' suggested in Kálmán and Kornai, 1985.

9

Applications

Contents

We started with Lewin's aphorism, "there is nothing as practical as a good theory". Vector semantics, the broad theory that was raised from a Firthian slogan to a computational theory by Schütze, 1993, has clearly proven its practicality on a wide range of tasks from Named Entity Recognition (see 8.1) to sentiment analysis. But the farther we move from basic labeling and classification tasks, the more indirect the impact becomes, until we reach a point where some conceptual model needs to be fitted to the text. Perhaps the best known such problem is time extraction and normalization, where our target model is the standard (Gregorian) calendar rather than the simple (naive) model we discussed in 3.2. In 9.1, based almost entirely on the work of Gábor Recski and his co-workers at TU Wien, we outline a system that probes for matches with a far more complex conceptual model, that of building codes and regulations in effect in the city of Vienna.

In 9.2 we turn to a problem well known to computational linguists: ungrammatical and fragmentary input. It is hard to deny that the (Nemeskey et al., 2013) system described here goes against the grain of contemporary computational linguistics. There are hard rules operating on discrete knowledge states, and there is no statistical component, yet the system escapes precisely the problem of ungrammaticality, the very problem that motivated the shift from discrete symbol manipulation to continuous optimization.

In 9.3 we survey the main steps we have taken towards the automatic building of representations. Instead of fully automatic acquisition of lexical entries, we leverage the considerable manual work that lexicographers have already done. We also describe how the technology can be put to work in three areas where word vectors brought very significant improvements: machine comprehension; computing the semantic similarity of words and sentences; and computing lexical entailment. In 9.4 we turn to one area where discrete, symbolic systems have considerable advantage over pure neural nets. This is the issue of explainability: how to justify the decisions a system makes, and how to make the entire formal system explainable (in simple words) to its future users.

Finally, in 9.5 we provide a summary of the program we outlined at the beginning for obtaining representations (vectors and matrices) by treating the dictionary as a system of equations and solving these. This section also serves as a change log for the current Version 2 of `4lang` and the improvements planned for V3. The entire V2 is published at https://github.com/kornai/4lang/blob/master/V2/700.tsv in the same timeframe as this volume, but with V3 we are obviously describing future work.

9.1 Fitting to the law

Unlike some Wild West cities where one can build pretty much any structure they wish to, the city of Vienna has highly specific building codes and zoning regulations, often down to the individual block or even lot level. When a builder wishes to raise a new structure, or alter an existing one, they need to submit a detailed plan to the Stadt Wien Baupolizei which will certify that the plan complies with all rules and regulations and issue a building permit, or provide feedback on the specific points where the plan fails.

The model of the BRISE project is one where builders submit their plans as they do today, electronically. After segmentation into sentences, the rules and regulations are translated into a system of deontic logic statements (Recski et al., 2021) via an intermediary structure of concept graphs that we called Algebraic Conceptual Representations in Kornai and Kracht, 2015 and Kornai et al., 2015 (see also 1.5 and S19 5.3–5). The concept graphs are intermediary structures that demarcate the boundary between ordinary language and expert knowledge, though in a manner somewhat different from Putnam's 1975 proposal discussed in 8.1. The goal of *semantic parsing*, computing the concept graphs from the sentences, is to deal with general language use, and provide a well-articulated formal representation for *everyday* language. The task-specific sensemaking effort aims at highly abstract models that are outside everyday linguistic competence, in this case formulas of a dyadic deontic logic (Ciabattoni and Lellmann, 2021).

This division of labor makes it possible to leverage preexisting software systems, in this case the Stanza NLP package from Stanford, which has state of the art facilities for German. Translating the Stanza output into concept graphs is done by Alto (Gontrum et al., 2017) which is sufficiently high level for permitting rapid prototyping. Readers of Recski et al., 2021 will be surprised how little 'glue' is required to go from German text to logically annotated concept graphs already sufficient for recognizing the Permitted, Forbidden, and Obligatory clauses and their conditions.

Equally important for the division of labor, `4lang` stands neutral on the precise choice of deontic logic that is to be deployed. As the reader familiar with the area knows only too well, there are many competing proposals (see Gabbay et al., 2013 for a summary that is rapidly becoming dated given the enormous progress in this area). `4lang`

offers only a skeletal theory (see 6.2) based on two simple definitions: *can* <do>, which   can/1246
is permissive only in the sense of physical permissibility; and *must* lack choose.   must
The discussion in 4.3 extends this to a more detailed theory of normative statements,
but only in a way that is characteristic of everyday permissions and prohibitions. The
building code that BRISE aims at goes far beyond this in sophistication. For a typical
example, consider *Für die mit BB5 bezeichneten Grundflächen wird bestimmt: Die bezeichneten Grundflächen sind mit Ausnahme von Vordächern von oberirdischer Bebauung
freizuhalten* 'For the areas designated with BB5 it is determined: With the exception of
canopies, the designated base areas are to be kept free of above-ground construction'.
From this Stanza creates the following UD parse:

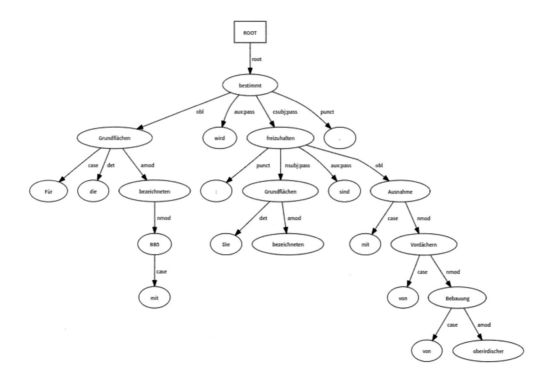

Fig. 9.1: Universal Dependencies analysis produced by Stanza

Based on this parse tree, BRISE creates the 4lang graph depicted in Fig. 10.2. This is
drastically simplified: notice the collapse of the two *mit*s in *mit BB5 bezeichneten* and
mit Ausnahme ... freizuhalten.

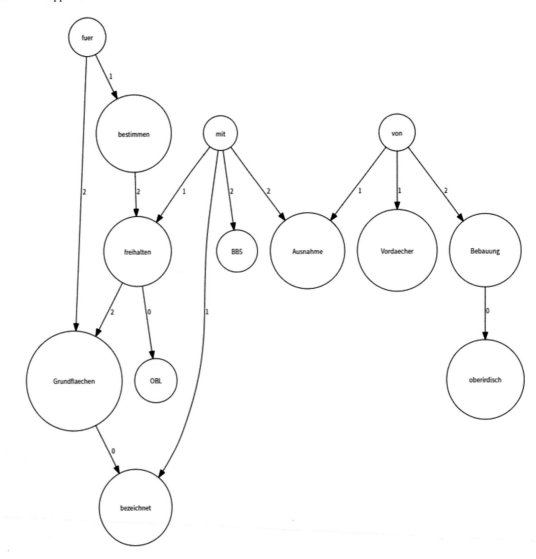

Fig. 9.2: 4lang analysis computed from UD analysis

Critically, after morphological reduction of *freizuhalten* to *zu freihalten* an OBL (obligation) node is created by lexical knowledge, that "ist zu" mark_ obligation, at least in the technical language these documents are written in. As a matter of fact, the language is not so strange: the dominant non-technical usage of *ist zu* will be imperative in everyday situations as well. A good portion of the words whose definition contains both before and after clauses invokes something of a normative element: consider *pay* money, before(=pat work), after(=pat has) or even *cause* before(=agt), after(=pat).

pay/237
cause_

Significant portions of the Building Regulation Information for Submission Envolvement (BRISE) have already been implemented. For an online demo, see https://ir-group.ec.tuwien.ac.at/brise-extract, and for source code, available under a permissive MIT license, see https://github.com/recski/brise-plandok.

Whether a narrowly targeted applied system provides the best context to study issues of such generality as imperatives/obligations is not a given. But the heuristic value of such systems for the theoretician is evident, and many of the `4lang` techniques in use are implemented in independent modules that can be deployed for other systems as well.

9.2 Pragmatic inferencing

One application area that we think has great potential is *pragmatic inferencing* as first deployed in Nemeskey et al. (2013). Unlike in BRISE, where both the regulations and the applications are formulated in a rigid, somewhat formal, stilted, and highly technical prose style, in ordinary situations people use fragmented, elliptic, and often grammatically incorrect language. Here we briefly describe how spreading activation deals with the example

Felsőgödre   kérek egy ilyen nyugdíjas
Felsőgöd.SBL please one such pensioner.

This appears in the MÁV (Hungarian State Rail) corpus, a small set of interactions between customers and a ticket clerk at the Western railroad station in Budapest. What the utterance actually means is "**please** give me a ticket from here **to Felsőgöd** with **a pensioner** discount". Even though only the words **in bold** were actually said, the clerk had no problem inferring the rest and providing the appropriate ticket.

In the early stages, the computational analysis proceeds in the manner described in S19:5.3: words undergo morphological analysis and NPs are built by a chunker. The bulk of the interpretative work is done by the data structures, attribute-value matrices (AVMs) controlled by spreading activation. Compared to ordinary `4lang` lexical entries such as *ticket* `1181 u N <paper>, <card>, before(pay/812),` `ticket for_ right/3122`, the AVMs that describe the technical vocabulary will also contain task-specific attribute slots such as Origin, FareClass, TimeRange, and Destination. In this case, no ticket can be issued until values are found for each of these slots.

The spreading activation algorithm was originally implemented using Eilenberg machines see https://github.com/kornai/pymachine. Here we present the idea using graphs (see 1.5) and ordinary finite state transducers. The key idea linking the two is that we think of the change from an 'unfilled' attribute to a 'filled' one, e.g. learning that Destination=Felsőgöd, both as extending the representation graph by the *Felsőgöd* node and as a change of knowledge state, a transition of an FST that keeps track of which values have already been filled and which still need to be found.

To control spreading we maintain two graphs: a *static* graph whose nodes are the machines corresponding to words and whose edges are the definitional links, and an *active* graph that keeps track of the active nodes pertaining to the currently analyzed

sentence. For every utterance, full or fragmentary alike, the active graph is initialized from the structures found by morphological analysis and chunk-level island building. In each iteration, the active graph is extended and transformed in three stages: expansion, activation, and linking. Expansion and activation are driven by the lexical definitions of words. In the expansion step we take every, as yet unexpanded, active word, and add to the graph the structures compiled from their definitions, each structure connected to the word it defines. This is also how AVMs are activated: each AVM is associated with a word (e.g. TicketAVM with the word *ticket*, GroundTransportAVM directly by the phrase *ground transport(ation)* or indirectly by *limo*), and becomes available only when the corresponding word is expanded. Activation works in the opposite direction: words whose definition structure is a subgraph of the currently active part are also added to the graph. Lastly, linking is responsible for filling the empty valency slots of verbs and AVMs. Linking is driven by a handful of explicitly designated linkers (Ostler, 1979), corresponding roughly to deep cases.

To see how this works on our example, notice that *Felsőgöd* is lexically specified as a town. It is of course absent from the core `4lang` lexicon, but will have to be present as Location (8.1) in any system capable of issuing tickets to Hungarian towns. We take the list of such towns from the Hungarian State Rail database, since even if we could find other towns in text (and we can, see 8.2), the background system would not be capable of issuing tickets to these. Next, the system knows, just by virtue of having entries for Hungarian case endings, that the ablative and delative cases are linkers for Origin, and the sublative is a linker for Destination. Here *Felsőgöd* appears with the sublative (a fact already detected at the initial morphological analysis stage), so will fill the Destination slot. As there are no ablative or delative marked elements, the Origin slot will have to be filled by the default `here`. More interesting is the case of *pensioner*, which is not a known fare class. But the morphology relates *pensioner* to *pension*, the lexicon relates *pension* to *old age*, and TicketAVM has an OldAgeSupplementary fare class.

The utterance has *kérek* 'ask,request 1sg' as a main verb, but the object of the asking, the (pensioner's) ticket, is not linked to this, since *pensioner* does not appear in the accusative case as it should. The indefinite article *egy* 'one' and the proadjectival *ilyen* 'such' also remain unlinked – the system in effect processed an even more fragmentary utterance, *Felsőgödre nyugdíjas*. Given the exponentially growing number of entries that can be activated in each turn, the key technical issue is to obtain a parse in a few steps, before activating the whole lexicon. In this system, we deployed heuristics that drive the activation towards nodes that are already active based on the structure of the static lexical graph, and in the systems discussed in 9.3 we simply limited the number of expansions to 2 or 3.

9.3 Representation building

Both the legal analysis and the ticket sales task depend on the ability to express in a semantic representation what the words mean. Word vectors are geared more towards semantic similarity than the discovery of meaning components, and can only be used in a support role for such highly conceptual undertakings. On the other hand, word vectors are quite easy to train automatically, while the `4lang` representations discussed here are obtained through a process of analysis performed by humans. In the case of `4lang` it is remarkably easy to teach the skill (in Nemeskey et al., 2013 we wrote "It takes only a few hours of training to teach the monosemic definitional style to undergraduates, comparing quite favorably to the effort it takes to explain e.g. the MUC named entity tagging guidelines") and in fact many current definitions originate with students (some are still marked by their initials in the comments column). Be it as it may, if there is one thing in contemporary NLP that is viewed with near-universal disdain it is 'manual' labor. There are, it must be admitted, very good reasons to steer systems away from manual labor, reasons that go well beyond the expense involved.

First, systems obtained this way are always suffering from the out of vocabulary (OOV) problem we discussed in 8.2. A complete dictionary is like the pot of gold at the end of the rainbow – no matter how much work we invest, we never quite get there. Even the task of getting to respectable coverage, say 95% in token frequency, will generally require a team of dictionary builders, and it is very hard to maintain consistency across members of such teams. While OOV is also a problem for the systems obtained entirely by automatic methods, because words with too few occurrences in the training corpus cannot be reliably assigned vectors, the problem can always be ameliorated by processing ever larger corpora. By doubling the corpus some of the words that were too rare will now be in scope (say, ten or more occurrences), but the end of the rainbow shifts. In large corpora over 50% of the word types appear only once (see Kornai, 2007 4.4 for discussion), so if we use a corpus twice as large as before, the number of word types we can't train to is more than doubled.

Second, manually created systems have a hard time shaking off the suspicion of cheating: what if the results are good only because the builders skimmed of the top, writing careful entries for the most frequent cases, but virtually guaranteeing failure for the more rare cases? The true measure of this is when we move the system from one domain to another. Besides encountering important words that were rare before, the whole logic of the system may prove to be brittle, requiring significant revisions in the new domain. In the pragmatic inferencing system described in 9.2, we tested this by moving from trains to planes, using the Air Travel Information System (ATIS) corpus (Hemphill, Godfrey, and Doddington, 1990). There are many obvious differences, as we wrote "there are no 'open jaw' railroad trips and no dining cars in the air. But the basic conceptual structure, as captured by the Ticket AVM, and the basic syntax of getting from A to B on day C, are shared across these domains. (. . .) adding new lexical entries to the network does not have deep ramifications for what may be taking place in some other corner. Changes are reasonably localized and debuggable. This is actually an advantage compared to systems with continuous weights, where bad effects (bugs) are impossible to attribute to specific causes", an important matter we shall return to in 9.4.

Finally, there is the deeper goal of learnability. Obviously, the sketch of learning presented in 5.3 does not amount to mature technology for the automatic acquisition of the extended lexical entries that do the work. Yet in a sense we actually have gold data: high quality teams of dictionary builders, lexicographers, have produced incredibly detailed and useful descriptions of word meaning. Our current approach is to leverage their work to the extent it is not encumbered by copyright restrictions. Recski, 2016 describes a system for creating `4lang` graphs based on English dictionary entries, and (Recski, Borbély, and Bolevácz, 2016) implemented a similar system for Hungarian. As we have argued in S19:1.3, the bulk of the information is in the words, not in the syntax. In fact, we consider the syntax acquisition problem largely solved, noting that modern transformers produce incredibly sophisticated English text of perfect grammaticality.

As its name suggests, the `dict_to_4lang` system is aimed at leveraging dictionary definitions of word senses. The overall mechanism goes even further in relying on externally supplied knowledge: the definition text is parsed by a state of the art deep learning-based parser, currently Stanford's Stanza, and can create `4lang` graphs from all sorts of running text by converting Stanza's output (a dependency parse tree) into the `4lang` format. Task-specificity is seen only in the fact that in the analysis of dictionary definitions the results improve markedly when we coerce the grammatical analysis into the maximum bar-level lexical category of the definiendum (NP for nouns, VP for verbs, etc). For source code, see https://github.com/recski/tuw-nlp, and for an online demo see https://ir-group.ec.tuwien.ac.at/fourlang.

Such hybrid use of neural and symbolic techniques may offend some readers' sense of purity. But our goal is to acquire structured meaning representations, and the neural systems by themselves are incapable of doing so *in principle*. Pure symbol-manipulation systems have been tried for decades, and they are clearly insufficient *in practice*. For systems such as BRISE or the ticket clerk to work and 'have legs' (be portable across domains), we need to first create an entire semantic world, the static graph. It is the reusability of the representations connected in this graph that guarantees portability and, at the same time, sets its limits. The TicketAVM will be very different for the plane, train, and theater domains, depending primarily on the peculiarities of the backend database system. But a good abstraction such as our definition of a ticket as something you pay for and something that confers rights, will work well for all these domains.

A particularly interesting question for the linguist and cognitive scientist (Pinker and Prince, 1994) is the extent to which such representations are precomputed. We all know that in the theater domain a ticket confers the rights to see a performance, in the tech support domain it confers the right to go directly to your issue without repeating all the effort that went into establishing what the issue was, and in the train/plane domain it confers the right to travel. Further, much of this knowledge clearly comes from the scripts we all store for attending a performance, calling tech support, or going to a ticket clerk. A spreading activation model can actually reach these scripts as long as the encyclopedia is built on lexical entries. In our case, a good portion of the Ticket-
`travel`   AVM can be obtained by using the sense so activated. Since *travel* `after(=agt at`

`place/1026[other,<city>])` implies a source and a destination, two of the four task-specific attribute slots Origin, FareClass, TimeRange, and Destination are already given.

A systematic investigation of how and under what circumstances `X is_a right` and `X has Y` leads from *(train) ticket* to *ticket request must have Origin, Destination* is well beyond the scope of the current volume. Clearly, the issue is not at all trivial: we have doubts that the existence of FareClass can be deduced from a generic (lexical) knowledge base that does not already contain this information. The TimeRange attribute, generally expressed via some free adverbial, is obligatory in the AVM, and we can well imagine the railway selling open tickets that can be used at any time (in fact, the Hungarian State Railway used to sell such tickets). However, the mere fact that such questions can be asked in the context of any system is a step forward compared to what Cabrera, 2001 calls the 'weak monist' position that there is no dictionary/encyclopedia distinction.

Remarkably, the representations we build are useful even for the core issue that vector semantics addresses, the similarity of words and sentences. One way to interpret word vectors is to say that our goal is to build an embedding where similar words are closer to one another than dissimilar ones. Recski et al., 2016 present a system that computes a similarity measure from similarity in the definitions, rather than (cosine) similarity of the vectors, see https://github.com/recski/wordsim. This idea extends to sentence similarity (Recski and Ács, 2015). The timely appearance of (Kovács et al., 2022a) makes it unnecessary to detail here how the `4lang` representations can combine with state of the art neural systems to improve result on the *lexical entailment* task (Schmitt and Schütze, 2019; Glavaš et al., 2020), an essential component for *machine comprehension* (Gémes, Kovács, and Recski, 2019) – for code see https://github.com/adaamko/wikt2def.

9.4 Explainability

One of the many goals of contemporary AI research is explainability. Here we will distinguish between the *narrow goal* of explaining individual decisions, and the *broad goal* of explaining global properties of the system. What gives particular importance to the narrow goal is that we want to make sure that decisions produced by the system can be explained to those affected by them. Consider credit approvals. Each year, applicants fill in millions of credit card applications. The banks approve or reject these based on guidelines similar in character to the building regulations we discussed in 9.1. For example, a bank may use the rules *credit will be extended for applicants with no history of bankruptcy who own their home and make over $30,000/year* and *credit will be denied to anyone with a history of bankruptcy*. (Actual rules are much more complex and generally not made public, so that applicants can't game the system.) Further, loan officers may have some discretion in applying these guidelines.

This situation fits the overall machine learning paradigm very well: thirty years ago, already plenty of gold (human-produced) training data was available, and then current computers were already sufficiently powerful for training neural nets for the problem. In fact, such NNs were known to perform better than human loan officers, yet the systems didn't gain traction. The main reason was that banks were afraid: what if they deny someone credit and the person sues them? It was clear that 'well, the neural net settled in an energy minimum with the output node at zero' did not, and at a jury trial would not, constitute an acceptable explanation. What is required is "judicial reasoning that builds from the bottom up, using case-by-case consideration of the facts to produce nuanced decisions" (Deeks, 2019). A notable step in this direction is the building of systems like POTATO (Kovács et al., 2022b) which provides "a task- and language- independent framework for human-in-the-loop learning of rule-based text classifiers using graph-based features". To the extent humans are acting in a decision-making role during the training phase (rule creation) we have good guarantees that the rules themselves are sensible *from a human standpoint*. (See https://github.com/adaamko/POTATO for code.)

From a more neutral 'philosophy of science' standpoint, perhaps weight-based explanations are already acceptable: after all, the weights the net trained to constitute a compact model of the training data, one that fits the seen instances best. Further, it was clearly self-consistent and gave good results on unseen (gold) data. It was obtained by mathematical optimization, and therefore had absolute guarantees that it was the best possible such model: messing with the connection weights would just make it worse. Yet people are, to this day, extremely reluctant to subject themselves to decisions by algorithms. It's not that they don't understand the algorithm: the NNs that perform well on the credit approval task are extremely simple, and can be explained to any high school student. Even the weights generally make a great deal of sense (income will have positive weight, bankruptcy negative) but somehow the set of weights that defines the model remains insufficient as a human-understandable explanation.

The major exception to this reluctance is in finance: more than half of the money invested these days is traded algorithmically, and the proportion is still growing. The increasing reliance on high frequency trading algorithms also makes clear that in certain contexts the narrow goal of explaining individual decisions is not very practical: by the time a human understands an explanation the system will have performed millions of trades. That said, the ability to furnish such explanations remains highly relevant for debugging the algorithms, but this is a rather different goal from explainable AI (XAI), as this goal is generally understood. We will use the finance domain to describe what we consider to be the *broad goal* of explainability: a human-understandable description of system parameters and behavior. This means both that we seek an explanation in terms of (possibly conflicting) goals and that such an explanation can only refer to a few, human-digestible parameters.

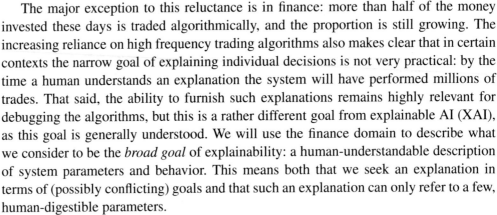

Readers will no doubt be familiar with cryptocurrencies. Besides the creation of trading infrastructure (increasingly trivial with Infuria, Alchemy, Moralis, and other services running on top of Ethereum and similar blockchains), minting a new coin is normally accompanied by a white paper that furnishes the broad justification: why should investors invest in this particular asset as opposed to those already on the market? What reason is

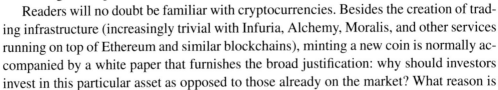

there to believe the new coin will keep its value? The most widespread justifications are the original proof of work (PoW) and the less energy-intensive proof of stake (PoS), but there are many others, such as proof of storage, proof of space-time, etc. that typically come with actual coins using the idea. The explanations of these concepts are rather technical, often requiring some mathematical proof that work is indeed being done, storage is indeed being used or made available, etc.

To some extent, the introduction of a new "proof of X" is sufficient for generating some investor excitement, with the general idea that can be summarized as 'proof of X is new, new things generate a new domain for doing business, and the best profits are always made in new domains'. This line of thinking, generally known as first-mover advantage, is typically explained in the business and marketing literature by means of historical examples. But such examples demonstrate only that first-mover advantage *can* exist, not that it actually is an exceptionless law of nature or, at the very least, a statistical regularity one can depend on.

This is one of the areas where likeliness-based reasoning of the sort we described in Chapter 5 plays a major role, since is very hard (expensive and prone to observer bias) to gather statistical data of the kind that a better, Kolmogorov-style probabilistic analysis would require. Here the individual steps are rather easy: first, it is easy to check whether for some X 'proof of X' is novel, or has already been proposed. Second, it is clear how to check whether X would indeed generate a new domain of business: all that is needed is a description of a business model of how to make money on X. Third, the notion that the best profits are always made in new domains is well entrenched, particularly as new domains are initially unregulated, and regulation is known as a significant cost to business.

Another way to see how broad explainability is connected to lexical semantics is to consider some hypothetical systems. In Proof of Prediction (PoP) what is on sale is a coin that is a full robotic trading algorithm, enabled by the smart contract system of the Ethereum infrastructure. By buying the coin, the owner tops it off with a certain amount of initial capital, and from this point onwards, the algorithm can make buy/sell decisions on its own unless it runs out of capital. As with other financial algorithms, the narrow justifications of individual program trading decisions are quite irrelevant to the owner: all that matters is the algorithm's ability to predict market moves. Based on some initial settings at the discretion of the first owner, some of the coins will thrive, others will decline in value: at any given moment, the size of assets under the management of the coin in and of itself constitutes proof of prediction ability. Given that a sophisticated trading algorithm will learn from its earlier experiences, the value of the coin may be more than the value of the assets it is managing.

This is actually less futuristic than it may sound. A non-hypothetical example would be a person, say in 1985, who invested in the Fidelity Magellan fund based on nothing else but the track record of the fund manager, Peter Lynch. In this particular case, Lynch took the trouble to outline the broad justification in a best-selling book (Lynch, 1989), but other widely successful fund managers, doctors, defense lawyers, and so on are quite incapable of putting into words what makes them successful. Clearly, global explainability is a bar that often proves too high for natural general intelligences (humans), so demanding it of AGIs may be unrealistic. However, our model situation is about introducing a new AGI, and it makes perfect sense to demand global explainability for these, even if we don't ask parents to publish a white paper each time they produce offspring.

Readers may see some sleight of hand in the definition of *prediction* used above, especially as 'past performance is no guarantee of future results'. But as with any other temporally fluctuating quality, say the temperature or humidity of a given location, the fertility of the soil, or the chances of sighting a whale at a particular location, past performance remains our best, albeit imperfect, indication of future results, and the above reasoning stands up well in the class of white papers in terms of offering a general explanation why anybody would want to invest in PoP coins.

Recall that our goal here is not so much to sell the actual product. Many readers will have problems with the business model 'invest in speculative assets so as to make more money' though many others will embrace it (and no doubt potential buyers would be recruited only from the second group). Rather, our goal is to link explainability in the broad sense to lexical semantics, and argue that the persuasive force of such explanations comes not so much from complex analytical models as from the meaning of the words they are composed of. In the explanations we rely only on discrete categories, (valuations in the sense of Definition 6 in 1.5, rather than in the financial sense), and elementary link-tracing logic and *kal va-chomer* (S19:9.4). Even continuous quantities, such as assets under the management of a PoP coin, are used only for trivial comparisons: if a coin instance is more predictive it is better, and worth more. For this, no sophisticated asset pricing method is required, and conversely, for humans no elaborate model confers the same explanatory value as such simple statements.

To see how explanations stem from the meaning of the words we provide another example, a Proof of Immortality coin. PoI will differ from PoP in one main main respect, a *faithfulness* parameter f that can be set anywhere between 0 and 1, representing the fraction of its current value that the sale of the coin would bring back to its owner. Our interest here is with the broad justification of why any investor would buy a coin that has $f < 1$. Perhaps a small management fee (going towards Ethereum gas fees etc.) could be justified for PoP already, but it would take extraordinary returns to justify $f = 0.2$ or similarly low values.

We start with the notion of immortality, 'the state of living for ever or being remembered for ever' (LDOCE). Let us start with the more modest goal of being remembered forever, which does not seem to contradict any commonsensical law such as *all men are mortal* that we discussed in 5.1. Even so, immortality in this weaker sense will already have its limits. LDOCE distinguishes between *forever*$_1$ 'for all future time' and *forever*$_2$ 'for a very long time' and it is clear that no guarantees can be made in the first sense: what if all intelligent life dies out? What if the universe itself has a finite lifespan? We assume here that investors will settle for a 'very long' time. How long is very long? Historical evidence fails to substantiate any person as non-mythical before the Bronze Age

– a good candidate for 'person longest remembered' may be Pharaoh Menes. This gives about 5,000 years as the record for 'very long'.

Let us next turn to the question: what would it take for someone to be remembered for 5,000 years? Clearly, one would need to create some sort of monument, or better yet, some active mechanism such as the clock built by the Long Now Foundation with a planned lifespan of 10,000 years. Let us simplify the problem of 'remembering' to the storage, and making available, of some records (text, images, etc.) that take up a terabyte compressed. This is vastly more than all the data we have on William Shakespeare, whose collected works are about 5.6 megabyte uncompressed. To store a terabyte costs about $4/month, and to maintain this in perpetuity would require an investment that is capable of yielding $50 for 5,000 years. A good traditional baseline is $1,000 conservatively invested to yield 5%.

To inflation-proof this would require an extra outlay (though storage prices are still falling dramatically, so this is less of an issue than it looks), and so would redundant storage. More complex is the issue of cultural drift, the format and the language of the original records slowly (in the case of storage formats, not so slowly) becoming hard to understand and deal with. Rather than investing in very well protected vaults for CD-ROMs and similar media, a more sensible strategy is to create a robot that monitors formats, and up-converts the data every time an old format is going out of fashion. With this, the problem is reduced to assuring the long-term viability of the robot, understood here not as a physical device but an algorithm operating on the blockchain, or several blockchains.

Over the 5,000 year time horizon several problems arise. First, there is no guarantee that industrial civilization, capable of sustaining the internet, let alone the current blockchain infrastructure, is preserved. Second, and perhaps more important, the legal system that currently guarantees contracts that contain 'in perpetuity' clauses as valid, may change. Besides the need to constantly monitor the software and hardware environment, the robot must adapt to the slow but unstoppable change of language, law, and the physical environment. In this situation, a generational system whereby an old and obsolescent robot is periodically replaced by one better adapted to current and future needs is not unreasonable.

Again, this is less futuristic than it sounds. Biological species survive, and slowly adapt to their environment often on a million-year timescale, though the payload they preserve, their genome, is typically less than 20GB, far below our 1TB target. Since the fossil record is ample proof of their immortality on a 5,000 year timescale, species provide a very strong model for PoI, even in the stronger sense 'living forever' not just the weaker 'being remembered forever' that we started out with. To implement them, we introduce the notion of *stirps*, the set of all coins inheriting from a single coin. The faithfulness parameter f determines how much a given coin is obliged to return to its parent. (To keep everything simple, we assume a single parent, as seen in prokaryotes.)

With this, we have arrived at a setup where artificial life is practically indistinguishable from Artificial Intelligence, except that the standard AI criterion for success, the

Turing test, can be performed at any time, while our criterion for immortality, a better than 5,000 year lifespan will, if taken literally, require at least this much time to achieve. A less literal demonstration, however, should still be feasible: we are quite convinced that Newtonian mechanics can predict celestial orbits 5,000 years out, even though the equations themselves have existed for less than 400 years and have already been supplanted by better ones. All we need are simulation models that show the basic inverse relation between stirps longevity l and the faithfulness parameter f. Once we know that a relation $f \sim 1/f$ holds, it is clear that by sufficiently low f we can obtain arbitrarily high l and we are done.

Our main thesis here is that global explainability is carried not so much by the exact mathematical model as by a verbal explanation of *why* this should be so. Here the explanation is easy to furnish: for any given amount of initial capital, the more a particular coin can spend on itself, the more able it will be to resist environmental changes or adapt to them. This guarantees a longer lifespan for the individual. By the same reasoning, the ability to spend more on offspring will guarantee the same for them, generation after generation. As soon as the blockchain becomes firmly embedded in the financial machinery of civilization, this may even be an experiment worth trying, but our point here is different, that only a small proto-logic carrying 'the more the merrier'-style arguments is required for the task that is of interest to us, namely to explain global behavior in terms that humans understand. There are, to be sure, many systems for which we lack such simple explanations, but XAI must be kept simple enough for humans to trust it.

9.5 Summary

The current (V2) version of 4lang differs from its predecessor, published in S19:4.8 and linked to Concepticon (List, Cysouw, and Forkel, 2016) in 2021, in several respects. The key advance is the change from the English definitions used in V1 to a formal language in V2. While the system is still human readable with a little practice, at this point our chief objective is to automatically translate the definitions to equations obtaining among polytopes and their distinguished points. Consider a d dimensional space where d is in the broadly used 300-800 range, and assume that each unary word corresponds to a vector in this space, and each binary to a transformation over the space L spanned by these vectors.

mark_   Before we turn to the unaries that dominate our data, let us briefly list all entries that are in some way exceptional. Chief among them is mark_ which denotes a relation between a form and its meaning. Since we don't have a theory of phonological form here, this relation remains a primitive. The same is true for wh, which would also require a theory of something outside L, namely the internal model that the speaker has of the knowledge state of the hearer. This is not to say that such theories could not be developed, at least at the naive level, for phonology or for modeling the hearer, but both would require a new volume, similar in size to the present one. As long as these volumes remain unwritten the only methodologically correct stance is to threat the key notions that would rely on them as primitives. When we state the equations, the mark_ clauses are simply omitted (such clauses appear in less than 5% of our definitions, chiefly for affixes). There are no definitions within 4lang that make reference to the interrogative morpheme wh. In 4.5 we rely on the traditional decomposition of *who, when, what, how,* ... as containing this morpheme, but for this treating wh itself as primitive is sufficient. Also exceptional are the unaries =agt and =pat, which trigger unification, and other which blocks unification.

wh

=agt
=pat

The use of vector semantics is particularly clear for gen, which gets mapped to the vector $\langle 1/d, 1/d, \ldots, 1/d|$ in all coordinate systems of interest; er_ which is reduced to arithmetic $>$ (see 7.2); is_a, which is reduced to set-theoretic containment; and lack, which we model by partial complementation rather than multiplication by -1. Logical *and* is built into the system (modeled by the ',' separating clauses within definitions) but logical *or* is largely external to it (see 4.6). Here we kept it binary, but in our opinion it is better treated as a unary predicate over plural entities than a true binary. Since no 4lang definition contains or, we leave the matter for further study, noting that omitting the standard and/or duality fits well with current notions of 'bilateralism' in logic.

This leaves only a handful of binaries to consider: each of these will be assigned a matrix B_j, and will be considered transformations over the linguistic space L. Our binaries are dominantly spatiotemporal at before between follow from in on under (see 3.1 and 3.2), or more conceptual case-like entities such as cause_ for_ has ins_ part_of. All other entries, very much including transitive and higher arity verbs, are treated as vectors, and our goal is to consider the lexicon as a system of equations that can be solved for the vectors and matrices. This is a *static* task in the sense that it pertains to the arrangement of lexical entries relative to one another, and must be distinguished from the *dynamic* task of building vector semantic representations for phrases, clauses, sentences, and even larger units on the fly.

To get a compact statement of the equation system, for each word vector w_i we need two associated notions, the polytope Y_i surrounding it (see 1.4); and the affine half-space H_i^b it defines by $\{x|\langle x, w_i\rangle > b\}$ with bias b (see 7.1). For example, our definition of *water*, repeated for convenience below, means that Y_2622, the polytope for water, is inside Y_846, the polytope for liquid; is outside the polytopes for taste, smell, and color; and Y_505, the polytope for life, is inside the water-need polytope.

```
water víz aqua woda mizu 水 shui3 水 2622 u N
        liquid, lack colour, lack taste, lack smell, life need
```

For humans, it is more convenient to work in the sparse and over-complete generating system given by the vectors w_i, which we call the *natural basis*. (We follow the literature and speak of a sparse overcomplete *basis* even though the vectors are not necessarily linearly independent, and speak of *scalar product* even when the matrix defining the bilinear form is not symmetrical.) If our goal was to get rid of the B_j matrices, which are also extremely hard to conceptualize in other bases, this could be accomplished by enriching the natural basis w_i by new vectors such as water-need, as we suggested for the dynamic case (recall fish-eat in 2.3). As the dynamic case suggest, we may simply consider `water-need` to be the intersection of the `water` and `need` polytopes (see 2.4).

Of particular importance is the coercion mechanism we introduced in 7.3. This is used almost all the time in the process of computing meaning representations during text understanding, e.g. when we identify Esau as the seller and Jacob as the buyer in the `exchange_` schema (Fig. 1.2). The same mechanism is used in about 7% of the

put definitions, for example *put* `=agt cause_ {=pat at place}, =agt move =pat, "locative" mark_ place`. The idea is simple: *X put Y at Z* (where we used *at* for locative, but other essives and inbound latives would work just as well) means the agent is causing the patient to be at the location marked by the locative. When there is no coercion, the defining clauses express generic truths, `cylinder is_a shape, elephant eat <grass>` but `=pat at place` is not generically true.

deep For an even simpler case, consider *deep* `has bottom[far]`, where `bottom[far]` is simply an abbreviation for `bottom is_a far`. Again, it is not at all the case that the bottom of things is far, this is only true of deep things.

This is the same relative clause problem we discussed in 8.3 using the example of *red*, and the overall solution is the same: we have a constraint mediated between `deep` and `far` by `bottom`, and it must be the same `bottom` that the deep thing has that is far. Let us consider how this plays out on Parsons-style examples comparing a *deep puddle*, which may be a couple of feet deep, to a *deep mine*, which may be

far   miles deep. Consulting `4lang` provides for *far* the definition `distance[great]`,
great for *great* the definition `big`, and for *big* the definition `er_ gen`. Applying these
big   we obtain `puddle has bottom is_a distance er_ gen` and `mine has bottom is_a distance er_ gen`. In both cases, we need to iteratively apply the relevant transformations, right to left. `gen` is a vector contained in the central region (polytope) of the positive octant. `er_ gen` is whatever the linear transformation $B_{er\_}$ maps this to. This is the 'set of all oversize things', not a concept we usually entertain, but one that makes perfect sense. Now `bottom` is a subset of this concept. What bottom? Well, the one that *puddle* or *mine* has, which is obtained by right-multiplying their respective polytopes by B_{has}. This means 'set of all things that puddles (resp. mines) have', and we simply apply 'oversize distances to bottom' relative to puddles (resp. mines).

To get this far, we need to coerce the `bottom` in the defining clause to the `place_` schema depicted in Fig. 3.2. This is one place where we don't need to solve any equation: coercion itself amounts to equating the slot with the filler. *Deep* simply assumes (lexically implies) a bottom that is far away. The exact same identification mechanism is often operative during sentence parsing. Consider *Tumithak threatened the destruc-*

threaten *tion of the city. threaten* is defined as `=agt express {after(=agt cause_ harm)}`. To make sense of this, we need to identify the `harm` that is being threatened with *the destruction of the city*. This is actually not hard: we know (from the lexicon) that *destroy* is 'damage something so badly that it no longer exists or cannot be used or

repaired' and that *damage* is 'physical harm', so `{destruction of the city}`
`is_a harm`' follows without any stipulation.

What is critical for all this to work is some pattern matching ability to recognize
specific instances as part of a general rule/pattern. In 1.2 we emphasized this for the
case of substituting English definitions in one another. "[LDOCE] defines PLANET as 'a
large body in space that moves around a star'. If we mechanically substitute this in the
definition of *Jupiter*, 'the largest __ of the Sun' we obtain 'the largest a large body in
space that moves around a star of the Sun'. It takes a great deal of sophistication for the
substitution algorithm to realize that *a large* is subsumed by *the largest* or that *a star* is
instantiated by *the Sun*. People perform these operations with ease, without conscious
effort, but for now we lack parsers of the requisite syntactic and semantic sophistication
to do this automatically." Part of our goal with `4lang` was to enable precisely this, a
mechanistic substitution syntax.

In principle, we need no additional mechanism for doing syntax than we already have
had to posit for describing the lexicon. In practice, a sophisticated treatment of grammatical constructions would need to be added, so as to leverage the pattern matching
ability of speakers and hearers in the production and analysis of blended constructions
first noted in BCG such as *let's not throw out the empirical baby with the theoretical
bathwater*. More elementary synechdoche, metonymy in particular, is already in scope,
given the lack of semantic subtyping (cf. our discussion of the 'institution' and 'building'
senses of *office* in 5.3). But blended constructions remain a great challenge for syntacticians, especially as the current generation of computational models, transformers such
as GPT-3, now produce fully grammatical and naturally flowing multi-paragraph texts.
This fact (besides putting the lie to all theories assuming a genetically defined 'language
organ') renders, in the eyes of many computational linguists, theoretical syntax obsolete.
Readers of this book will have seen that the author is not at all dismissive of linguistic
theory, but he certainly shares in the urgency of producing grammars that work.

Using the current (V2) version of `4lang` means that we work in an overcomplete
basis composed of 760 vectors w_i plus 16 matrices B_j. We have as many equations
as there are definitions, and solving this system of equations can be approached in
many ways. Here we begin with the simplest, which is successive elimination of un-
knowns. At every step we have the option of eliminating a vector, say *atmosphere*   atmosphere
based on its definition `air, Earth has`. This is an ordinary step in solving a system of equations by substitution, but one that has a bad effect on sparseness. Some
of the definitions affected by this change are not affected much, e.g. *Earth* orig-   Earth
inally defined as `planet, in space/2509, life on, ocean on, land
on, has atmosphere` would now become `planet, in space/2509, life
on, ocean on, land on, has air`. But others, such as *rain* `water, from`   rain
`atmosphere, fall/2694, many(drop), weather` are getting perceptibly
more complicated. Instead of the elementary clause `rain from atmosphere` we
now require two clauses `rain from air, Earth has air` which together model

the subordinate clause structure *from the air that Earth has* by automatic unification of the two instances of `air`.

Altogether, substitution affects the sparseness: in the limiting case, with a truly uroboros set of maybe 120 elements, we expect the definitions to become much longer and more convoluted. Whether 62, 120, or 200 primitives remain in the V3 uroboros core, several hundred substitutions can be carried out in the current dictionary, leaving us with p primitives. Ultimately, we will arrive at a dense, p-dimensional basis where we still have p equations, but not all of these are useful. For example, we have `for_/2782` 'dativus finalis' for which we have no definition better than itself. There is an equation `for_ = for_` to be sure, but it adds nothing to the rank r of the equation system. What we have in the normal basis is a very sparse matrix $B_{for\_}$ that is hot only at certain coordinates, e.g. *company = organization, for_ business* means that the `company` row has a high value (near 1) at the `business` column, but not conversely, there is no statement that the raison d'être of business is a company. Perhaps the second clause

business   of our definition of *business* `organization, make money` can be construed as purposive, but we will not pursue this line of thought here – be it as it may, the matrix $B_{for\_}$ is not symmetrical.

The lack of symmetry in these matrices actually points to a much larger issue, namely that the attention matrix P will not give rise to a well-formed scalar product during the dynamic computation of semantic representations. While this does not make actual computations harder (after all, the B_i are explicitly listable very sparse matrices in the natural basis), it does muddy the geometric interpretation a bit, since norm and distance are ordinarily conceived of in term of a symmetrical scalar product matrix. At this point, the hypergraph view introduced in 1.5 is actually easier to visualize. Here the schemas, subordinate clauses, and all material collected between { } as ComplexCaluse (Rule 8) is viewed as a hypernode. In vector semantic terms, the { } are the usual set-forming notation: we treat complex clauses, including schemas, as sets of their component vectors.

Some of the vectors that appear in schemas are easy to name, e.g. `seller` and `buyer` in the 'commercial exchange' schema we introduced in 1.4 (see Fig. 1.2), while others are more abstract. But all are subject to coercion: the seller must be equated to some (agentive) participant, and the goods to some patient. This is not to say that all slots must be filled: many may remain underdetermined. But the relations that are prescribed for the slots are always preserved: for example, the implication that after the exchange the goods become the property of the buyer is true for any particular fillers for these slots. Finally, one area where the vector model can be leveraged in novel ways is temporal modeling. This requires three copies of the natural basis: one for the past, one for the present, and one for the future. Needless to say, the same method of creating separate worlds is equally applicable in any modal situation.

Version news

Perhaps the most perceptible difference between the previous (V1) and the current (V2) version of `4lang` is the manual addition of bindings in two important languages,

Japanese by László Cseresnyési; and Chinese (Mandarin) by Huba Bartos. These required the addition of a few minor lexicographic principles. In Japanese, when two equally good, or equally bad, possibilities exist, we selected the less scholarly/learned, stylistically less marked, and shorter one. In Chinese, where speakers choose between a monosyllabic and a disyllabic, or between a disyllabic and a trisyllabic form based on rhythmic considerations, we always chose the shorter one.

While not exactly a proof of universality in the sense we discussed in 1.2, the relative ease of creating these new bindings goes some way toward ameliorating concerns of eurocentricity. However, it should be emphasized that `4lang` is not a polyglot dictionary, the different bindings are not intended as faithful translations of one another, it is the concepts that are at the center of attention, not the words of one language or another. Less reliable (machine extracted) binding exist for 40 languages (Ács, Pajkossy, and Kornai, 2013), and (Hamerlik, 2022) extends this range further.

V2 is available at GitHub under kornai/4lang/tree/master/V2/700.tsv. The Japanese and Chinese bindings are added after the original fourth column as fifth and sixth respectively. Both begin with Latin transliteration (Hepburn and PinYin respectively) followed by utf8 codepoints that attempt to go around difficulties of Han unification. With this, our first example from 1.2 now looks

```
self önmaga ipse sam jibun 自分 zi4- 自; zi4ji3 自己 1851 e N
     =pat[=agt], =agt[=pat]
```

The English material in `700.tsv` appears in the Appendix. The definitions have been checked for syntactic correctness by Ádám Kovács's `def_ply_parser.py` and are known to contain errors. First, some entries which have no proper graph definition (`=agt` and `=pat` in particular) fail to parse. Since these are primitives (are defined by themselves) this will not be fixed. Second, number tokens (included as defaults e.g. in *hour*) `time, unit, day has <24>, has <60>`(minute) throw errors, and **hour** this again is unlikely to get fixed. Third, and most important, the parser's treatment of U/V alternation (2.5) leaves something to be desired, in that it forces several typically intransitive entries such as *fly* or *work* to the *binaries* list. While it is true that such entries have transitive readings *fly a kite, work the fields*, this will likely get revamped in V3 as part of the planned move to parallel synchronous rewriting. For now, the parser remains a debugging tool, particularly valuable as it already has the ability to undo the anuvṛtti.

References

Abend, Omri and Ari Rappoport (2013). "UCCA: A semantics-based grammatical annotation scheme". In: *IWCS'13*, pp. 1–12.

Abney, Steven (1991). "Parsing by chunks". In: *Principle-based parsing*. Ed. by Robert Berwick, Steven Abney, and Carol Tenny. Kluwer Academic Publishers, pp. 257–278.

Ács, Evelin and Gábor Recski (2018). "Semantic parsing with Interpreted Regular Tree Grammars". In: *Proceedings of the Automation and Applied Computer Science Workshop 2018 : AACS'18*. Ed. by Dmitriy Dunaev and István Vajk. Budapest University of Technology and Economics, pp. 43–53.

Ács, Judit and András Kornai (2020). "The Role of Interpretable Patterns in Deep Learning for Morphology". In: *XVI. Magyar Számítógépes Nyelvészeti Konferencia (MSZNY2020)*. Szeged, pp. 171–179.

Ács, Judit, Dávid Márk Nemeskey, and Gábor Recski (2019). "Building word embeddings from dictionary definitions". In: *K + K = 120: Papers dedicated to László Kálmán and András Kornai on the occasion of their 60th birthdays*. Ed. by Katalin Mády Beáta Gyuris and Gábor Recski. Research Institute for Linguistics, Hungarian Academy of Sciences (RIL HAS).

Ács, Judit, Katalin Pajkossy, and András Kornai (2013). "Building basic vocabulary across 40 languages". In: *Proceedings of the Sixth Workshop on Building and Using Comparable Corpora*. Sofia, Bulgaria: Association for Computational Linguistics, pp. 52–58.

Ács, Judit and Géza Velkey (2017). "Comparing word segmentation algorithms". In: *Proceedings of the Automation and Applied Computer Science Workshop 2017 : AACS'17*. Budapest University of Technology and Economics.

Agirre, Eneko and Philip Edmonds, eds. (2007). *Word Sense Disambiguation*. Springer.

Aho, Alfred V. and Jeffrey D. Ullman (1971). "Translations on a context-free grammar". In: *Information and Control* 19, pp. 439–475.

Ajdukiewicz, Kazimierz (1935). "Die Syntaktische Konnexität". In: *Studia Philosophica* 1, pp. 1–27.

Ajtai, Miklós (1994). "The complexity of the Pigeonhole Principle". In: *Combinatorica* 14.4, pp. 417–433. DOI: 10.1007/BF01302964.

Akbik, Alan, Tanja Bergmann, and Roland Vollgraf (June 2019). "Pooled Contextualized Embeddings for Named Entity Recognition". In: *Proceedings of the 2019 Conference of the North American Chapter of the Association for Computational Linguistics: Human Language Technologies, Volume 1 (Long and Short Papers)*. Minneapolis, Minnesota: Association for Computational Linguistics, pp. 724–728. DOI: 10.18653/v1/N19-1078. URL: https://www.aclweb.org/anthology/N19-1078.

Akbik, Alan, Duncan Blythe, and Roland Vollgraf (Aug. 2018). "Contextual String Embeddings for Sequence Labeling". In: *Proceedings of the 27th International Conference on Computational Linguistics*. Santa Fe, New Mexico, USA: Association for

Computational Linguistics, pp. 1638–1649. URL: `https://www.aclweb.org/anthology/C18-1139`.

Allauzen, Alexandre et al., eds. (2013). *Proceedings of the Workshop on Continuous Vector Space Models and their Compositionality*. Association for Computational Linguistics. URL: `http://aclweb.org/anthology/W13-32`.

Anderson, J.M. (2006). *Modern grammars of case: a retrospective*. Oxford University Press, USA.

Anderson, Stephen R. (1982). "Where Is Morphology?" In: *Linguistic Inquiry* 13, pp. 571–612.

Angluin, Dana (1981). "A Note on the Number of Queries Needed to Identify Regular Languages". In: *Information and Control* 51, pp. 76–87.

— (1987). "Learning Regular Sets From Queries and Counterexamples". In: *Information and Computation* 75, pp. 87–106.

Angluin, Dana and Leslie Valiant (1979). "Fast Probabilistic Algorithms for Hamiltonian circuits and Matchings". In: *Journal of Computer and System Sciences* 18, pp. 155–193.

Arora, Sanjeev et al. (2015). "Random Walks on Context Spaces: Towards an Explanation of the Mysteries of Semantic Word Embeddings". In: *arXiv:1502.03520v1* 4, pp. 385–399.

Atiyah, Michael (2001). *Geometry vs Algebra. An excerpt from Mathematics in the 20th century*. URL: `https://people.math.umass.edu/~hacking/461F19/handouts/atiyah.pdf`.

Ayadi, Ali et al. (2019). "Ontology population with deep learning-based NLP: a case study on the Biomolecular Network Ontology". In: *Procedia Computer Science*. Elsevier, pp. 572–581. URL: `https://hal.archives-ouvertes.fr/hal-02317227/file/li.pdf`.

Bach, Emmon (1977). *An extension of classical transformational grammar*. UMASS.

Badia, Antonio (2009). *Quantifiers in Action: Generalized Quantification in Query, Logical and Natural Languages*. Springer. ISBN: 9780387095639.

Bahdanau, Dzmitry, Kyunghyun Cho, and Yoshua Bengio (2015). "Neural machine translation by jointly learning to align and translate". In: *International Conference on Learning Representations (ICLR 2015)*.

Baker, Collin, Michael Ellsworth, and Katrin Erk (June 2007). "SemEval-2007 Task 19: Frame Semantic Structure Extraction". In: *Proceedings of the Fourth International Workshop on Semantic Evaluations (SemEval-2007)*. Prague, Czech Republic: Association for Computational Linguistics, pp. 99–104. URL: `https://aclanthology.org/S07-1018`.

Baker, Mark C. (2003). *Lexical Categories*. Oxford University Press.

Banarescu, Laura et al. (2013). "Abstract Meaning Representation for Sembanking". In: *Proceedings of the 7th Linguistic Annotation Workshop and Interoperability with Discourse*. Sofia, Bulgaria: Association for Computational Linguistics, pp. 178–186. URL: `https://www.aclweb.org/anthology/W13-2322`.

Barwise, Jon and John Perry (1983). *Situations and Attitudes*. MIT Press.

Bazzi, Issam (2002). "Modelling out-of-vocabulary words for robust speech recognition". PhD thesis. Massachusetts Institute of Technology.

Belinkov, Yonatan et al. (2017a). "Evaluating layers of representation in neural machine translation on part-of-speech and semantic tagging tasks". In: *IJCNLP*. arXiv preprint arXiv:1801.07772.

Belinkov, Yonatan et al. (2017b). "What do neural machine translation models learn about morphology?" In: *ACL*. arXiv preprint arXiv:1704.03471.

Bell, John Lane (1988). "Infinitesimals". In: *Synthese* 75, pp. 285–315.

— (2008). *A primer of infinitesimal analysis*. Cambridge University Press.

Bellard, Fabrice (2019). *Lossless Data Compression with Neural Networks*. URL: https://bellard.org/nncp/nncp.pdf.

Belnap, Nuel D. (1977). "How a computer should think". In: *Contemporary Aspects of Philosophy*. Ed. by G. Ryle. Newcastle upon Tyne: Oriel Press, pp. 30–56.

Benacerraf, Paul (1973). "Mathematical Truth". In: *The Journal of Philosophy* 70.19, pp. 661–679.

Bengio, Yoshua (2008). "Neural net language models". In: *Scholarpedia* 3.1, p. 3881. URL: http://www.scholarpedia.org/article/Neural_net_language_models.

Bergelson, Elika and Daniel Swingley (2013). "The acquisition of abstract words by young infants". In: *Cognition* 127, pp. 391–397.

Berko, Jean (1958). "The child's learning of English morphology". In: *Word* 14, pp. 150–177.

Berlin, Brent and Paul Kay (1969). *Basic Color Terms: Their Universality and Evolution*. Berkeley: University of California Press.

Blackburn, Patrick, Maarten de Rijke, and Yde Venema (2001). *Modal logic*. Cambridge University Press.

Bloomfield, Leonard (1926). "A set of postulates for the science of language". In: *Language* 2, pp. 153–164.

— (1933). *Language*. London: George Allen and Unwin.

Boguraev, Branimir K. and Edward J. Briscoe (1989). *Computational Lexicography for Natural Language Processing*. Longman.

Bohnstingl, Thomas et al. (2021). "Towards efficient end-to-end speech recognition with biologically-inspired neural networks". In: *arXiv preprint arXiv:2110.02743*.

Bojanowski, Piotr et al. (2017). "Enriching Word Vectors with Subword Information". In: *Transactions of the Association for Computational Linguistics* 5, pp. 135–146. ISSN: 2307-387X. URL: https://transacl.org/ojs/index.php/tacl/article/view/999.

Bommasani, Rishi, Kelly Davis, and Claire Cardie (July 2020). "Interpreting pretrained Contextualized Representations via Reductions to Static Embeddings". In: *Proceedings of the 58th Annual Meeting of the Association for Computational Linguistics*. Online: Association for Computational Linguistics, pp. 4758–4781. DOI: 10 .

18653/v1/2020.acl-main.431. URL: https://aclanthology.org/
2020.acl-main.431.

Boole, George (1854). *An Investigation of the Laws of Thought on Which are Founded the Mathematical Theories of Logic and Probabilities*. Macmillan.

Borbély, Gábor et al. (2016). "Denoising composition in distributional semantics". In: *DSALT: Distributional Semantics and Linguistic Theory*. poster.

Borschev, Vladimir and Barbara Partee (2014). "Ontology and integration of formal and lexical semantics". In: *In Computational Linguistics and Intellectual Technologies: Annual Conference "Dialogue"*. Ed. by V.P. Selegey, pp. 114–127. URL: http://www.dialog-21.ru/digests/dialog2014/materials/pdf/BorschevVBParteeBH.pdf.

Böttner, Michael (2001). "Peirce Grammar". In: *Grammars* 4.1, pp. 1–19.

Brachman, R.J. and H. Levesque (1985). *Readings in knowledge representation*. Morgan Kaufmann Publishers Inc., Los Altos, CA.

— (2004). *Knowledge Representation and reasoning*. Morgan Kaufmann Elsevier, Los Altos, CA.

Brants, Thorsten and Alex Franz (2006). *Web 1T 5-gram Version 1*. Philadelphia: Linguistic Data Consortium.

Bresnan, Joan (1982). "The passive in lexical theory". In: *The mental representation of grammatical relations*. Ed. by Joan Bresnan. MIT Press, pp. 3–86.

Brill, Eric (1994). "Some Advances in Transformation-Based Part of Speech Tagging". In: *National Conference on Artificial Intelligence*, pp. 722–727.

Brown, Peter F et al. (1993). "The mathematics of statistical machine translation: Parameter estimation". In: *Computational linguistics* 19.2, pp. 263–311.

Bruening, Benjamin (2018). "The lexicalist hypothesis: Both wrong and superfluous". In: *Language* 94.1, pp. 1–42. DOI: 10.1353/lan.2018.0000.

Buck, Carl Darling (1949). *A Dictionary of Selected Synonyms in the Principal Indo-European Languages*. University of Chicago Press.

Bullon, Stephen (2003). *Longman Dictionary of Contemporary English*. 4th ed. Longman.

Burnard, Lou and Guy Aston (1998). *The BNC handbook: exploring the British National Corpus*. Edinburgh University Press.

Butt, Miriam et al. (2002). "The Parallel Grammar Project". In: *COLING-02: Grammar Engineering and Evaluation*. URL: https://aclanthology.org/W02-1503.

Cabrera, Julio (2001). ""The Lexicon-Encyclopedia Interface" by Bert Peeters (ed.)" In: *Pragmatics and Cognition* 9.2, pp. 313–327. DOI: https://doi.org/10.1075/pc.9.2.09cab.

Campos, Joseph J., Alan Langer, and Alice Krowitz (1970). "Cardiac Responses on the Visual Cliff in Prelocomotor Human Infants". In: *Science* 170.3954, pp. 196–197. ISSN: 0036-8075. DOI: 10.1126/science.170.3954.196. eprint: https:

//science.sciencemag.org/content/170/3954/196.full.pdf.
URL: https://science.sciencemag.org/content/170/3954/196.

Carlson, G.N. (1977). "A unified analysis of the English bare plural". In: *Linguistics and Philosophy* 1, pp. 413–457. DOI: https://doi.org/10.1007/BF00353456.

Carroll, John A. (1983). *An island parsing interpreter for the full augmented transition network formalism.* ACL Proceedings, First European Conference, pp. 101–105.

Cawdrey, Robert (1604). *A table alphabetical of hard usual English words.*

Chandlee, Jane and Adam Jardine (Mar. 2019). "Autosegmental Input Strictly Local Functions". In: *Transactions of the Association for Computational Linguistics* 7, pp. 157–168. DOI: 10.1162/tacl_a_00260. URL: https://aclanthology.org/Q19-1010.

Chang, Angel X. and Christopher Manning (May 2012). "SUTime: A library for recognizing and normalizing time expressions". In: *Proceedings of the Eighth International Conference on Language Resources and Evaluation (LREC'12)*. Istanbul, Turkey: European Language Resources Association (ELRA), pp. 3735–3740. URL: http://www.lrec-conf.org/proceedings/lrec2012/pdf/284_Paper.pdf.

Chen, Conrad and Hsi-Jian Lee (2004). "A three-phase system for Chinese names entity recognition". In: *Proceedings of RO-CLING.*

Chomsky, Noam (1959). "Review of Skinner 1957". In: *Language* 35.1, pp. 26–58.

— (1966). *Cartesian Linguistics.* Harper and Row.

— (1970). "Remarks on nominalization". In: *Readings in English Transformational Grammar.* Ed. by R. Jacobs and P. Rosenbaum. Waltham, MA: Blaisdell, pp. 184–221.

— (1973). "Conditions on Transformations". In: *A festschrift for Morris Halle.* Ed. by S.R. Anderson and P. Kiparsky. New York: Holt, Rinehart and Winston.

Chung, Sandra (2012). "Are lexical categories universal? The view from Chamorro". In: *Theoretical Linguistics* 38, pp. 1–56.

Church, Alonzo (1936). "An Unsolvable Problem of Elementary Number Theory". In: *American Journal of Mathematics* 58.2, pp. 345–363. URL: http://www.jstor.org/stable/2371045.

Ciabattoni, A. and B. Lellmann (2021). "Sequent rules for reasoning and conflict resolution in conditional norms". In: *DEON 2020/2021.* Ed. by F. Liu et al. College Publications.

Clark, Kevin et al. (Aug. 2019). "What Does BERT Look at? An Analysis of BERT's Attention". In: *Proceedings of the 2019 ACL Workshop BlackboxNLP: Analyzing and Interpreting Neural Networks for NLP*. Florence, Italy: Association for Computational Linguistics, pp. 276–286. DOI: 10.18653/v1/W19-4828. URL: https://www.aclweb.org/anthology/W19-4828.

Collins, A.M. and E.F. Loftus (1975). "A spreading-activation theory of semantic processing". In: *Psychological Review* 82, pp. 407–428.

Collins, A.M. and M.R. Quillian (1969). "Retrieval time from semantic memory". In: *Journal of Verbal Learning and Verbal Behavior* 8, pp. 240–247.

Collins, Michael and Yoram Singer (1999). *Unsupervised Models for Named Entity Classification*.

Collobert, R. et al. (2011). "Natural Language Processing (Almost) from Scratch". In: *Journal of Machine Learning Research (JMLR)*.

Collobert, Ronan and Jason Weston (2008). "A Unified Architecture for Natural Language Processing: Deep Neural Networks with Multitask Learning". In: *Proceedings of the 25th International Conference on Machine Learning*. ICML '08. Helsinki, Finland: ACM, pp. 160–167.

Comrie, Bernard (1976). "The syntax of causative constructions: cross-linguistic similarities and divergences". In: *Syntax and semantics 6*. Ed. by Masayoshi Shibatani. Academic Press.

Couprie, Dirk (2004). "How Thales Was Able to "Predict" a Solar Eclipse Without the Help of Alleged Mesopotamian Wisdom". In: *Early Science and Medicine* 9 (4), pp. 321–337.

Courcelle, Bruno and Joost Engelfriet (2012). *Graph structure and monadic second-order logic*. Cambridge University Press.

Cresswell, Max J. (1975). "Hyperintensional Logic". In: *Studia Logica* 34.1, pp. 25–38. URL: http://www.jstor.org/stable/20014742.

— (1976). "The semantics of degree". In: *Montague Grammar*. Ed. by Barbara Partee. Academic Press, pp. 263–292.

Creutz, Mathias and Krista Lagus (2007). "Unsupervised models for morpheme segmentation and morphology learning". In: *ACM Transactions on Speech and Language Processing (TSLP)* 4.1, p. 3.

da Costa, Newton C.A. (1980). "A Model-Theoretical Approach to Variable Binding Term Operators". In: *Mathematical Logic in Latin America*. Ed. by A.I. Arruda, R. Chuaqui, and N.C.A. Da Costa. Vol. 99. Studies in Logic and the Foundations of Mathematics. Elsevier, pp. 133–162. DOI: https://doi.org/10.1016/S0049-237X(09)70484-7. URL: https://www.sciencedirect.com/science/article/pii/S0049237X09704847.

Dahlgren, Kathleen (1988). *Naive Semantics for Natural Language Understanding*. Kluwer. ISBN: 978-0-89838-287-7.

— (1995). "A linguistic ontology". In: *International Journal of Human-Computer Studies* 43, pp. 809–818.

Davidson, Donald (1967). "The logical form of action sentences". In: *The Logic of Decision and Action*. Ed. by N. Rescher. University of Pittsburgh Press, pp. 81–95.

— (1980). *Essays on Actions and Events*. Oxford: Clarendon Press.

De Mey, M. (1972). "The Psychology of Negation and Attention". In: *Logique et Analyse* 15, pp. 137–153.

Deeks, Ashley (2019). "The judicial demand for explainable artificial intelligence". In: *Columbia Law Review* 119.7, pp. 1829–1850. URL: `https://www.jstor.org/stable/26810851`.

Deerwester, Scott C., Susan T Dumais, and Richard A. Harshman (1990). "Indexing by latent semantic analysis". In: *Journal of the American Society for Information Science* 41.6, pp. 391–407.

Dehaene, Stanislas (1997). *The number sense*. Oxford University Press.

Desprès, Sylvie et al. (2020). "Enhancing a Biomedical Ontology with Knowledge from Discharge Summaries". In:

Devlin, Jacob et al. (2019). "BERT: Pre-training of Deep Bidirectional Transformers for Language Understanding". In: *Proc. of NAACL*.

Devlin, Keith (1991). *Logic and Information*. Cambridge University Press.

— (2008). *The Unfinished Game: Pascal, Fermat, and the Seventeenth-Century Letter that Made the World Modern*. Basic Books.

Diederich, Paul Bernard (1939). *The frequency of Latin words and their endings*. The University of Chicago Press.

Doherty, Patrick, Witold Lukaszewicz, and Andrzej Szalas (2000). "Efficient reasoning using the local closed world assumption". In: *Proc 9th International Conference on AI: Methodology, Systems, Applications (AIMSA 2000)*.

Dowty, David (1986). *On the semantic content of the notion thematic role*.

— (1989). "On the semantic content of the notion thematic role". In: *Property theory, type theory and natural language semantics*. Ed. by G. Chierchia, B. Partee, and R. Turner. Dordrecht: D. Reidel.

Drewes, Frank, Hans-Jörg Kreowski, and Annegret Habel (1997). "Hyperedge replacement graph grammars". In: *Handbook of Graph Grammars and Computing by Graph Transformation*. Ed. by Grzegorz Rozenberg. World Scientific, pp. 95–162.

Dufter, Philipp, Nora Kassner, and Hinrich Schütze (June 2021). "Static Embeddings as Efficient Knowledge Bases?" In: *Proceedings of the 2021 Conference of the North American Chapter of the Association for Computational Linguistics: Human Language Technologies*. Online: Association for Computational Linguistics, pp. 2353–2363. DOI: `10.18653/v1/2021.naacl-main.186`. URL: `https://aclanthology.org/2021.naacl-main.186`.

Elman, Jeffrey L (1990). "Finding structure in time". In: *Cognitive science* 14.2, pp. 179–211.

Endresen, Anna (2013). *Morphological Intensifiers Beyond Adjectives: Evidence from productive patterns of Russian prefixation*. URL: `https://sites.ualberta.ca/~iclc2013/PRESENTATIONS/Endresen/Intensifiers%20handout.pdf`.

Etherington, D.W. (1987). "Formalising Nonmonotonic Reasoning Systems". In: *Artificial Intelligence* 31, pp. 41–85.

Fauconnier, Gilles (1985). *Mental Spaces*. MIT Press.

Fedorenko, Evelina et al. (2020). "Lack of selectivity for syntax relative to word meanings throughout the language network". In: *Cognition* 203, p. 104348. ISSN: 0010-0277. DOI: https://doi.org/10.1016/j.cognition.2020.104348. URL: https://www.sciencedirect.com/science/article/pii/S0010027720301670.

Fillmore, Charles (1968). "The case for case". In: *Universals in Linguistic Theory*. Ed. by E. Bach and R. Harms. New York: Holt and Rinehart, pp. 1–90.

Fillmore, Charles and Sue Atkins (1998). "FrameNet and lexicographic relevance." In: *Proceedings of the First International Conference on Language Resources and Evaluation*. Granada, Spain.

Firbas, J. (1971). *On the Concept of Communicative Dynamism in the Theory of Functional Sentence Perspective*. Brno University.

Firth, John R. (1957). "A synopsis of linguistic theory". In: *Studies in linguistic analysis*. Blackwell, pp. 1–32.

Fodor, Jerry A. (1983). *The Modularity of Mind*. MIT Press.

— (1998). *Concepts*. Clarendon Press.

Fossati, Davide et al. (2006). "The problem of ontology alignment on the Web: A first report". In: *Proceedings of the 2nd International Workshop on Web as Corpus*, pp. 51–58. URL: https://aclanthology.org/W06-1708.

Frege, Gottlob (1879). *Begriffsschrift: eine der arithmetischen nachgebildete Formelsprache des reinen Denkens*. Halle: L. Nebert.

— (1892). "On sense and reference". In: *The Philosophy of Language*. Ed. by A.P. Martinich. New York: Oxford University Press (4th ed, 2000), pp. 36–56.

Fuenmayor, David and Christoph Benzmüller (2019). "Harnessing Higher-Order (Meta-)Logic to Represent and Reason with Complex Ethical Theories". In: *PRICAI 2019: Trends in Artificial Intelligence*. Ed. by Abhaya C. Nayak and Alok Sharma. Springer International Publishing, pp. 418–432. ISBN: 978-3-030-29908-8.

Gabbay, Dov et al., eds. (2013). *Handbook of Deontic Logic and Normative Systems*. College Publications.

Gallese, Vittorio and George Lakoff (2005). "The Brain's concepts: the role of the Sensory-motor system in conceptual knowledge". In: *Cognitive Neuropsychology* 22.3–4, pp. 455–479. DOI: 10.1080/02643290442000310. URL: https://doi.org/10.1080/02643290442000310.

Gärdenfors, Peter (2000). *Conceptual Spaces: The Geometry of Thought*. MIT Press.

— ed. (2007). *Generalized quantifiers*. Reidel.

Gazdar, Gerald et al. (1985). *Generalized Phrase Structure Grammar*. Oxford: Blackwell.

Gémes, Kinga, Ádám Kovács, and Gábor Recski (2019). "Machine comprehension using semantic graphs". In: *Proceedings of the Automation and Applied Computer Science Workshop 2019 : AACS'19*. Ed. by Dmitriy Dunaev and István Vajk. Budapest University of Technology and Economics, pp. 90–98.

Genabith, Josef Van and Richard Crouch (1999). In: *Semantics and syntax in Lexical Functional Grammar: The resource logic approach*. Ed. by Mary Dalrymple. MIT Press, pp. 209–260.

Gewirth, A. (1978). *Reason and morality*. University of Chicago Press.

Ghosh, Swarnendu et al. (2018). "The journey of graph kernels through two decades". In: *Computer Science Review* 27, pp. 88–111. ISSN: 1574-0137. DOI: https://doi.org/10.1016/j.cosrev.2017.11.002. URL: https://www.sciencedirect.com/science/article/pii/S1574013717301429.

Giannakidou, Anastasia (1997). "The Landscape of Polarity Items". PhD thesis. University of Groningen.

Gittens, Alex, Dimitris Achlioptas, and Michael W. Mahoney (2017). "Skip-Gram − Zipf + Uniform = Vector Additivity". In: *Proceedings of the 55th Annual Meeting of the Association for Computational Linguistics (Volume 1: Long Papers)*. Vancouver, Canada: Association for Computational Linguistics, pp. 69–76. DOI: 10.18653/v1/P17-1007. URL: http://aclweb.org/anthology/P17-1007.

Givón, Talmy (1979). *On understanding grammar*. Academic Press.

Glavaš, Goran et al. (2020). "SemEval-2020 Task 2: Predicting Multilingual and Cross-Lingual (Graded) Lexical Entailment". In: *Proceedings of the 13th International Workshop on Semantic Evaluation*. Association for Computational Linguistics.

Goddard, Cliff (2002). "The search for the shared semantic core of all languages". In: *Meaning and Universal Grammar − Theory and Empirical Findings*. Ed. by Cliff Goddard and Anna Wierzbicka. Vol. 1. Benjamins, pp. 5–40.

Goddard, Cliff and Anna Wierzbicka (2014). *Words and Meanings: Lexical Semantics across Domains, Languages and Cultures*. Oxford University Press. ISBN: 978-0-19-966843-4.

Goldberg, Yoav (2017). *Neural Network Methods for Natural Language Processing*. Morgan Claypool.

Gontrum, Johannes et al. (2017). "Alto: Rapid Prototyping for Parsing and Translation". In: *Proceedings of the Software Demonstrations of the 15th Conference of the European Chapter of the Association for Computational Linguistics*. Valencia, Spain: Association for Computational Linguistics, pp. 29–32. URL: https://www.aclweb.org/anthology/E17-3008.

Gordon, Andrew and Jerry Hobbs (2017). *A Formal Theory of Commonsense Psychology: How People Think People Think*. Cambridge University Press.

Gove, Philip Babcock, ed. (1961). *Webster's Third New International Dictionary of the English Language, Unabridged*. G. & C. Merriam.

Graham, AC (1958). *Two Chinese Philosophers*. Lund Humphries.

Graves, Alex (2012). "Offline Arabic handwriting recognition with multidimensional recurrent neural networks". In: *Guide to OCR for Arabic scripts*. Springer, pp. 297–313.

Graves, Alex, Abdel-rahman Mohamed, and Geoffrey Hinton (2013). "Speech recognition with deep recurrent neural networks". In: *2013 IEEE international conference on acoustics, speech and signal processing*. IEEE, pp. 6645–6649.

Graves, Alex and Jürgen Schmidhuber (2009). "Offline handwriting recognition with multidimensional recurrent neural networks". In: *Advances in neural information processing systems*, pp. 545–552.

Greff, Klaus et al. (2015). "LSTM: A search space odyssey". In: arXiv: `1503.04069 [cs.NE]`.

Grice, H.P. (1975). "Logic and Conversation". In: *Syntax and Pragmatics 3: Speech Acts*. Ed. by P. Cole and J. Morgan. Academic Press, pp. 41–58.

Grishman, Ralph and Beth Sundheim (May 1996). "Design of the MUC-6 Evaluation". In: *TIPSTER TEXT PROGRAM PHASE II: Proceedings of a Workshop held at Vienna, Virginia, May 6-8, 1996*. Vienna, Virginia, USA: Association for Computational Linguistics, pp. 413–422. DOI: `10.3115/1119018.1119072`. URL: `https://www.aclweb.org/anthology/X96-1047`.

Groschwitz, Jonas, Alexander Koller, and Christoph Teichmann (2015). "Graph parsing with s-graph grammars". In: *Proceedings of the 53rd ACL and 7th IJCNLP*. Beijing.

Guberman, Nitzan (2016). *On Complex Valued Convolutional Neural Networks*. arXiv: `1602.09046 [cs.NE]`.

Gumma, Krishna Murali et al. (2011). "Expansion of urban area and wastewater irrigated rice area in Hyderabad, India". In: *Irrigation and Drainage Systems* 25 (3), pp. 135–149. DOI: `10.1007/s10795-011-9117-y`. URL: `https://doi.org/10.1007/s10795-011-9117-y`.

Guralnik, David B., ed. (1958). *Webster's New World Dictionary of the American Language*. The World Publishing Company.

Gurevich, Yuri (June 1988). "On Kolmogorov Machines And Related Issues". In: *Bulletin of EATCS* 35, pp. 71–82.

Gutzmann, Daniel et al., eds. (2021). *The Wiley Blackwell Companion to Semantics*. Wiley-Blackwell. ISBN: 978-1-118-78831-8.

Gyenis, Zalán and András Kornai (2019). "Naive probability". In: *ArXiv*, p. 1905.10924.

Halácsy, Péter et al. (2008). "Parallel Creation of Gigaword Corpora for Medium Density Languages-an Interim Report." In: *LREC*.

Hamawand, Zeki (2011). *Morphology in English: Word Formation in Cognitive Grammar*. Continuum. ISBN: 9781441111371.

Hamerlik, Endre (2022). "Polyglot core vocabulary". In: *MSc Thesis, Budapest University of Technology and Economics*.

Hammond, Michael (1995). "Metrical phonology". In: *Annual Review of Anthropology* 24, pp. 313–342.

Hanks, Patrick (2000). "Do word meanings exist". In: *Computers and the Humanities*, pp. 171–177.

Hanne, Sandra, Frank Burchert, and Shravan Vasishth (2016). "On the nature of the subject–object asymmetry in wh-question comprehension in aphasia: evidence from eye tracking". In: *Aphasiology* 30.4, pp. 435–462. DOI: 10.1080/02687038.2015.1065469. URL: https://doi.org/10.1080/02687038.2015.1065469.

Harris, Zellig (1951). *Methods in Structural Linguistics*. University of Chicago Press.

Harris, Zellig S. (1954). "Distributional structure". In: *Word* 10.23, pp. 146–162.

Hasnul, Muhammad Anas et al. (2021). "Electrocardiogram-Based Emotion Recognition Systems and Their Applications in Healthcare – A Review". In: *Sensors* 21.15. ISSN: 1424-8220. DOI: 10.3390/s21155015. URL: https://www.mdpi.com/1424-8220/21/15/5015.

Haspelmath, Martin (2021). "Word class universals and language-particular analysis". In: *ms, Max Planck Institute for Evolutionary Anthropology*.

Haugen, Einar (1957). "The Semantics of Icelandic Orientation". In: *Word* 13.3, pp. 447–459. DOI: 10.1080/00437956.1957.11659646.

Haviland, John B. (2000). "Pointing, gesture spaces, and mental maps". In: *Language and gesture*. Cambridge University Press, pp. 13–46.

Hayes, Patrick J. (1978). *The Naive Physics Manifesto*. Geneva: Institut Dalle Molle.

— (1979). "The naive physics manifesto". In: *Expert Systems in the Micro-Electronic Age*. Ed. by D. Michie. Edinburgh University Press, pp. 242–270.

Head, Henry and Gordon Holmes (Nov. 1911). "Sensory disturbances from cerebral lesions". In: *Brain* 34.2-3, pp. 102–254. ISSN: 0006-8950. DOI: 10.1093/brain/34.2-3.102. eprint: http://oup.prod.sis.lan/brain/article-pdf/34/2-3/102/933215/34-2-3-102.pdf. URL: https://doi.org/10.1093/brain/34.2-3.102.

Heider, Fritz and Marianne Simmel (1944). "An Experimental Study of Apparent Behavior". In: *The American Journal of Psychology* 57.2, pp. 243–259. DOI: 10.2307/1416950.

Heim, Irene (1982). *The Semantics of Definite and Indefinite Noun Phrases*. University of Massachusetts, Amherst, MA: PhD thesis.

Hellwig, P. (1993). "Extended Dependency Unification Grammar". In: ed. by V. Agel et al., pp. 593–635.

Hemphill, Charles T, John J Godfrey, and George R Doddington (1990). "The ATIS spoken language systems pilot corpus". In: *Proc. DARPA speech and NL workshop*, pp. 96–101.

Hertz, John A, Anders S Krogh, and Richard G Palmer (1991). *Introduction to the Theory of Neural Computation*. Vol. 1. Redwood City, CA: Addison-Wesley.

Hewitt, John and Christopher D Manning (2019). "A structural probe for finding syntax in word representations". In: *Proceedings of the 2019 Conference of the North American Chapter of the Association for Computational Linguistics: Human Language Technologies, Volume 1 (Long and Short Papers)*, pp. 4129–4138.

Hirose, Akira, ed. (2003). *Complex-valued Neural Networks*. World Scientific.

Hobbs, Jerry R. and Feng Pan (2004). "An Ontology of Time for the Semantic Web". In: *ACM Transactions on Asian Language Processing (TALIP)* 3.1, pp. 66–85.

Höche, Silke (2009). *Cognate Object Constructions in English: A Cognitive-linguistic Account*. Gunter Narr Verlag. ISBN: 978 3 8233 6489 4.

Hochreiter, Sepp and Jürgen Schmidhuber (Nov. 1997). "Long Short-Term Memory". In: *Neural Computation* 9.8, pp. 1735–1780.

Horn, Larry (1989). *The Natural History of Negation*. Chicago: University of Chicago Press.

Hovav, Malka Rappaport and Beth Levin (2008). "The English dative alternation: The case for verb sensitivity". In: *Journal of Linguistics* 44.1, pp. 129–167. DOI: 10.1017/S0022226707004975.

Huffman, David A. (1952). "A method for the construction of minimum redundancy codes". In: *Proceedings of the IRE*. Vol. 40, pp. 1098–1101.

Ioannidis, John P. A. (2005). "Why Most Published Research Findings Are False". In: *PLoS Medicine*. URL: https://doi.org/10.1371/journal.pmed.0020124.

Jackendoff, Ray S. (1972). *Semantic Interpretation in Generative Grammar*. MIT Press.

— (1977). *X-bar Syntax: A Study of Phrase Structure*. MIT Press.

— (1983). *Semantics and Cognition*. MIT Press.

— (1990). *Semantic Structures*. MIT Press.

Jackendoff, Ray and Jenny Audring (2020). *The texture of the lexicon*. Oxford University Press.

Jaynes, E.T. (2003). *Probability theory*. Cambridge University Press.

Jespersen, Otto (1965). *A Modern English grammar on historical principles*. Vol. VI. London: Allen and Unwin.

Johnson, Kent (2015). "Notational Variants and Invariance in Linguistics". In: *Mind & Language* 30.2, pp. 162–186. DOI: 10.1111/mila.12076. eprint: https://onlinelibrary.wiley.com/doi/pdf/10.1111/mila.12076. URL: https://onlinelibrary.wiley.com/doi/abs/10.1111/mila.12076.

Jones, Gary, Fernand Gobet, and Julian M. Pine (2000). "A Process Model of Children's Early Verb Use". In: *Proceedings of the 22nd Annual Meeting of the Cognitive Science Society*. Ed. by L.R. Gleitman and A.K. Joshi. Lawrence Erlbaum, pp. 723–728.

Jordan, Michael I. (May 1986). *Serial order: a parallel distributed processing approach*. Tech. rep. ICS 8604. San Diego, California: Institute for Cognitive Science, University of California.

Joshi, S.D. and Saroja Bhate (1984). *Fundamentals of Anuvrtti*. Poona University Press.

Jozefowicz, Rafal et al. (2016). "Exploring the limits of language modeling". In: *arXiv preprint arXiv:1602.02410*.

Jurafsky, Daniel and James H. Martin (2022). *Speech and Language Processing*. 3rd edition. URL: https://web.stanford.edu/~jurafsky/slp3/.

Kadmon, Nirit and Fred Landman (1993). "Any". In: *Linguistics and Philosophy* 16 (4), pp. 353–422.

Kahneman, Daniel (2011). *Thinking, fast and slow*. Farrar, Straus, and Giroux.

Kálmán, László (1990). "Deferred information: The semantics of commitment". In: *Papers from the Second Symposium on Logic and Language*, pp. 125–157.

Kálmán, László and András Kornai (1985). *Pattern matching: a finite state approach to generation and parsing*.

Kamp, Hans (1981). "A Theory of Truth and Semantic Representation". In: *Formal Methods in the Study of Language*. Ed. by J.A.G. Groenendijk, T.M.V. Jansen, and M.B.J. Stokhof. Amsterdam: Mathematisch Centrum, pp. 277–322.

Kapitula, Todd (2015). *Ordinary Differential Equations and Linear Algebra: A Systems Approach*. SIAM. ISBN: 978-1-611974-08-9.

Kaplan, David (1978). "On the logic of demonstratives". In: *Journal of Philosophical Logic* 8, pp. 81–98.

Karpathy, Andrej, Justin Johnson, and Li Fei-Fei (2015). "Visualizing and understanding recurrent networks".

Karpathy, Andrej and Fei-Fei Li (2014). "Deep Visual-Semantic Alignments for Generating Image Descriptions". In: *CoRR* abs/1412.2306. arXiv: `1412.2306`. URL: `http://arxiv.org/abs/1412.2306`.

Karttunen, Lauri (1989). "Radical lexicalism". In: *Alternative Conceptions of Phrase Structure*. Ed. by Mark Baltin and Anthony Kroch. University of Chicago Press, pp. 43–65.

— (2014). *Three ways of not being lucky*. URL: `http://web.stanford.edu/~laurik/presentations/LuckyAtSALTwithNotes.pdf`.

Katz, Jerrold J. and Paul M. Postal (1964). *An Integrated Theory of Linguistic Descriptions*. Cambridge: MIT Press.

Kaufman, Daniel (2009). "Austronesian nominalism and its consequences: A Tagalog case study". In: *Theoretical Linguistics* 35, pp. 1–49.

Kiparsky, Paul (1982). "From cyclic phonology to lexical phonology". In: *The structure of phonological representations, I*. Ed. by H. van der Hulst and N. Smith. Dordrecht: Foris, pp. 131–175.

— (2016). "Stems". In: *Oxford Research Encyclopedia in Linguistics*. Ed. by Mark Aronoff. URL: `https://global.oup.com/academic/product/oxford-research-encyclopedias-linguistics-9780199384655`.

Kipper, Karin, Hoa Trang Dang, and Martha Palmer (2000). "Class Based Construction of a Verb Lexicon". In: *AAAI-2000 Seventeenth National Conference on Artificial Intelligence*. Austin, TX.

Kiss, George (1973). "Grammatical Word Classes: A Learning Process and its Simulation". In: *Psychology of Learning and Motivation*. Ed. by Gordon Bower. Vol. 7. Academic Press, pp. 1–41.

Knuth, Donald E. (1968). "Semantics of context-free languages". In: *Mathematical Systems Theory* 2, pp. 127–145.

Koller, Alexander (2015). "Semantic construction with graph grammars". In: *Proceedings of the 11th International Conference on Computational Semantics*. London, UK:

Association for Computational Linguistics, pp. 228–238. URL: `https://www.aclweb.org/anthology/W15-0127`.

Koller, Alexander and Marco Kuhlmann (2011). "A generalized view on parsing and translation". In: *Proceedings of the 12th International Conference on Parsing Technologies (IWPT)*. Dublin.

Kolmogorov, Andrei N. (1933). *Grundbegriffe der Wahrscheinlichkeitsrechnung*. Springer.

— (1953). "O ponyatii algoritma". In: *Uspehi matematicheskih nauk* 8.4, pp. 175–176.

Kornai, András (2006). "Evaluating geographic information retrieval". In: *Accessing Multilingual Information Repositories*. Springer, pp. 928–938.

— (2007). *Mathematical linguistics*. Springer.

— (2008). "On the proper definition of information". In: *Living, Working and Learning beyond Technology: Conference Proceedings of ETHICOMP 2008*. Ed. by T. Bynum et al. Tipographia Commerciale, pp. 488–495.

— (2010a). "The algebra of lexical semantics". In: *Proceedings of the 11th Mathematics of Language Workshop*. Ed. by Christian Ebert, Gerhard Jäger, and Jens Michaelis. LNAI 6149. Springer, pp. 174–199. DOI: `10.5555/1886644.1886658`.

— (2010b). "The treatment of ordinary quantification in English proper". In: *Hungarian Review of Philosophy* 54.4, pp. 150–162.

— (2012). "Eliminating ditransitives". In: *Revised and Selected Papers from the 15th and 16th Formal Grammar Conferences*. Ed. by Ph. de Groote and M-J Nederhof. LNCS 7395. Springer, pp. 243–261.

— (2014). "Bounding the impact of AGI". In: *Journal of Experimental and Theoretical Artificial Intelligence* 26.3, pp. 417–438.

— (2019). *Semantics*. Springer Verlag. ISBN: 978-3-319-65644-1. URL: `http://kornai.com/Drafts/sem.pdf`.

— (2021). "Vocabulary: Common or Basic?" In: *Frontiers in Psychology*. DOI: `10.3389/fpsyg.2021.730112`. URL: `https://www.frontiersin.org/articles/10.3389/fpsyg.2021.730112/full`.

Kornai, András and Péter Halácsy (2008). "Google for the linguist on a budget". In: *Proceedings of the 4th Web as Corpus Workshop*. Ed. by S. Evert, A. Kilgarriff, and S. Sharoff. LREC WAC-4, pp. 8–11.

Kornai, András and Marcus Kracht (July 2015). "Lexical Semantics and Model Theory: Together at Last?" In: *Proceedings of the 14th Meeting on the Mathematics of Language (MoL 14)*. Chicago, IL: Association for Computational Linguistics, pp. 51–61.

Kornai, András and Lisa Stone (2004). "Automatic translation to controlled medical vocabularies". In: *Innovations in intelligent systems and applications*. Ed. by Anil K. Jain. Springer, pp. 413–434.

Kornai, András et al. (2006). "Web-based frequency dictionaries for medium density languages". In: *Proc. 2nd Web as Corpus Workshop (EACL 2006 WS01)*. Ed. by A. Kilgariff and M. Baroni, pp. 1–8.

Kornai, András et al. (2015). "Competence in lexical semantics". In: *Proceedings of the Fourth Joint Conference on Lexical and Computational Semantics*. Denver, Colorado: Association for Computational Linguistics, pp. 165–175. DOI: `10.18653/v1/S15-1019`. URL: `https://www.aclweb.org/anthology/S15-1019`.

Kovács, Ádám et al. (2022a). "Explainable lexical entailment with semantic graphs". In: *Natural Language Engineering*. DOI: `https://www.doi.org/10.1017/S1351324922000092`.

Kovács, Ádám et al. (2022b). *POTATO: exPlainable infOrmation exTrAcTion frameWOrk*. arXiv: `2201.13230 [cs.CL]`. URL: `https://arxiv.org/pdf/2201.13230.pdf`.

Kracht, Marcus (2011a). "Gnosis". In: *Journal of Philosophical Logic* 40.3, pp. 397–420.

— (2011b). *Interpreted Languages and Compositionality*. Vol. 89. Studies in Linguistics and Philosophy. Berlin: Springer.

Kratzer, Angelika (1995). "Stage Level and Individual Level Predicates". In: *The Generic Book*. Ed. by G. Carlson and F.J. Pelletier. University of Chicago Press.

Kripke, Saul A. (1972). "Naming and necessity". In: *Semantics of Natural Language*. Ed. by D. Davidson. D. Reidel, Dordrecht, pp. 253–355.

Kushman, Nate et al. (2014). "Learning to Automatically Solve Algebra Word Problems". In: *Proc. ACL 2014*.

Labov, William (1984). "Intensifiers". In: *Proc. GURT*. Ed. by Deborah Schiffrin. Washington, DC: Georgetown University Press, pp. 43–70.

Lakoff, George (1970). *Irregularity in Syntax*. Holt, Rinehart, and Winston.

— (1987). *Women, Fire, and Dangerous Things: What Categories Reveal About the Mind*. University of Chicago Press. ISBN: 978-0-226-46803-7.

Landman, Fred (1986). *Towards a Theory of Information*. Dordrecht: Foris.

— (2004). *Indefinites and the Type of Sets*. Blackwell Publishing.

Langacker, Ronald (1987). *Foundations of Cognitive Grammar*. Vol. 1. Stanford University Press.

— (2001). "What *wh* means". In: *Conceptual and Discourse Factors in Linguistic Structure*. Ed. by Alan Cienki, Barbara Luka, and Michael B. Smith. CSLI Publications, pp. 137–152.

Laparra, Egoitz et al. (2018). "SemEval 2018 Task 6: Parsing Time Normalizations". In: *SemEval@NAACL-HLT*.

Lappin, Shalom (1996). "Generalized Quantifiers, Exception Phrases, and Logicality". In: *Journal of Semantics* 13, pp. 197–220.

Lazaridou, Angeliki et al. (2013). "Compositionally Derived Representations of Morphologically Complex Words in Distributional Semantics". In: *ACL (1)*, pp. 1517–1526. URL: `http://aclweb.org/anthology/P/P13/P13-1149.pdf`.

LeCun, Yann, Yoshua Bengio, and Geoffrey Hinton (2015). "Deep learning". In: *Nature* 521, pp. 436–444.

Lehrer, Adrienne (1985). "Markedness and Antonymy". In: *Journal of Linguistics* 21.2, pp. 397–429.

Lenat, Douglas B. and R.V. Guha (1990). *Building Large Knowledge-Based Systems*. Addison-Wesley.

Lévai, Dániel and András Kornai (Jan. 2019). "The impact of inflection on word vectors". In: *XV. Magyar Számítógépes Nyelvészeti Konferencia*.

Lewin, Kurt (1943). "Psychology and the Process of Group Living". In: *The Journal of Social Psychology* 17.1, pp. 113–131. DOI: 10.1080/00224545.1943.9712269.

Lewis, D. (1970). "General semantics". In: *Synthese* 22.1, pp. 18–67.

Li, Yujia et al. (2022). *Competition-Level Code Generation with AlphaCode*. URL: https://storage.googleapis.com/deepmind-media/AlphaCode/competition_level_code_generation_with_alphacode.pdf.

Lieber, Rochelle (1992). *Deconstructing Morphology: Word Formation in Syntactic Theory*. Chicago: University of Chicago Press.

Lipshits, Mark and Joseph McIntyre (1999). "Gravity affects the preferred vertical and horizontal in visual perception of orientation". In: *NeuroReport* 10, pp. 1085–1089. URL: https://journals.lww.com/neuroreport/Fulltext/1999/04060/Gravity_affects_the_preferred_vertical_and.33.aspx.

List, Johann-Mattis, Michael Cysouw, and Robert Forkel (May 2016). "Concepticon: A Resource for the Linking of Concept Lists". In: *Proceedings of the Tenth International Conference on Language Resources and Evaluation (LREC'16)*. Portorož, Slovenia: European Language Resources Association (ELRA), pp. 2393–2400. URL: https://www.aclweb.org/anthology/L16-1379.

Little, W. A. (1974). "The existence of persistent states in the brain". In: *Mathematical Biosciences* 19, pp. 101–120.

Liu, Yinhan et al. (2019). *RoBERTa: A robustly optimized bert pretraining approach*. arXiv: 1907.11692 [cs.CL].

Luong, Minh-Thang, Hieu Pham, and Christopher D Manning (2015). "Bilingual Word Representations with Monolingual Quality in Mind". In: *Proceedings of NAACL-HLT*, pp. 151–159.

Lynch, Peter (1989). *One Up on Wall Street*. Simon and Schuster. ISBN: 0671661035.

Maclagan, Diane and Bernd Sturmfels (2015). *Introduction to Tropical Geometry*. AMS.

Mager, Manuel et al. (2022). "BPE vs. Morphological Segmentation: A Case Study on Machine Translation of Four Polysynthetic Languages". In: *arXiv:2203.08954*.

Makrai, Márton (2014). "Deep cases in the 4lang concept lexicon". In: *X. Magyar Számítógépes Nyelvészeti Konferencia (MSZNY 2014)*. Ed. by Attila Tanács, Viktor Varga, and Veronika Vincze, 50–57 (in Hungarian), 387 (English abstract). ISBN: 978-963-306-246-3.

Maler, Oded and Amir Pnueli (1994). "On the cascaded decomposition of automata, its complexity, and its application to logic". In: *ACTS Mobile Communication*.

Manning, Christopher D. (2011). "Part-of-Speech Tagging from 97% to 100%: Is It Time for Some Linguistics?" In: *Computational Linguistics and Intelligent Text Processing*. Ed. by Alexander F. Gelbukh. Berlin, Heidelberg: Springer Berlin Heidelberg, pp. 171–189. ISBN: 978-3-642-19400-9. DOI: 10.1007/978-3-642-19400-9_14.

Matsuzaki, Takuya et al. (July 2017). "Semantic Parsing of Pre-university Math Problems". In: *Proceedings of the 55th Annual Meeting of the Association for Computational Linguistics (Volume 1: Long Papers)*. Vancouver, Canada: Association for Computational Linguistics, pp. 2131–2141. DOI: 10.18653/v1/P17-1195. URL: https://www.aclweb.org/anthology/P17-1195.

McCann, Bryan et al. (2017). "Learned in translation: Contextualized word vectors". In: *Advances in Neural Information Processing Systems*, pp. 6294–6305.

McCarthy, John (1963). "A Basis For A Mathematical Theory Of Computation". In: *Computer Programming and Formal Systems*. North-Holland, pp. 33–70.

McClelland, James L and Jeffrey L Elman (1986). "The TRACE model of speech perception". In: *Cognitive Psychology* 18.1, pp. 1–86. ISSN: 0010-0285. DOI: https://doi.org/10.1016/0010-0285(86)90015-0. URL: http://www.sciencedirect.com/science/article/pii/0010028586900150.

McIntosh, E., ed. (1951). *The Concise Oxford Dictionary of Current English*. 4th ed. Oxford University Press.

McKeown, Margaret G. and Mary E. Curtis (1987). *The nature of vocabulary acquisition*. Lawrence Erlbaum Associates.

Meillet, Antoine (1912). "L'évolution des formes gramaticales". In: *Scientia* 12.26.

Mihalcea, Rada F. (May 2002). "Bootstrapping Large Sense Tagged Corpora". In: *Proceedings of the Third International Conference on Language Resources and Evaluation (LREC'02)*. Las Palmas, Canary Islands - Spain: European Language Resources Association (ELRA). URL: http://www.lrec-conf.org/proceedings/lrec2002/pdf/310.pdf.

Mikolov, Tomas, Wen-tau Yih, and Geoffrey Zweig (2013). "Linguistic Regularities in Continuous Space Word Representations". In: *Proceedings of the 2013 Conference of the North American Chapter of the Association for Computational Linguistics: Human Language Technologies (NAACL-HLT 2013)*. Atlanta, Georgia: Association for Computational Linguistics, pp. 746–751.

Mikolov, Tomas et al. (May 2013). "Efficient Estimation of Word Representations in Vector Space". In: *1st International Conference on Learning Representations, ICLR 2013, Workshop Track Proceedings*. Ed. by Y. Bengio and Y. LeCun. arXiv: 1301.3781 [cs.CL]. URL: http://arxiv.org/abs/1301.3781.

Miller, George A. (1956). "The magical number seven, plus or minus two: some limits on our capacity for processing information". In: *Psychological Review* 63, pp. 81–97.

— (1995). "WordNet: a lexical database for English". In: *Communications of the ACM* 38.11, pp. 39–41.

Miller, George A. and Noam Chomsky (1963). "Finitary models of language users". In: *Handbook of Mathematical Psychology*. Ed. by R.D. Luce, R.R. Bush, and E. Galanter. Wiley, pp. 419–491.

Minsky, Marvin (1975). "A framework for representing knowledge". In: *The Psychology of Computer Vision*. Ed. by P.H. Winston. McGraw-Hill, pp. 211–277.

Mitra, Arindam and Chitta Baral (Aug. 2016). "Learning To Use Formulas To Solve Simple Arithmetic Problems". In: *Proceedings of the 54th Annual Meeting of the Association for Computational Linguistics (Volume 1: Long Papers)*. Berlin, Germany: Association for Computational Linguistics, pp. 2144–2153. DOI: 10.18653/v1/P16-1202. URL: https://www.aclweb.org/anthology/P16-1202.

Moerdijk, Ieke and Gonzalo Reyes (1991). *Models for smooth infinitesimal analysis*. Springer-Verlag.

Mohammadshahi, Alireza and James Henderson (Nov. 2020). "Graph-to-Graph Transformer for Transition-based Dependency Parsing". In: *Findings of the Association for Computational Linguistics: EMNLP 2020*. Online: Association for Computational Linguistics, pp. 3278–3289. DOI: 10.18653/v1/2020.findings-emnlp.294. URL: https://aclanthology.org/2020.findings-emnlp.294.

Mohr, P.J., D.B. Newell, and B.N. Taylor (2016). "CODATA recommended values of the fundamental physical constants: 2014". In: *Journal of Physical and Chemical Reference Data* 45.4, p. 043102.

Mollica, Francis and Steven T. Piantadosi (2019). "Humans store about 1.5 megabytes of information during language acquisition". In: *Royal Society Open Science*. URL: https://doi.org/10.1098/rsos.181393.

Moltmann, Frederike (2013). *Abstract Objects and the Semantics of Natural Language*. Pxford University Press.

Moltmann, Friederike (1995). "Exception Phrases and Polyadic Quantification". In: *Linguistics and Philosophy* 18, pp. 223–280.

Montague, Richard (1970). "Universal Grammar". In: *Theoria* 36, pp. 373–398.

— (1973). "The proper treatment of quantification in ordinary English". In: *Formal Philosophy*. Ed. by R. Thomason. Yale University Press, pp. 247–270.

Moon, Rosamund (1987). "The Analysis of Meaning". In: *Looking up: An account of the COBUILD Project in Lexical Computing*. Ed. by J. M. Sincair. London, Glasgow: Collins ELT, pp. 86–103.

Moradshahi, Mehrad et al. (2020). *{HUBERT} Untangles {BERT} to Improve Transfer across {NLP} Tasks*. URL: https://openreview.net/forum?id=HJxnM1rFvr.

Nemeskey, Dávid Márk (2017). "emLam – a Hungarian Language Modeling baseline". In: *XIII. Magyar Számítógépes Nyelvészeti Konferencia (MSZNY2017)*. Szeged, pp. 91–102. arXiv: 1701.07880 [cs.CL].

— (2020). "Natural Language Processing Methods for Language Modeling". PhD thesis. Eötvös Loránd University.

Nemeskey, Dávid Márk and András Kornai (2018). "Emergency Vocabulary". In: *Information Systems Frontiers (ISF)* 20.5, pp. 909–923. ISSN: 1387-3326. DOI: 10 . 1007 / s10796 - 018 - 9843 - x. URL: https : / / link . springer . com / content/pdf/10.1007/s10796-018-9843-x.pdf.

Nemeskey, Dávid et al. (2013). "Spreading activation in language understanding". In: *Proceedings of the 9th International Conference on Computer Science and Information Technologies (CSIT 2013)*. Yerevan, Armenia: Springer, pp. 140–143. URL: https://hlt.bme.hu/media/pdf/nemeskey_2013.pdf.

Newman, Paul (1968). "The Reality of Morphophonemes". In: *Language* 44, pp. 507–515.

Nivre, Joakim, Mitchell Abrams, Željko Agić, et al. (2018). *Universal Dependencies 2.3*. LINDAT/CLARIN digital library at the Institute of Formal and Applied Linguistics (ÚFAL), Faculty of Mathematics and Physics, Charles University. URL: http:// hdl.handle.net/11234/1-2895.

Nivre, Joakim et al. (May 2016). "Universal Dependencies v1: A Multilingual Treebank Collection". In: *Proc. LREC 2016*, pp. 1659–1666.

Núñez, Rafael E and Eve Sweetser (2006). "With the future behind them: Convergent evidence from Aymara language and gesture in the crosslinguistic comparison of spatial construals of time". In: *Cognitive science* 30.3, pp. 401–450.

Nzeyimana, Antoine and Andre Niyongabo Rubungo (2022). "KinyaBERT: a Morphology-aware Kinyarwanda Language Model". In: *arXiv preprint arXiv:2203.08459*.

Onea, Edgar (2009). "Exhaustiveness of Hungarian Focus. Experimental Evidence from Hungarian and German". In: *Focus at the Syntax-Semantics Interface*. Ed. by Arndt Riester and Edgar Onea. Vol. 3. Working Papers of the SFB 732. University of Stuttgart.

Osgood, Charles E., William S. May, and Murray S. Miron (1975). *Cross Cultural Universals of Affective Meaning*. University of Illinois Press.

Osgood, Charles E., George Suci, and Percy Tannenbaum (1957). *The measurement of meaning*. University of Illinois Press.

Ostler, Nicholas (1979). *Case-Linking: a Theory of Case and Verb Diathesis Applied to Classical Sanskrit*. MIT: PhD thesis.

Park, Jaehui (2019). "Selectively Connected Self-Attentions for Semantic Role Labeling". In: *Applied Sciences* 9, p. 1716.

Parsons, Terence (1970). "Some problems concerning the logic of grammatical modifiers". In: *Synthese* 21.3–4, pp. 320–334.

— (1974). "A Prolegomenon to Meinongian Semantics". In: *The Journal of Philosophy* 71.16, pp. 561–580.

— (1980). *Nonexistent Objects*. New Haven: Yale University Press.

— (2017). "The Traditional Square of Opposition". In: *The Stanford Encyclopedia of Philosophy*. Ed. by Edward N. Zalta. Summer 2017. Metaphysics Research Lab, Stanford University.

Partee, Barbara (1984). "Nominal and temporal anaphora". In: *Linguistics and Philosophy* 7, pp. 243–286.

— (1985). "Situations, Worlds and Contexts". In: *Linguistics and Philosophy* 8.1, pp. 53–58.

Pearl, Judea (2009). *Causality: Models, Reasoning, and Inference*. 2nd. Cambridge University Press.

Peeters, Bert, ed. (2000). *The Lexicon-Encyclopedia Interface*. Elsevier.

Pennington, Jeffrey, Richard Socher, and Christopher Manning (2014). "Glove: Global Vectors for Word Representation". In: *Proceedings of the 2014 Conference on Empirical Methods in Natural Language Processing (EMNLP)*. Doha, Qatar: Association for Computational Linguistics, pp. 1532–1543. DOI: 10.3115/v1/D14-1162. URL: http://www.aclweb.org/anthology/D14-1162.

Pereira, Fernando (2012). "Low-Pass Semantics". In: http://videolectures.net/metaforum2012_pereira_semantic/.

Perlmutter, D.M. (1978). "Impersonal passives and the unaccusative hypothesis". In: *BLS* 4, pp. 128–139.

Perlmutter, David M. (1980). "Relational grammar". In: *Current approaches to syntax*. Ed. by Wirth and Moravcsik. Academic Press, pp. 195–229.

Peters, Matthew et al. (2018). "Deep Contextualized Word Representations". In: *Proceedings of the 2018 Conference of the North American Chapter of the Association for Computational Linguistics: Human Language Technologies, Volume 1 (Long Papers)*. New Orleans, Louisiana: Association for Computational Linguistics, pp. 2227–2237. DOI: 10.18653/v1/N18-1202. URL: http://aclweb.org/anthology/N18-1202.

Pike, Kenneth L. (1982). *Linguistic concepts: an introduction to tagmemics*. University of Nebraska Press.

Pinker, S. and A. Prince (1994). "Regular and irregular morphology and the psychological status of rules of grammar". In: *The reality of linguistic rules*. Ed. by S. D. Lima, R. Corrigan, and G. Iverson. John Benjamins Publishing Co, pp. 321–351.

Pinker, Steven and Alan S. Prince (1988). "On language and connectionism: Analysis of a parallel distributed processing model of language acquisition". In: *Cognition* 28, pp. 73–194.

Plag, Ingo (1998). "The polysemy of -ize derivatives: On the role of semantics in word formation". In: *Yearbook of Morphology 1997*. Ed. by Geert Booij and Jaap Van Marle. Dordrecht: Springer, pp. 219–242. ISBN: 978-94-011-4998-3. DOI: 10.1007/978-94-011-4998-3_8. URL: https://doi.org/10.1007/978-94-011-4998-3_8.

— (2003). *Word-Formation in English*. Cambridge Textbooks in Linguistics. Cambridge University Press. ISBN: 9780521525633.

Pollard, Carl and Ivan Sag (1987). *Information-based Syntax and Semantics. Volume 1. Fundamentals*. Stanford, CA: CSLI.

Pozdniakov, Konstantin (2018). *The numeral system of Proto-Niger-Congo: A step-by-step reconstruction*. Language Science Press.

Premack, David and Guy Woodruff (1978). "Does the chimpanzee have a theory of mind?" In: *Behavioral and Brain Sciences* 1.4, pp. 515–526. DOI: 10.1017/S0140525X00076512.

Procter, Paul (1978). *Longman Dictionary of Contemporary English*. 1st ed. Longman.

Purver, Matthew et al. (2021). "Incremental Composition in Distributional Semantics". In: *Journal of Logic, Language and Information* 30, pp. 379–406. DOI: https://doi.org/10.1007/s10849-021-09337-8.

Pustejovsky, James (1995). *The Generative Lexicon*. MIT Press.

Pustejovsky, James et al. (2003). "TimeML: Robust specification of event and temporal expressions in text". In: *New Directions in Question Answering, Papers from 2003 AAAI Spring Symposium*, pp. 28–34.

Putnam, Hilary (1975). "The Meaning of "Meaning"". In: *Language, mind, and knowledge* 7, pp. 131–193.

Qi, Peng et al. (2020). "Stanza: A Python Natural Language Processing Toolkit for Many Human Languages". In: *Proceedings of the 58th Annual Meeting of the Association for Computational Linguistics: System Demonstrations*. Online: Association for Computational Linguistics, pp. 101–108. DOI: 10.18653/v1/2020.acl-demos.14. URL: https://aclanthology.org/2020.acl-demos.14.

Quillian, M. Ross (1967). "Semantic memory". In: *Semantic information processing*. Ed. by Minsky. Cambridge: MIT Press, pp. 227–270.

— (1969). "The teachable language comprehender". In: *Communications of the ACM* 12, pp. 459–476.

Quine, Willard van Orman (1947). "On universals". In: *Journal of Symbolic Logic* 12 (3), pp. 74–84.

Quirk, Randolph et al. (1985). *A Comprehensive Grammar of the English Language*. 2nd Revised edition. Longman. ISBN: 9780582965027.

Radford, Alec et al. (2019). "Language Models are Unsupervised Multitask Learners". https://github.com/openai/gpt-2. URL: https://d4mucfpksywv.cloudfront.net/better-language-models/language-models.pdf.

Rambow, Owen and Giorgio Satta (1994). *A Two-Dimensional Hierarchy for Parallel Rewriting Systems*. Tech. rep. URL: http://repository.upenn.edu/ircs_reports/148.

Rauch, Erik, Michael Bukatin, and Kenneth Baker (2003). "A confidence-based framework for disambiguating geographic terms". In: *HLT-NAACL 2003 Workshop: Analysis of Geographic References*. Ed. by András Kornai and Beth Sundheim. Edmonton, Alberta, Canada: Association for Computational Linguistics, pp. 50–54.

Rawski, Jonathan and Hossep Dolatian (2020). "Multi-Input Strict Local Functions for Tonal Phonology". In: *Proceedings of the Society for Computation in Linguistics*. Vol. 3. 25.

Recski, Gábor (2016). "Building Concept Graphs from Monolingual Dictionary Entries". In: *Proceedings of the Tenth International Conference on Language Resources and Evaluation (LREC 2016)*. Ed. by Nicoletta Calzolari et al. Portorož, Slovenia: European Language Resources Association (ELRA). ISBN: 978-2-9517408-9-1.

— (2018). "Building concept definitions from explanatory dictionaries". In: *International Journal of Lexicography* 31 (3), pp. 274–311. DOI: 10.1093/ijl/ecx007.

Recski, Gábor and Judit Ács (2015). "MathLingBudapest: Concept Networks for Semantic Similarity". In: *Proceedings of the 9th International Workshop on Semantic Evaluation (SemEval 2015)*. Denver, Colorado: Association for Computational Linguistics, pp. 138–142. DOI: 10.18653/v1/S15-2025. URL: https://www.aclweb.org/anthology/S15-2025.

Recski, Gábor, Gábor Borbély, and Attila Bolevácz (2016). "Building definition graphs using monolingual dictionaries of Hungarian". In: *XI. Magyar Számítógépes Nyelvészeti Konferencia [11th Hungarian Conference on Computational Linguistics*. Ed. by Attila Tanács, Viktor Varga, and Veronika Vincze.

Recski, Gabor et al. (2021). "Explainable Rule Extraction via Semantic Graphs". In: *Proceedings of the Fifth Workshop on Automated Semantic Analysis of Information in Legal Text (ASAIL 2021)*. São Paulo, Brazil: CEUR Workshop Proceedings, pp. 24–35. URL: http://ceur-ws.org/Vol-2888/paper3.pdf.

Recski, Gábor et al. (2016). "Measuring Semantic Similarity of Words Using Concept Networks". In: *Proceedings of the 1st Workshop on Representation Learning for NLP*. Berlin, Germany: Association for Computational Linguistics, pp. 193–200. DOI: 10.18653/v1/W16-1622. URL: https://www.aclweb.org/anthology/W16-1622.

Redmon, Joseph et al. (2016). "You only look once: Unified, real-time object detection". In: *Proceedings of the IEEE conference on computer vision and pattern recognition*, pp. 779–788.

Řehůřek, Radim and Petr Sojka (May 2010). "Software Framework for Topic Modelling with Large Corpora". English. In: *Proceedings of the LREC 2010 Workshop on New Challenges for NLP Frameworks*. Valletta, Malta: ELRA, pp. 45–50. URL: http://is.muni.cz/publication/884893/en.

Reiter, Raymond and Giovanni Criscuolo (1983). "Some representational issues in default reasoning". In: *Computers and Mathematics with Applications* 9.1, pp. 15–27.

Rényi, Alfréd (1972). *Letters on Probability*. Wayne State University Press.

Rescorla, Leslie A. (1980). "Overextension in Early Language Development". In: *Journal of Child Language* 7.2, pp. 321–335. DOI: 10.1017/S0305000900002658.

Rogers, James et al. (2013). "Cognitive and sub-regular complexity". In: *Formal Grammar*. Vol. 8036. Lecture Notes in Computer Science. Springer, pp. 90–108.

Rosch, Eleanor (1975). "Cognitive Representations of Semantic Categories". In: *Journal of Experimental Psychology* 104.3, pp. 192–233.

Rosenblatt, Frank (1957). *The Perceptron: a perceiving and recognizing automaton*. Tech. rep. 85-460-1.

Ross, Alf (1941). "Imperatives and Logic". In: *Theoria* 7, pp. 53–71.

Rothe, Sascha, Sebastian Ebert, and Hinrich Schütze (June 2016). "Ultradense Word Embeddings by Orthogonal Transformation". In: *Proceedings of the 2016 Conference of the North American Chapter of the Association for Computational Linguistics: Human Language Technologies*. San Diego, California: Association for Computational Linguistics, pp. 767–777. arXiv: `1602.07572 [cs.CL]`. URL: `http://www.aclweb.org/anthology/N16-1091`.

Roy, Subhro and Dan Roth (2016). "Solving General Arithmetic Word Problems". In: *CoRR* abs/1608.01413. arXiv: `1608.01413`. URL: `http://arxiv.org/abs/1608.01413`.

— (2017). "Unit Dependency Graph and its Application to Arithmetic Word Problem Solving". In: *Proc. of the Conference on Artificial Intelligence (AAAI)*. URL: `http://cogcomp.org/papers/14764-64645-1-SM.pdf`.

Russell, Bertrand (1905). "On denoting". In: *Mind* 14, pp. 441–478.

Sadock, Jerrold M. (Oct. 1999). "The Nominalist Theory of Eskimo: A Case Study in Scientific Self-Deception". In: *International Journal of American Linguistics* 65.4, pp. 383–406.

Scha, R. (1981). "Distributive, Collective and Cumulative Quantification". In: *Formal Methods in the Study of Language, Part 2*. Ed. by J. A. G. Groenendijk, T. M. V. Janssen, and M. B. J. Stokhof. Mathematisch Centrum, pp. 483–512.

Schmitt, Martin and Hinrich Schütze (2019). "SherLIiC: A Typed Event-Focused Lexical Inference Benchmark for Evaluating Natural Language Inference". In: *Proceedings of the 57th Annual Meeting of the Association for Computational Linguistics*. Florence, Italy: Association for Computational Linguistics, pp. 902–914. DOI: `10.18653/v1/P19-1086`. URL: `https://www.aclweb.org/anthology/P19-1086`.

Schönhage, A. (1980). "Storage Modification Machine". In: *SIAM Journal on Computing* 9.3, pp. 490–508.

Schulte, Marion (2015). *The Semantics of Derivational Morphology: a Synchronic and Diachronic Investigation of the Suffixes -age and -ery in English*. Language in Performance (LIP). Narr. ISBN: 9783823379638.

Schütze, Hinrich (1993). "Word Space". In: *Advances in Neural Information Processing Systems 5*. Ed. by SJ Hanson, JD Cowan, and CL Giles. Morgan Kaufmann, pp. 895–902.

— (1998). "Automatic Word Sense Discrimination". In: *Computational Linguistics Special-Issue-on-Word Sense Disambiguation* 24.1. URL: `http://www.aclweb.org/anthology/J98-1004`.

Schwartz, Roy, Sam Thomson, and Noah A. Smith (2018). "SoPa: Bridging CNNs, RNNs, and Weighted Finite-State Machines". In: *Proc. 56th ACL Annual Meeting*. Melbourne, Australia, pp. 295–305.

Shao, Yan, Christian Hardmeier, and Joakim Nivre (2018). "Universal Word Segmentation: Implementation and Interpretation". In: *Transactions of the Association for*

Computational Linguistics 6, pp. 421–435. DOI: 10.1162/tacl_a_00033. URL: https://aclanthology.org/Q18-1030.

Shi, Yunzi (2017). *Sarah's Interactive Voronoi Diagram*. URL: http://yunzhishi.github.io/voronoi.html.

Shieber, Stuart M. (1986). *An Introduction to Unification-Based Approaches to Grammar*. Stanford, CA: CSLI.

— (May 2004). "Synchronous Grammars as Tree Transducers". In: *Proceedings of the Seventh International Workshop on Tree Adjoining Grammar and Related Formalisms (TAG+ 7)*. Vancouver, Canada. URL: http://nrs.harvard.edu/urn-3:HUL.InstRepos:2019322.

Shvachko, Konstantin V. (1991). "Different modifications of Pointer Machines and their Computational Power". In: *Proc. 16th Mathematical Foundations of Computer Science*. Springer, pp. 426–441.

Sinclair, John M. (1987). *Looking up: an account of the COBUILD project in lexical computing*. Collins ELT.

Smit, Peter et al. (Apr. 2014). "Morfessor 2.0: Toolkit for statistical morphological segmentation". In: *Proceedings of the Demonstrations at the 14th Conference of the European Chapter of the Association for Computational Linguistics*. Gothenburg, Sweden: Association for Computational Linguistics, pp. 21–24. DOI: 10.3115/v1/E14-2006. URL: https://www.aclweb.org/anthology/E14-2006.

Smolensky, Paul (1990). "Tensor product variable binding and the representation of symbolic structures in connectionist systems". In: *Artificial intelligence* 46.1, pp. 159–216.

Somers, Harold L (1987). *Valency and case in computational linguistics*. Edinburgh University Press.

Sondheimer, Norman K., Ralph M. Weischedel, and Robert J. Bobrow (1984). "Semantic Interpretation Using KL-ONE". In: *Proceedings of the 10th International Conference on Computational Linguistics and 22nd Annual Meeting of the Association for Computational Linguistics*. Stanford, California, USA: Association for Computational Linguistics, pp. 101–107.

Song, Xinying et al. (Nov. 2021). "Fast WordPiece Tokenization". In: *Proceedings of the 2021 Conference on Empirical Methods in Natural Language Processing*. Online and Punta Cana, Dominican Republic: Association for Computational Linguistics, pp. 2089–2103. DOI: 10.18653/v1/2021.emnlp-main.160. URL: https://aclanthology.org/2021.emnlp-main.160.

Szabolcsi, Anna (1981). *Compositionality in Focus*. Vol. 15. 1–2. Folia Linguistica, pp. 141–161.

— (2015). "What do quantifier particles do?" In: *Linguistics and Philosophy* 38.2, pp. 159–204. ISSN: 1573-0549. DOI: 10.1007/s10988-015-9166-z. URL: https://doi.org/10.1007/s10988-015-9166-z.

Tallerman, Maggie (2011). *Understanding syntax*. Hodder Education.

Talmy, L. (1983). "How Language Structures Space". In: *Spatial Orientation: Theory, Research, and Application*. Ed. by H. Pick and L. Acredolo. Plenum Press, pp. 225–282.

— (1988). "Force dynamics in language and cognition". In: *Cognitive science* 12.1, pp. 49–100.

Talmy, Leonard (2000). *Toward a cognitive semantics*. MIT Press.

Thalbitzer, William (1911). "Eskimo: an illustrative sketch". In: *Handbook of the American languages*. Vol. 1. US Government Printing Office.

Tjong Kim Sang, Erik F. and Fien De Meulder (2003). "Introduction to the CoNLL-2002 shared task: Language-Independent Named Entity Recognition". In: *Proceedings of the Seventh Conference on Natural Language Learning at HLT-NAACL 2003*, pp. 142–147. URL: https://aclanthology.org/W03-0419.

Tomasello, Michael (1992). *First verbs: A case study of early grammatical development*. Cambridge University Press. DOI: https://doi.org/10.1017/CBO9780511527678.

— (2003). *Constructing a language: A usage-based theory of language acquisition*. Harvard University Press.

Trabelsi, Chiheb et al. (2017). *Deep Complex Networks*. arXiv: 1705.09792 [cs.NE].

Tshitoyan, Vahe et al. (2019). "Unsupervised word embeddings capture latent knowledge from materials science literature". In: *Nature* 571.7763, pp. 95–98.

Tulving, E. (1972). "Episodic and Semantic Memory". In: *Organization of memory*. Ed. by E. Tulving and W. Donaldson. Academic Press, pp. 381–403.

Uspensky, Vladimir and Alexei Semenov (1993). *Algorithms: Main Ideas and Applications*. Springer.

Valiant, Leslie G. (1984). "A theory of the learnable". In: *Communications of the ACM* 27.11, pp. 1134–1142.

Vaswani, Ashish et al. (2017). "Attention is All you Need". In: *Advances in Neural Information Processing Systems 30*. Ed. by I. Guyon et al. Curran Associates, Inc., pp. 5998–6008. arXiv: 1706.03762 [cs.CL]. URL: http://papers.nips.cc/paper/7181-attention-is-all-you-need.pdf.

Waliński, Jacek Tadeusz (2018). *Verbs in fictive motion*. Łódź University Press.

Wang, Youkai, Huinan Wei, and Zhuangwen Li (2018). "Effect of magnetic field on the physical properties of water". In: *Results in Physics* 8, pp. 262–267. ISSN: 2211-3797. DOI: https://doi.org/10.1016/j.rinp.2017.12.022. URL: https://www.sciencedirect.com/science/article/pii/S2211379717317230.

Wells, Roulon S. (1947). "Immediate constituents". In: *Language* 23, pp. 321–343.

Whitney, William Dwight (1885). "The roots of the Sanskrit language". In: *Transactions of the American Philological Association (1869–1896)* 16, pp. 5–29.

Wierzbicka, Anna (1985). *Lexicography and conceptual analysis*. Ann Arbor: Karoma.

— (1992). *Semantics, culture, and cognition: Universal human concepts in culture-specific configurations*. Oxford University Press.

Wierzbicka, Anna (1996). *Semantics: Primes and universals*. Vol. 26. Oxford University Press Oxford.

— (2000). "Lexical prototypes as a universal basis for cross-linguistic identification of "parts of speech"". In: *Approaches to the typology of word classes*. Ed. by Petra M. Vogel and Bernard Comrie. Berlin: Mouton de Gruyter, pp. 285–320.

Wieting, John et al. (2015). "From Paraphrase Database to Compositional Paraphrase Model and Back". In: *TACL* 3, pp. 345–358.

Woods, William A. (1975). "What's in a link: Foundations for semantic networks". In: *Representation and Understanding: Studies in Cognitive Science*, pp. 35–82.

Xiao, Shiyuan et al. (Nov. 2019). "Similarity Based Auxiliary Classifier for Named Entity Recognition". In: *Proceedings of the 2019 Conference on Empirical Methods in Natural Language Processing and the 9th International Joint Conference on Natural Language Processing (EMNLP-IJCNLP)*. Hong Kong, China: Association for Computational Linguistics, pp. 1140–1149. DOI: 10.18653/v1/D19-1105. URL: https://aclanthology.org/D19-1105.

Yli-Jyrä, Anssi (2015). "Three Equivalent Codes for Autosegmental Representations". In: *Proceedings of the 12th International Conference on Finite-State Methods and Natural Language Processing 2015 FSMNLP 2015 Düsseldorf*.

Yngve, Victor H. (1961). "The depth hypothesis". In: *Structure of Language and its Mathematical Aspects*. Ed. by R. Jakobson. Providence, RI: American Mathematical Society, pp. 130–138.

Zalta, Edward N. (1983). *Abstract objects*. D. Reidel.

Zimmermann, Thomas E. (1999). "Meaning Postulates and the Model-Theoretic Approach to Natural Language Semantics". In: *Linguistics and Philosophy* 22, pp. 529–561.

Zipf, George K. (1949). *Human Behavior and the Principle of Least Effort*. Addison-Wesley.

Index

A. Kornai, *Vector Semantics*, Cognitive Technologies, https://doi.org/10.1007/978-981-19-5607-2

External index

4lang
ALU
Alto
Apache OpenNLP
Artificial General Intelligence, AGI
Artificial Intelligence, AI
Berkeley Construction Grammar, BCG
Bessel functions
British National Corpus, BNC
Buridan
Cancer Genome Atlas
Categorial Grammar, CG
Colt's Manifacturing Company
Combinatory Categorial Grammar, CCG
Common Crawl
CommonCrawl
De Veritate
DeepDive
Dependency Grammar, DG
Discourse Representation Theory, DRT
Doc Martens
Dogger Bank
Dun and Bradstreet
Dunbar's number
Dynamic Semantics

Edgar Codd
Edward Herbert
Eilenberg machines
Energy (Illinois)
Ethereum
F-measure
FastText
Getty thesaurus
Goodstein's theorem
Han unification
Hatfields and McCoys
Head-driven Phrase Structure Grammar, HPSG
Heider-Simmel experiment
Hidden Markov model, HMM
Hutter prize
Hyderabad (India)
Hyderabad (Pakistan)
I can eat glass
ISO TimeML
Jena
Kalam cosmological argument
Kock-Lawvere Axiom
Law of Total Probability
Lexical Functional Grammar, LFG
Long Now Foundation

A. Kornai, *Vector Semantics*, Cognitive Technologies, https://doi.org/10.1007/978-981-19-5607-2

Appendix: `4lang`

This Appendix shows only the English binding, concept number, status, lexical category, and definition fields of the `700.tsv 4lang` file – for the full column structure see 1.2. In addition to the Hungarian, Latin, and Polish bindings available since the beginning, Release V2 also provides Chinese and Japanese bindings as well as a mapping to Concepticon (List, Cysouw, and Forkel, 2016) concept numbers.

en	num	s	pos	def
-able	21	e	G	gen allow {gen stem_ =agt}, "_-able" mark_ stem
-er	14	e	G	er_, =agt has quality, "_-er" mark_ stem_[quality], "than _" mark_ =pat, =pat has quality
-er	3627	e	G	stem_-er is_a =agt, "_ -er" mark_ stem_
-est	3625	e	G	er_ all
-est	1513	e	G	er_ other
-ing	2	e	G	stem_-ing is_a event, "_-ing" mark_ stem_
-ist	29	e	G	person[<profession>], think {stem_[important]}, "_-ist" mark_ stem_
-ize	17	e	G	cause_ after({=pat has property, stem_ has property}), "_-ize" mark_ stem
-th	4	e	G	part, in whole, before(divide)
-th	5	e	G	position, in sequence
=agt	3225	p	G	/=agt
=pat	3376	p	G	/=pat
acid	2064	e	N	substance, <burn>, has taste[sharp,sour], lack kind
act	2373	e	V	do
act	2379	e	N	act/2373
action	399	u	N	person do 29
activity	2383	e	N	act
add	1859	e	V	=agt cause_ {=pat in place}, "to/2743 _" mark_ place
after	2533	c	G	follow, in order/2739 88
aggressive	3338	e	A	angry, threaten, want [fight,attack,defeat]

© The Author(s) 2023
A. Kornai, *Vector Semantics*, Cognitive Technologies, https://doi.org/10.1007/978-981-19-5607-2

agriculture	3603 e N	practice, raise/1788 crop, raise/1788 animal
aim	363 e N	purpose 11
air	1540 c N	gas, life need, Earth has
alcohol	158 e N	liquid, <drink>, <cause_ person[drunk]> 73
all	1695 u N	gen, whole 94,96,160
allow	670 e V	=agt lack {=agt stop =pat} 146
amount	1666 u N	quantity 159
anger	488 c N	feeling, bad, strong, aggressive 141,149
angry	999 u A	anger
animal	78 u N	live, move
area	2366 e N	place/2326, in country, <has border>
arm	1231 c N	organ, long, human has body, body has, limb, hand at, wrist at, shoulder at 83
around	1388 u D	at, =agt[round], "around _" mark_ =pat
artefact	3141 u N	object, human make
ash	991 e N	powder[<grey>,<white>,<black>], {<wood> burn} make 153
Asia	3146 e N	land, @Asia
at	2744 u G	=agt has place, =pat[place], "at _" mark_ =pat
atmosphere	3379 u N	air, Earth has 215
atom	181 e N	particle, lack part, small 103
attack	2034 u V	violent, <ins_ weapon>, want hurt, want damage
attract	2664 e V	=agt cause_ {=pat want {=pat near =agt}} 31,130,195
authority	2811 c N	power, command, determine, judge
autumn	1882 u N	season/548, follow summer, winter follow, rain in, cool/1103, leaf change colour
awake	494 e A	conscious
axis	3377 e N	line, round around, turn around, at middle, shape has
bad	2043 c A	cause_ hurt 29,102,122,
bake	2130 e V	cook/825, =pat[<bread>, <cake>], =agt cause_ =pat[hard] 152
bark	2517 e U	sound/993[short,loud], <dog> make
base	146 e N	part_of whole, at bottom, whole has bottom, cause_ whole[fix] 82
beam	2722 e N	line, light/739, from <sun>
beautiful	2170 u A	attract gen
bed	68 e N	furniture, rest in
bee	1601 e N	insect, has wing, sting, make honey 35
before	2768 p G	before 88
begin	1312 e V	after(=pat)
belief	1063 e N	believe
believe	1062 u V	=agt think =pat[true,real]
below	1534 e A	under 35
bend	1112 e N	line, lack straight/563 4
bend	975 e V	has form[change], after(lack straight/563) 4

best	1515 e A	good, -est 162
between	1409 e G	=agt at =pat[place/2326], =agt separate =pat, "between _" mark_ =pat
big	1744 e A	er_ gen 96,214
bird	1576 c N	animal, make egg, has feather, has two(wing)
bite	1001 u V	cut, ins_ <tooth> 63
black	761 e A	colour, dark, night has colour, coal has colour 64
blade	1954 e N	cut, has edge, flat, <sword> has
blame	1052 e V	think =pat[responsible]
blood	2599 e N	liquid, in body, red 24
blue	1237 e A	colour, sky has colour, cold
body	2370 u N	object, animal has
bone	431 e N	material, rigid, frame, part_of body
book	1384 e N	artefact, text in, has more(page), has cover, gen read
border	1011 e N	line, official/1065, separate <two(country)>
bottom	787 c N	part_of whole, position, deep(er_ whole)
bowl	1462 e N	round, wide, food in
box	478 e N	container, has lid, cube
brain	122 e N	organ, control body, feel, conscious, thought in, memory in, emotion in
branch	66 u N	part, long, from trunk, <tree has>
breath	1500 e N	air, in lung, from lung
breathe	1501 c U	breath
bright	2629 e A	shine
broadcast	2135 e V	signal, <radio,television> receive 153
brush	1277 e N	device, has hair/3359, has handle, clean ins_, polish ins_, paint ins_
building	3125 c N	artefact, structure, has roof, has more(wall), <house>
bullet	901 e N	artefact, metal, from gun
burn	497 c U	fire, <=agt[wood]>, <after(ash)> 88
bus	356 e N	vehicle, more(passenger) in, has engine, has regular(way)
business	2974 e N	organization, make money 216
buttocks	3363 e N	organ, sit on
buy	2609 e V	=agt receive =pat, =agt pay seller, "from _" mark_ seller 31,63,71
buyer	3628 u N	=agt, buy, -er/3627 41
calculate	2141 e V	=agt cause_ {=agt know =pat[number]}
calm	1827 c A	quiet, lack motion, lack angry, lack nervous, lack upset
camera	1221 c N	machine, make photograph, has lens
can	1246 u V	<do> 146
can	1427 u N	cylinder, metal, contain [<food>,<drink>] 146
car	184 e N	vehicle, has four(wheel), <has engine>
carbon	3426 e N	material, element, coal is_a, diamond is_a, graphite is_a

care	82	e V	help, =pat[sick], =agt want {=pat feel good}, =agt think =pat[important]
carry	1080	e V	hold, transport
cause	1891	u N	reason 67
cause_	3290	u V	before(=agt), after(=pat) 64,67,74,
centre	1412	c N	middle 84
cereal	3340	e N	plant, <at farm>, make food, wheat is_a, rice is_a
chair	2163	e N	furniture, sit ins_, has seat, has leg(four), has <back>, support person
change	2554	u V	after(=pat[different]) 74
characteristic	3352	e N	is_a
chew	1983	e N	bite, cause_ {=pat[liquid]}
chicken	412	e N	bird, make meat, make egg
child	931	u N	person, young, parent make
chin	73	e N	part_of face, at centre, under mouth 83
choose	2546	e V	=agt cause_ {=pat for_ =agt}
circle	1386	e N	shape, round, close/3381, curve
circular	1294	u A	shape, resemble circle, curve
clean	2389	u A	lack dirt
clear	2390	u A	can/1246(see through)
close	3381	u V	move, after(part at other(part)), after(gen lack through)
cloth	2232	e N	material, sheet, has thread, cotton is_a
cloud	770	e N	object, visible, water, drop, in sky, <white>, <grey>
coal	2169	e N	mineral, hard, black, burn, make heat
cold	1053	u N	temperature, low
colour	2207	c N	sensation, light/739, red is_a, green is_a, blue is_a 9
colour	2219	c V	add colour/2207 9
column	3633	e N	shape, tall, vertical, solid
command	1941	e N	speak, has authority, cause_ {person do =pat}, "dative_" mark_ person 71
common	1415	e A	at many, more has, public
communicate	3145	u U	give information
company	2549	u N	organization, for_ business 61
compete	2608	c U	want win, =agt has opponent, opponent want win
complex	3341	e A	has many(part), difficult(understand)
conduct	3353	e V	=agt cause_ {=pat move}, "to/2743 _" mark_ place, <energy[flow] in> 71
confident	3027	e A	think after(self[good])
conform	3375	u N	similar, <gen expect =pat>, "to _" mark_ =pat 71
connect	1227	e V	after(connected)
connected	3637	e A	move(together)
conscious	2459	c A	notice, realize, awake, understand, deliberate, think, know

constant	3365 u A	lack change
contact	3366 c N	touch, =agt at =pat
contain	2313 e V	=pat in =agt
container	2801 u N	artefact, contain
control	253 u N	=agt want {=pat do}
cook	2152 e N	person, <profession>, make food 4
cook	822 e U	get heat 4
cook	825 e V	=agt make <=pat[food]>, ins_ heat 4
cool	1101 e V	after(cold)
cool	1103 e A	temperature, normal er_, er_ cold
cord	1395 e N	rope, thin/2598
corner	2062 e N	point, turn at, sharp, two(side) at
correct	1029 e A	conform norm
cotton	1936 e N	plant, soft, white, fibre
count	2142 u V	calculate, other(number) follow number
country	1913 u N	state/76
court	3124 e N	place/1026, law ins_, judge at
cover	750 u V	=agt on =pat, protect, cause_[lack{gen see =pat}] 129
cow	2335 e N	cattle, female, make milk, <at farm>
crop	2361 c N	plant, at agriculture, product
cube	2701 u N	shape, has four(side), has flat(top), has flat(bottom)
cupboard	2180 e N	furniture, has door, has shelf, store
curve	898 u N	line, lack straight/563, lack plane/2090, bend/1112, direction lack constant
cut	2542 u V	separate, ins_ edge
cutlery	3354 c N	knife is_a, fork is_a, spoon is_a, for_ eat 61
cylinder	2704 u N	shape, round(column), has flat(bottom), has flat(top)
damage	1209 u N	bad(situation)
danger	2610 u N	can/1246(harm) 166
dark	2110 u A	lack light/739
day	1754 c N	period, week has, month has, has more(hour), sun on sky, work at
death	981 e N	end, life has
decide	471 u U	want, before(=agt has more(possibility)), after(=pat)
decorate	2736 e V	=agt cause_ {=pat[beautiful]}
deep	1602 e N	has bottom[far] 214
defeat	1520 u V	before(compete), =agt has success, =agt cause_ {=pat lose}, after(=agt control =pat)
defend	2592 e V	=agt cause_ {=pat[safe]} 32
deliberate	2814 e A	do, conscious, =agt want effect/1014
desire	2544 e V	feeling, want 146,149
detail	1999 e N	part_of information, exact

determine	1627 e V	decide
device	3142 u N	artefact, machine
diamond	925 e N	hard, mineral, lack colour, <jewellery>
different	1566 u A	=pat has quality, =agt lack quality, "from _" mark_ =pat 71,102,161
difficult	1771 u A	act need large(effort), "to/3600 _" mark_ act 71
dimension	3355 e N	quantity, size, place/2326 has 96
direction	1148 u N	relation, has end, in space/2327
dirt	1438 e N	soil is_a, mud is_a, dust is_a
disease	1370 e N	bad(situation), organ in
distance	2290 u N	space/2327 has size, space/2327 between
disturb	2673 e V	interrupt, =agt cause_ {=pat lack calm}
divide	1919 u V	split
do	2372 u V	cause, =agt[animal], =pat[happen] 67
dog	1465 e N	animal, has leg(four), bark, bite, faithful
door	128 u N	artefact, at entrance, open/1814, close/3381
dot	1968 e N	mark, small, round 35
down	1498 e D	vertical(gen er_) 80
drink	1161 e V	=pat[liquid,<alcoholic>] in mouth, =agt has mouth, =agt swallow
drink	1164 e N	liquid, in mouth, swallow, <alcoholic>
drive	2614 e V	=agt cause_ {=pat move}, <=pat[car]>, control
drop	588 u V	fall/2694
drunk	1165 c A	quality, person has quality, alcohol cause_, lack control 4
dry	2145 e A	lack wet
dust	1970 u N	substance, fine, dry, particle, powder, <dirt> 54
ear	870 e N	organ, hear ins_
early	1431 e A	time, gen er_ time
earth	815 c N	planet, in space/2509, life on, ocean on, land on, has atmosphere
easy	1380 e A	lack difficult
eat	700 e V	=agt cause_ {=pat in mouth}, swallow, <=pat[food]>, <bite/1001>, <chew>, =agt has mouth 163
edge	503 c N	part, thin/2598, <sharp>, blade has, cut, instrument has
effect	1014 u N	<event> cause_
effort	687 u N	use energy, try
egg	2422 c N	food, round, animal make, animal[female], <animal[chicken]>, has thin/2598(shell)
electric	2633 u A	has electricity
electricity	88 e N	liquid, has power, move<device>, <in wire>
element	3452 u N	material, has atom(same)
elephant	605 c N	animal, large, eat <grass>, lack hair/3359, has long(nose), has large(ear), has tusk[ivory]

emotion	3010	u N	state/77, in mind, feeling 149
empty	2501	e A	lack {gen in}
enclose	285	e V	around, =agt cause_ {=pat[shut/2668]}
end	2596	u N	part, after(lack exist)
energy	2804	e N	work[physical] 35
engine	893	u N	machine, cause_ move
enter	258	e V	go, after(=agt in =pat)
entrance	256	u N	place/1026, {<person>[enter]} through, <door>
equipment	2788	e N	tool, gen has
er_	3272	p G	er_ 11
event	692	u N	activity, at place/1026, <important>
exact	705	e A	correct, lack different
exchange	405	u N	before(=pat at person), after(=pat at other(person)) 59,88
exist	2587	u V	real 143
expect	2557	e V	=agt think {=pat[real] at future}
experience	2308	e V	after(=agt remember =pat), =pat[real]
express	2757	e V	=agt cause_ {gen know =pat}
extreme	2786	e A	er_ gen
eye	2182	u N	organ, see ins_, animal has, on face, <two>
face	177	c N	organ, surface, front, part_of head, forehead part_of, chin part_of, ear part_of, jaw part_of 83
fact	2323	e N	has proof[exist] 143
faith	1064	c N	belief, religious
faithful	1099	e A	has faith
fall	1883	e N	autumn 80
fall	2694	e U	move, after(down) 80
family	383	e N	group, <two>(parent) part_of, <more>(child) part_of
far	1678	u A	distance[great] 214
fast	940	e A	quick 35
fat	3337	e N	substance, soft, <in food>, <under skin>
fear	734	u N	sensation, danger cause_, <anxiety> 167
feather	2427	e N	organ, soft, at bird
feel	521	u V	=pat in mind, =pat at body, =agt has body, =agt has mind 147,149
feeling	533	u N	liquid, in mind, joy is_a, sorrow is_a, fear is_a, anger is_a 149
female	1794	u N	sex
fibre	3357	c N	in food, cause_ health, thread, in rope, in cloth, thin/2598, natural, in wood
fight	1002	c N	person want {harm at other(person)}, ins_ weapon 28
fine	809	e A	small, light/1381, thin/2598
finger	2522	e N	part_of hand, long, thin/2598
fire	2454	c N	substance, cause_ heat, cause_ light/739, has flame, burn, <cause_ smoke>

firm	2215 e A	rigid, lack soft
firm	362 e N	company/2549[<small>]
first	649 c A	-th/5[one], lack before, second/1569 follow 94
five	2989 e A	number, follow four
fix	2026 u V	=agt cause_ {=pat[stable]}
flame	1473 e N	air, hot, light/739 from, fire has
flat	1493 c A	has surface, horizontal, lack slope, lack curve
flesh	1093 u N	material, soft, muscle
flexible	3371 e A	can/1246(change), can/1246(bend/975)
floor	1932 e N	surface, room/2235 has, stand on
flour	1545 e N	food, from/2742 grain
flow	847 u V	<liquid> move[smooth]
flower	2637 e N	plant has, <has colour>
fly	2018 u U	move, =agt in air, control
follow	1400 u V	=agt has direction, =pat has direction, after(=agt), before(=pat)
food	542 u N	material, gen eat 36,135
foot	1466 e N	organ, leg has, at ground 129
for	2824 p G	exchange_, "for _" mark_ money 71
for_	2782 p G	for_ 62
force	683 u N	power 167
forehead	1077 e N	part_of face, front, eye under, hair at, at temple/982 83
fork	2630 e N	cutlery, <metal>, has {<four>(branch)}, has handle, {move food} ins_
form	141 e N	object has
four	2988 u A	number, follow three
frame	1303 u N	artefact, has part[more, together], structure, give shape, fix, border
frequent	919 e A	often
fright	2812 e N	sudden, intense, fear
frighten	2916 e V	cause_ fear
from	2742 u G	before(in =pat), after(far)
front	608 u N	part, first 83
fruit	945 e N	part_of plant, seed part_of, food, sweet, has flesh
fur	1974 e N	hair/3359, cover skin, mammal has 130
furniture	343 c N	object, in room/2235, chair is_a, table is_a, bed is_a, cupboard is_a
further	2435 u A	more
future	1197 u N	time, follow now
gas	885 u N	substance, thin/1038, air is_a
gen	3635 p N	gen 58,105
get	1206 e V	after(has)
get	1223 e V	after(=agt has =pat), after(has)
give	113 e V	=agt cause_ {person has =pat}, dative_ mark_ person 63
go	1654 u U	move, ins_ leg, =agt has leg

good	1189 u A	gen want 102	
govern	2615 e V	control system[<country>], =agt has power	
government	1433 e N	authority, country has, people in	
grain	2183 c N	seed, small, dry, cereal has	
graphite	3601 e N	material, black, soft, is_a carbon	
great	1746 e N	big 214	
green	2679 e A	colour, plant has colour	
grey	2268 e A	colour, dark(cloud) has	
ground	2297 e N	surface, solid, at Earth	
group	432 c N	member[several], together	
grow	1796 e U	after(size(er_ gen))	
gun	1980 c N	weapon, has[metal(tube)], cause_ bullet[move], shot ins_	
hair	3359 u N	organ, fine, thread, grow, long, on body, animal has body	
hand	1264 c N	organ, part_of arm, human has arm, for_ [move gen], wrist part_of, palm part_of, five(finger) part_of, thumb part_of 61	
handle	834 u N	part_of object, for_ hold(object in hand) 12	
happen	2418 e V	change 74	
hard	1291 u A	gen lack bend/975, gen lack soft	
harm	2067 u N	bad	
has	288 p V	=agt control =pat, =agt has =pat 45,64,168	
head	756 u N	organ, brain in, face on, top	
health	554 e N	healthy	
healthy	555 u A	has body, body in good(situation)	
hear	987 c V	=agt perceive =pat[sound/993], ins_ ear	
heat	1070 u N	energy, warm	
heavy	1772 e A	weight(er_ gen) 105	
height	1583 e N	distance, vertical 82	
help	2072 u V	=agt cause_ {=pat succeed/2718}, =agt together =pat	
high	1582 u A	top er_ gen, has top	
hill	482 e N	on land, high, mountain er_ 145	
hold	2309 u V	=pat in hand, =agt has hand	
hole	1557 u N	empty(place/2326), in solid	
hole	1557 u N	empty(place/2326), in solid	
hollow	2500 e A	hole in	
home	1924 e N	place/1026, =agt at, "_'s" mark_ =agt	
honey	3342 c N	food, sweet, sticky, bee make	
horizontal	3144 u A	direction, flat(ground) has, still(water) has 80	
horse	1547 e N	animal, animal, has[four (leg)], ride ins_, pull	
hot	862 u A	temperature, high	
hour	1834 c N	time, unit, day has <24>, has <60>(minute) 217	
house	963 e N	building, home	
human	658 e N	man/659	

hurt	200	u V	cause_ {=pat has pain}, offend 141
husband	745	e N	male, has wife, "to/of _" mark_ wife
ice	1178	e N	water, cold, hard
idea	703	e N	in mind, think make 147
ill	273	u A	has disease, lack healthy
image	1242	u N	has form, resemble object, gen see, represent object, "of _" mark_ object
imagination	725	e N	make mental(image), lack see, \<picture\>
important	853	u A	cause, has value
in	2758	c G	place, =agt at place, =pat[place] contain =agt, "in _" mark_ place 58,82
information	1141	c N	liquid, study give, experience give, gen know
injury	3360	e N	damage, body has
ins_	702	u G	=pat make =agt[easy] 62,147
insect	2048	u N	small(animal), has leg[six], has head, \<has wing\>
institution	3372	e N	organize at, work at, has purpose, system, society/2285 has, has long(past), building, people in, conform norm 125
instrument	701	u N	object, work ins_, gen use, has purpose, at hand 146
intense	3369	e A	er_ gen, extreme
interrupt	1645	u V	=agt cause_[pause in =pat]
iron	2589	e N	hard(metal)
is_a	2585	p V	is_a 58
item	3343	c N	one, in list, in group, in set/2746
ivory	3606	u N	bone, elephant has
jaw	72	e N	organ, animal has, at mouth, tooth at, part_of face
join	392	e V	after(together)
joint	1169	u N	part, join at
joy	1856	u N	sensation, good 34
judge	289	c N	human, part_of court/3124, decide, make official(opinion)
kind	1274	u A	like/3382, help
king	1350	u N	monarch, man/744, lead/2617 country, part_of royal(family)
knife	1256	e N	instrument, for_ cut, has blade[\<metal\>], has handle 61,64
know	2455	u V	=agt has information, information connect =pat 164
lack	3306	p V	lack 101
land	816	u N	solid, ground, area/2366
large	1745	e A	big 96
law	1200	u N	rule, system, society/2285 has, official/1065, norm
lead	1832	e N	metal, soft
lead	2617	e V	=agt cause_ {=pat[change]}
leaf	723	c N	organ, green, flat, at stem, part_of plant
leg	1467	c N	limb, animal has, move ins_, support, low 83
legal	2806	e A	law

lens	3344 e N	shape, part_of camera, light/739 through, for_ clear(image), <glass>[curve], image has different(size), <look ins_> 61
letter	1539 e N	message, gen write
letter	278 e N	symbol, small, mark_ sound, gen write
level	2781 e N	position, at scale
lid	751 u N	cover, hollow has
life	505 u N	live
light	1381 e A	weight(gen er_)
light	739 c N	material, wave, cause_[animal see thing], beam/2722
light	944 e V	after(=pat burn)
like	3382 e V	feel {=pat[good], good for_ =agt} 44
limb	3345 e N	part_of body, leg is_a, arm is_a 83
limit	1012 e N	lack further
line	2118 u N	shape, long, has position, has direction, <straight/563>
liquid	846 u A	substance, flow, has shape[change]
list	1544 e N	series, item member, written
little	1355 e A	small 96
live	504 u U	exist, breathe, grow, reproduce, eat, act, change
long	1086 c A	gen er_ size[horizontal], has axis
lose	656 e V	after(lack) 106
loud	995 e A	sound/993, intense
love	2200 u V	emotion, good, =pat[person], care/82
low	139 u A	height(gen er_)
lung	2441 u N	organ, breathe ins_
machine	894 u N	object, work
main	818 e A	er_ other, rank, lead/2617 162
make	409 u V	=agt cause_ {=pat[exist]} 36,74
male	1039 e N	sex
mammal	2729 c N	animal, has fur, has milk
man	659 e N	animal, has two(leg), has two(hand), think, can/1246(speak), can/1246(work), has society/2285 5
man	744 e N	person, male 5
many	2113 u A	quantity, er_ gen 94,159
mark	1182 u N	sign, visible 31
mark_	3331 u V	=agt[sign], =pat[meaning], represent 11,31,212
marriage	1487 c N	legal(union), husband part_of, wife part_of
mass	2410 e N	amount, object has
material	2798 u N	object, real, <build> use
mean	1186 e V	=agt represent =pat, =agt[sign]
meaning	528 e N	information in mind, sign represent
measure	1608 e N	cause_[person know quantity]
meat	1094 u N	flesh, food, animal[<mammal>] has

member	2293 u N	group has, in group 161
memory	666 u N	information, human has, connect past
mental	3039 u A	in mind
message	2508 u V	information, <written>
metal	738 u N	material, solid, shine, conduct electricity
middle	1410 u N	part, place/1026, near centre 84
milk	2337 c N	liquid, white, cow make, fat/3337 in
mind	2457 c N	human has, in brain, human has brain, think ins_, perceive ins_, emotion ins_, will ins_, memory ins_, imagination ins_ 147
mineral	97 u N	substance, solid, natural, in Earth
minute	1956 e N	time, unit, has second/1570, hour has
monarch	3370 c N	king is_a, queen is_a, lead/2617 country
money	1952 u N	artefact, for_ exchange, has value, official/1065 61
month	1068 e N	unit, time, <twelve>, part_of year
more	2404 u A	quantity, er_ gen
most	1518 e A	all er_
motion	1729 e N	move
mountain	1024 c N	object, on Earth, natural, high, has <steep/1673>(side), high(er_ hill)
mouth	2137 c N	organ, on face, food in, speak ins_, can/1246(open/1814)
move	1731 u U	before(=agt at place/1026), after(=agt at other(place/1026)) 67,88,164
much	2114 e A	many 160
mud	2056 u N	wet, soil, soft, sticky, water in 24
muscle	1168 e N	material, animal has, move
must	1286 u G	lack choose 201
natural	2972 e A	normal
near	1414 u A	distance, gen er_ distance
neck	1803 u N	part_of body, cause_[head at]
need	2259 e N	=agt want =pat 146
needle	2448 e N	artefact, long, thin/2598, steel, pierce, has hole, <sew ins_> 130
nervous	2721 u A	feel fright
nice	1275 e A	cause_ joy
night	500 u N	period, follow sunset, sunrise follow, dark, lack sun, <sleep at> 106
noise	2671 u N	sound/993, <bad>, loud, frighten gen
norm	3361 u N	good for_ society 61
normal	2799 u A	resemble other
nose	1912 e N	organ, part_of face, animal has face, front, at centre, smell, air[move] in 83
notice	540 e V	know, see 36
now	1726 c A	time, this
number	2138 u N	quantity

object	2705 c N	thing, \<has colour\>, has shape, has weight, \<has surface\>, has position, \<lack life\> 19
ocean	1829 u N	water, salt in, cover most(Earth)
offend	201 e V	=agt cause_[harm at =pat]
official	1065 e A	has authority
official	2398 e N	person, has authority
often	921 u D	many in time, little(distance) between
on	33 u G	at, =agt touch =pat, \<high(=agt er_ =pat)\> 84
one	559 u N	number, lack other
open	1814 e A	move[can/1246], move through, lack shut/2668
open	1815 e V	after(=pat open/1814)
opinion	1768 u N	thought, person has, person[confident], person lack proof 106
opponent	631 e N	person, oppose, \<compete\>, \<in battle\> 153
oppose	630 u V	fight
or	2568 p G	"_ or _" mark_ choose 112,149
order	2739 e N	relation, more(item) has, first part_of
organ	2203 u N	object, part_of body, has purpose
organization	2204 u N	group, person member, has purpose, structure
organize	2949 u V	=agt cause_ {=pat has structure}
other	1567 p N	different 140,161,196
out	1316 e D	lack in 82
outdoor	1455 e A	place lack {in building}
outer	1456 e A	part, other(part) in
owner	3610 e N	=agt, =agt has
page	1491 e N	artefact, surface, paper has, write on, in \<book\>
pain	1318 u N	bad, sensation, injury cause_ 67,144
paint	795 c V	=agt cause_ {=pat[liquid] cover surface, =pat[picture] on surface}, liquid has colour, decorate, ins_ brush
palm	3630 e N	surface, hand has, hold
paper	1940 e N	material, from wood, flexible, sheet, has two(side)
parallel	1931 e A	constant(distance) between, lack contact
parent	2266 c N	make child 144
part	1997 u N	in, connected 58
part_of	3368 p G	part_of 64
particle	3373 u N	piece, separate, small
passenger	2534 u N	person, person[travel], person in vehicle, other(person) drive vehicle 140
past	1732 u N	period, part_of time, now follow
pause	2267 e N	lack action, before(action), after(action) 89,135
pay	237 e N	money, before(=pat work), after(=pat has)
pay	812 e V	=agt give money
people	1762 u N	human(group)

perceive	531	u V	know, ins_ sense, hear is_a, smell is_a, see is_a, touch is_a 164
period	1123	u N	time, has start, has end
person	2185	c N	man/659
photograph	1243	e N	image, camera make
phrase	1721	e N	sign, at text, has word
physical	2809	u A	body has
picture	1244	u N	image
piece	449	e N	thing, small, part_of thing[large]
pierce	2256	e V	=agt cause_[hole in =pat], ins_ sharp
pipe	418	u N	artefact, space/2327 in, cylinder, liquid[move] in
place	2326	c N	thing in/2758 28,138,139
place_	1026	c N	point, gen at
plan	2369	e N	after(structure)
plane	2807	e N	vehicle, fly, has two(wing), <has engine>
planet	3599	e N	object, in space/2509, size(er_ gen)
plant	2792	c N	live, lack move, has leaf[many], has root, at soil
pleasant	1288	u A	nice
point	1969	u N	place, lack part_of 139
polish	740	u V	=agt cause_ surface[smooth, shine], =pat has surface 36
political	1965	e A	politics
politics	1964	u N	activity, get power
position	1027	u N	place/1026
possibility	1524	e N	possible
possible	1525	u A	gen allow, can/1246 146
powder	1971	u N	substance, more(particle) 35
power	684	e N	cause_ change 167
power	979	e V	cause_ change<move> 167
practice	2512	u V	do, frequent
pressure	3132	e N	force, gas has
price	86	e N	amount, gen pay/812, at exchange
problem	2785	e N	situation, difficult, after(solve)
product	2359	e N	artefact, for_ sell 66
programme	2948	u N	plan, action, in television
proof	298	u N	prove 143
protect	2593	e V	=agt cause_ {=pat[safe]} 11
prove	1127	e V	after(other(people) know =pat[true]), real ins_ 143
public	1407	u A	lack owner 106
pull	1096	e V	=agt cause_ {=pat at =agt}
purpose	365	u N	gen want
put	2374	e V	=agt cause_ {=pat at place}, =agt move =pat, "locative" mark_ place 214
quality	1699	u N	gen has, characteristic, <good>

quantity	1667 c N	gen count, gen measure, <much> 159
queen	1353 e N	woman, monarch
quick	941 e A	act in short(time) 90
quiet	986 e A	lack noise
radio	1982 e N	device, wave in air, communicate, device make sound/993, programme in, broadcast
rain	698 e N	water, from atmosphere, fall/2694, many(drop), weather 215
raise	1788 e V	help {=pat grow}
raise	661 e V	after(=pat[high])
range	2367 e N	many, between gen , different
rank	1992 u N	position, in organization, official/2398, in <army>, in <police>, in <navy>
read	1908 e V	=agt cause_ {meaning in mind}, =pat[written] has meaning, =agt has mind
real	1126 u A	exist 143
realize	2956 e V	after(know)
reason	1892 e N	cause_ thing, gen understand thing
receive	1225 u V	get/1223
recent	1692 e A	time, before now, near
rectangular	3133 u A	has side[four, parallel], has four(corner)
red	2658 c N	colour, warm, fire has colour, blood has colour, resemble anger 9,196
regular	2147 u A	conform rule
relation	2646 u N	one at, other at
religion	2580 e N	system, faith has system
religious	2581 u A	has religion, has faith
remember	667 u V	=pat in mind, =agt has mind
report	1042 e N	information, connect recent(event), detail in
represent	1741 c V	sign has meaning, =agt[sign], =pat[meaning]
reproduce	3138 e U	=agt make other[similar] 196
resemble	3397 e V	=agt has quality, =pat has quality 196
responsible	766 c A	has control, has authority, has blame 130
rest	1959 c U	quiet, calm, before(tired), after(has energy)
rice	2021 e N	plant, food, grain, in Asia
ride	1555 e U	travel, =agt on <horse>, ins_ <horse> 13
rigid	3131 u A	has shape, shape lack change
road	2481 e N	way, has hard (surface), vehicle on
roof	2376 u N	top, part_of house
room	2235 c N	place/1026, has more(wall), part_of building, has floor, has ceiling
root	936 u N	under ground, part_of plant, support, at base/146 82
rope	1396 u N	artefact, long, flexible

round	1295 c A	circular/1294
royal	1352 e A	monarch
rule	2530 u N	govern
run	882 e U	move, fast/940, ins_ foot
sad	2248 e A	emotion, bad 150
safe	303 u A	lack danger
salt	2108 e N	mineral, white, has taste, powder
same	192 e A	lack different 102
say	1719 e V	communicate, ins_ sound/993, person hear sound/993, "dative" mark_ person
scale	2107 c N	range, level at, measure ins_, regular
screen	2977 e N	artefact, <part_of electric(machine)>, picture on
season	546 c N	period, time, part_of year, activity in 181
season	548 c N	period[<four>], part_of year, spring/2318 is_a, summer is_a, fall/1883 is_a, winter is_a, has weather 181
seat	2494 u N	sit in
second	1569 u A	-th/5[two], follow first
see	1476 c V	perceive, ins_ eye 4,164
seed	1577 u N	organ, part_of plant, make other(plant)
self	1851 e N	=pat[=agt], =agt[=pat] 3,102,217
sell	595 c V	=agt cause_ {buyer has =pat}, buyer cause_ {=agt has money_}, dative_ mark_ buyer 63
seller	3629 u N	=agt sell, -er/3627 41
sensation	534 u N	sense ins_, <touch>
sense	2458 e N	animal has, hear is_a, see is_a, smell is_a, touch is_a, taste is_a
sentence	1722 c N	in text, has phrase
separate	1450 u V	=agt cause_ {=pat at other(place/1026)}, "from _" mark_ place/1026
sequence	3137 u N	many(thing) part_of, thing follow other, has order/2739
series	2951 u N	structure, has item, item follow other(item)
set	2375 e V	=agt cause_ {=pat at position[<stable>,<proper>]}
set	2746 e N	group, has many(item), together, unit, item has common(characteristic)
seven	2996 e A	number, follow six
several	2116 e A	many
sew	2588 e V	=agt cause_[cloth[fix]], ins_ needle, ins_ thread
sex	1780 c N	male is_a, female is_a
shape	142 u N	form
sharp	611 u A	has [<edge>,<point>]
sheep	1204 e N	mammal, <at farm>, has wool
sheet	1492 u N	rectangular, flat
shelf	1962 u N	surface, vertical, hold, store, <rectangular>, rigid

shell	1216 u N	cover[hard, outer], <animal> in
shine	742 u V	light/739
shoot	1551 e V	after(bullet fly, bullet[fast]), ins_ gun
shop	329 u N	institution, sell in
short	2029 e A	size[horizontal] er_ gen
shot	1550 e N	shoot
shoulder	2548 e N	organ, part_of body, neck at, arm at
show	1742 e V	=agt cause_[gen see =pat]
shut	2668 u A	gen lack [move through]
sick	274 u A	ill, <vomit>
side	1903 u N	part, <two>, centre[far], oppose, object has 84
sign	1183 c N	gen perceive, information, show, has meaning 67
signal	1184 e N	communicate, people see
similar	2794 u A	=agt has quality, =pat has quality, "to _" mark_ =pat 44,45
sit	2493 c U	=agt at surface[<seat>], =agt has buttocks, buttocks at surface, =agt has trunk/2759[vertical]
situation	2784 u N	around
six	2990 u A	number, follow five
size	1605 c N	dimension 96
skin	318 u N	organ, part_of body, cover
sky	496 c N	high(er_ air), animal see, cloud on, sun on, star on
slide	434 e N	move, on surface, has constant(contact)
slope	1529 e U	=agt has direction, has end[high], =agt has end[other,low]
small	1356 u A	gen er_ 96
smell	2151 c N	feel ins_ nose, =pat in air
smooth	2092 u A	surface, easy(slide on)
snow	1066 u N	ice, fall/2694, soft, white
society	2285 u N	group, <people> member, has rule
soft	1979 c A	hard(gen er_)
soil	2298 u N	ground, plant in
solid	2216 u A	firm/2215
solve	760 c V	work, =pat problem, =agt want {gen lack problem}
sorrow	341 u N	emotion, er_ sad 150
sound	512 e A	whole, firm/2215
sound	993 e N	wave, human hear, in air
soup	1541 e N	food, liquid
sour	680 c A	taste, resemble acid, <bad>
space	2327 c N	thing in, empty, three(dimension) in
space	2509 c N	sun in, star in, atmosphere under
speak	270 u V	talk
speech	268 u N	sound, =agt say
split	1007 e V	after(separate), <break>

spoon	1222 u N	cutlery, {eat soup} ins_, has bowl, has handle
spring	2318 e N	first(season/548), warm, plant[live], love in, summer follow, follow winter
stable	3130 e A	lack move
stand	74 u U	=agt[vertical], <=agt on foot[two]> 81
star	408 u N	planet, shine, dot, at sky, at night
start	1313 e N	after(act)
state	76 u N	land, political(unit), has government, control self
steel	112 e N	metal, hard, strong, contain iron, contain carbon
stem	2421 u N	part_of plant, long, leaf on, flower on, fruit on 66
stem_	3280 u N	part_of word, stable 66
stick	338 e N	object, long, <wood>, <gen use>
sticky	1987 u A	stick
still	1828 e A	lack move
sting	2257 u V	pierce, <insect>
stomach	939 e N	organ, animal has, tube, food in
stop	1615 e V	after(=agt lack move) 88
store	330 u N	shop
straight	563 e A	has constant(direction)
strong	688 e A	has force[great] 72
structure	2944 u N	has more(part), connected
student	462 e N	person, study, in <school>
study	2305 e N	work, want know =pat
substance	172 u N	has mass, in space/2327, physical 196
succeed	1401 e V	=pat before =agt
succeed	2718 e V	after(aim[real]), =agt has aim
success	2969 u N	real, good, before(desire)
sudden	1061 u A	lack warn, before(lack (gen know)) 88
sugar	440 u N	material, sweet, <white>, in food, in drink 9,31
summer	1802 c N	season/548, follow spring/2318, autumn follow, warm, fruit at, much(life) at, long(day) at
sun	1755 c N	planet, give light/739, give heat, bright, yellow, at sky, at day
sunrise	3136 e N	after(sun at sky)
sunset	3135 e N	after(sky lack sun)
support	2310 u V	=agt cause_ =pat[stable], =agt[below]
surface	781 u N	part, separate, object has part, object in
swallow	1805 c V	=agt cause_ {=pat[move]}, after(=pat in stomach), =pat in mouth, =pat in throat, =agt has stomach, =agt has mouth, =agt has throat 164
sweet	495 c A	taste, good, pleasant, sugar has taste, honey has taste 9
symbol	2976 u N	mean/1186, represent
system	2015 u N	group, complex, relation between part
table	180 e N	furniture, has leg[<more>], has surface[flat,horizontal]

talk	269	e U	communicate, ins_ sentence
tall	1581	u A	height(er_ gen)
taste	1113	c N	\<food\> has, person feel, ins_ tongue
television	2343	c N	electric(equipment), box, has screen, programme on screen, man/659 see programme
temperature	1071	c N	physical, quality, hot
temple	982	u N	flat, side, part_of head
text	3127	u N	information in, sentence part_of, \<written\>
thick	2134	e A	lack thin/1038
thick	2752	e A	{distance between surface} er_ gen, has more(surface)
thin	1038	e A	flow(er_ gen)
thin	2598	e A	gen er_ {distance between surface}
thing	481	u N	exist, \<object\>
think	907	u U	=pat in mind, =agt has mind 148
this	706	u N	now, near, before(speak)
thought	908	u N	idea, in mind 147
thread	366	u N	fine(cord), \<sew ins_\>, \<in cloth\>
threaten	789	u V	=agt express {after(=agt cause_ harm)} 214
three	2970	e A	number, follow two 4
throat	2432	e N	organ, pipe, in neck, at mouth
through	100	u G	before(=agt on side), =pat has side, in =pat, after(=agt on other(side)), =pat has side[other] 135
thumb	1098	u N	part_of hand, human has hand, short, thick/2752
time	1120	c N	event in, has direction, past part_of, now part_of, future part_of
tired	3634	e A	want rest
to	12	u G	after(=agt in =pat)
to	2743	u G	after(=agt at/2744 =pat)
to	3600	u G	is_a thing, "to/3600 _" mark_ thing
together	586	u D	similar \<place, intent\>
tongue	1808	e N	part_of body, at mouth, taste ins_, speak ins_
tool	2202	u N	object, work ins_
tooth	827	c N	organ, animal has, hard, in jaw, bite/1001 ins_, chew ins_, attack ins_, defend ins_ 63
top	2377	e N	part, at position, vertical(position er_ part[other]) 82
touch	522	u V	\<hand\> at =pat[surface], \<=agt has hand\>, contact, feel surface
transport	3057	u N	move
travel	2537	e U	after(=agt at place/1026[other,\<city\>]) 206
tree	709	e N	plant, has material[wood], has trunk/2759, has many(branch) 12,29
true	1125	c A	fact 143
trunk	2759	u N	main(part), long, stable
try	1976	e V	=agt want =agt[=pat]
tube	419	u N	pipe

turn	860 u V	move, change(direction), \<has axis>
tusk	3605 e N	tooth, long, elephant has
two	2967 u A	number, one in, other in, follow one 139
under	136 u D	high(=pat er_ =agt)
understand	525 u V	meaning in mind, =pat has meaning, =agt has mind
union	2525 e N	together, public
unit	3038 u N	amount, measure
upset	1166 u N	disturb
use	1008 u V	=agt has purpose, =pat help purpose, "for _" mark_ purpose, "to/3600 _" mark_ purpose 71
useful	3134 e A	for_ gen 61
value	526 e N	amount, < gen pay/812>
vehicle	1172 c N	machine, has wheel(many), {move people} ins_, {move object} ins_, car is_a, truck is_a, bus is_a
vertical	869 c N	direction, has top, has middle, has bottom, Earth pull in direction 80
violent	690 e A	ins_ physical(force)
visible	3128 e A	can/1246(gen see)
wall	721 u N	object, vertical, enclose, divide, protect, building has, long, high
want	131 c V	=agt feel {=agt need =pat} 102,144,146
warm	1655 e A	temperature(er_ gen)
warm	878 e V	after(warm/1655)
warn	803 e V	cause_ {=pat know danger}
water	2622 u N	liquid, lack colour, lack taste, lack smell, life need 4,101,213
wave	1104 e N	in sequence, move, on surface, liquid move[vertical], has surface
way	2484 u N	artefact, gen move on, has direction 64
weapon	754 e N	instrument, fight ins_
weather	1121 u N	state/77, atmosphere has, at time, at place/1026, temperature, water in air, wind, air has pressure
week	1021 u N	period, time, seven(day) in
weight	2127 c N	physical(quantity), heavy
wet	1769 u A	liquid cover
wh	3636 p G	wh 4,213
wheat	344 u N	plant, has grain, make flour
wheel	1293 u N	artefact, part_of \<vehicle>, circular/1294, turn, \<has spoke>, \<has hub
white	755 c N	colour, light/739, snow has colour, empty, clear
whole	553 c A	all member 161
wide	2166 u A	distance[great, horizontal] between side, has more(side)
wife	767 e N	in marriage, female
will	132 e V	want
win	937 u U	best, succeed/2718, before(compete), before(effort), get/1223 \<prize> 88
wind	2164 e N	air, move[horizontal]

wing	2146 u N	object, fly ins_, part_of body
winter	2322 c N	season/548, follow autumn, spring/2318 follow, cold, snow in, death in
woman	1795 u N	female, person
wood	710 c N	material, hard, tree has
wool	924 c N	material, soft, sheep has 130
word	2224 u N	sign, speech 67
work	1740 e N	useful
wrist	438 e N	organ, joint, at hand, at end, arm has end 83
write	1109 u V	put {<letter/278>,<more(word)>} on surface<paper>, ins_ {<pen>,<pencil>}
written	3126 u A	letter/278 on surface[<paper>]
year	545 c N	period, time, month part_of
yellow	2057 e N	colour, sun has colour
young	799 u N	early, in life

Printed in the United States
by Baker & Taylor Publisher Services